Handbook of Theoretical Activity Diagrams Depicting Chemical Equilibria in Geologic Systems Involving an Aqueous Phase at One Atm and 0° to 300°C

Harold C. Helgeson, Thomas H. Brown, and Robert H. Leeper

Northwestern University

Freeman, Cooper & Company

1736 Stockton Street, San Francisco, California 94133

Copyright © 1969 Ⓕ Freeman, Cooper & Company

All rights to reproduce this book in whole or in part are reserved, with the exception of the right to use short quotations for review of the book.

Printed in the United States of America

Library of Congress Catalogue Card Number 73-97467

Preface

THIS handbook is an outgrowth of an attempt in this laboratory over the past several years to compile an internally consistent set of thermodynamic data for geologic systems involving aqueous solutions. Part of this effort has been directed toward development of computer techniques and theoretical equations, and prediction of thermodynamic values to fill in gaps in the experimental record. The numerical results of this work (Helgeson, 1969) together with the outstanding compilation of thermodynamic data contributed recently by Robie and Waldbaum (1968) make it possible to construct a comprehensive set of *theoretical* equilibrium diagrams for geologic systems involving an aqueous phase at elevated temperatures. This handbook represents a beginning in this task. Because of incomplete thermodynamic data and numerical uncertainties it will no doubt require repeated revision, correction, and expansion in the future. In the meantime we can only hope that it will serve as a useful (albeit provisional) supplement to such publications as *Phase Diagrams for Ceramists* (Levin, Robbins, and McMurdie, 1964) and *Equilibrium Diagrams for Minerals at Low Temperature and Pressure* (Schmitt, 1962) on the geologist's reference shelf.

We should like to express our appreciation to W. H. Freeman for his willing cooperation in undertaking publication of this book. We are also grateful to C. L. Christ, R. M. Garrels, and F. T. Mackenzie for many helpful discussions during preparation of the handbook, and to H. J. Greenwood and P. B. Barton for their constructive suggestions and critical reviews of the manuscript; however, we hasten to absolve these individuals of responsibility for errors and omissions in the pages that follow. Finally, and most importantly, we are indebted to the many scientists who have endeavoured through the years to produce reliable thermodynamic data for the many minerals and aqueous species found in the earth. Although we are far short of realizing a complete set of such data, we seem to have reached the threshold required for comprehensive and quantitative theoretical interpretation of phase relations in multicomponent geologic systems.

H. C. Helgeson

June, 1969, Evanston

Table of Contents

	Page
Introduction	5
References	11
List of minerals	12
Activity diagrams	15
Fugacity diagrams	227
Index of diagrams	245

Introduction

Activity diagrams depicting chemical equilibrium among minerals and aqueous solutions are important references for the modern geologist. Such diagrams facilitate prediction and interpretation of the chemical environments in which mineral assemblages form in geochemical processes. They also enable the experimental geochemist to define critical experiments and interpret his results in terms of the composition of the aqueous phase. Diagrams of this kind are presented below for a variety of geologic systems at one or more of the following temperatures: 0°, 25°, 60°, 100°, 150°, 200°, 250°, and 300° C.

Thermodynamic relations and methods used in the construction and interpretation of equilibrium activity diagrams have been summarized in a number of publications (c.f., Garrels and Christ, 1965; Helgeson, 1968, 1970). Accordingly, we shall limit our discussion to a brief review of a few of the more fundamental relations.

The activities of individual ions, complexes, or both may be used as descriptive variables in equilibrium activity diagrams, but the activities of cations common to minerals are perhaps the most generally suitable. The compositions of the bulk of the rock-forming minerals in the earth's crust can be described in the system MgO—CaO—FeO—Na_2O—K_2O—Al_2O_3—SiO_2—CO_2—H_2O. If the component HCl is added to the system, the major compositional characteristics of natural aqueous solutions can also be described in terms of these components. Provision for ore deposits and oxidation-reduction equilibria can be incorporated by adding the components H_2S, H_2SO_4, Cu_2S, PbS, ZnS, Ag_2S, etc. This set of components is one of several alternate sets that could be chosen to describe the same geologic system (Helgeson, 1970).

Three of the oxide components listed above do not dissociate to an appreciable extent in aqueous solution. These are the components SiO_2, CO_2, and H_2O, which occur primarily as neutral molecules or as complexes with other ions in solution. The remaining six dissociate almost completely according to

$$M_{\nu_+}O_{\nu_-} + \nu_+ Z_+ H^+ \rightleftharpoons \nu_+ M^{Z_+} + \frac{\nu_+ Z_+}{2} H_2O \tag{1}$$

where M refers to the cation in the oxide component, O stands for oxygen, H represents hydrogen, ν_+ and ν_- are the number of moles of the cation and anion, respectively, in one mole of the oxide component, and Z_+ is the charge on the cation. Below $\sim 300°$ C, HCl and the metal-sulfide components also exhibit nearly complete dissociation in aqueous solution (Helgeson, 1969, 1970). In contrast, H_2S does not dissociate significantly in acid solutions and H_2SO_4 is commonly represented by HSO_4^-.

The Law of Mass Action for reaction (1) can be written (with i representing a given oxide component) as

$$\frac{a_{M(i)}^{\nu_+(i)} a_{H_2O}^{(\nu_+(i)Z_+(i)/2)}}{a_i a_{H^+}^{(\nu_+(i)Z_+(i))}} = K_{1(i)} \tag{2}$$

where a_i stands for the activity of the subscripted species in the aqueous phase, and $K_{1(i)}$ is the equilibrium constant for a statement of reaction (1) involving the ith oxide component. It can be shown (c.f., Helgeson, 1967, 1968) that the logarithmic derivatives of equivalent statements of Equation (2) for two oxide components designated by the subscripts (1) and (2) can be combined and expressed (with $a_{H_2O} = 1$—see below) as

$$\frac{d \log (a_{M(1)}/a_{H^+}^{Z_+(1)})}{d \log (a_{M(2)}/a_{H^+}^{Z_+(2)})} = \frac{\nu_{+(2)}}{\nu_{+(1)}} \frac{d \log a_1}{d \log a_2} = \frac{\nu_{+(2)}}{\nu_{+(1)}} \frac{d \log f_1}{d \log f_2} = \frac{\nu_{+(2)}}{\nu_{+(1)}} \frac{d\mu_1}{d\mu_2} \tag{3}$$

where μ_1 and f_1 refer to the chemical potential and fugacity, respectively, of the subscripted component.

The logarithms of the activity ratios shown on the left side of Equation (3) may be used to construct equilibrium activity diagrams. It can be deduced from this equation that the slope of a stability field boundary on such a diagram is proportional to that of the boundary on the corresponding chemical potential or logarithmic fugacity diagram. Similar relations can be derived for the other components listed above (Helgeson, 1970).

A stability field boundary on a logarithmic activity diagram can be represented by a reversible chemical reaction such as[1]

[1] Note that reaction (4), which is written to conserve aluminum among the solid phases, is only one of several chemical reactions that can be written to describe equilibrium among K-feldspar, kaolinite, and the aqueous phase. Other reactions may be balanced to conserve other components among the solids. The mass action equations for any of these may be used to represent equilibrium among the phases on logarithmic activity diagrams, but the descriptive variables and/or the conditions specified for each diagram will be different. Alternate diagrams of this kind are presented below for a number of geologic systems.

Introduction

$$2KAlSi_3O_8 + 2H^+ + 9H_2O$$
(K-feldspar)

$$\rightleftharpoons Al_2Si_2O_5(OH)_4 + 2K^+ + 4H_4SiO_{4(aq)} \quad (4)$$
(kaolinite)

for which we can write (with a_{H_2O} again equal to one)

$$\log (a_{K^+}/a_{H^+}) = \frac{\log K_4 - 4 \log a_{H_4SiO_{4(aq)}}}{2} \quad (5)$$

where K_4 is the equilibrium constant for reaction (4) at the temperature and pressure of interest. The value of log K_4 defines the intercept of the stability field boundary corresponding to the coexistence of kaolinite, K-feldspar, and an aqueous phase on a diagram in which $\log (a_{K^+}/a_{H^+})$ is plotted against $\log a_{H_4SiO_4}$. The slope of the field boundary is -2. If components other than those represented in reaction (5) are present in the system, either kaolinite, K-feldspar, or both may not be stable with respect to other minerals that might coexist with the aqueous phase. Determination of the correct stability fields can be accomplished quickly with the aid of a computer.

The activity and fugacity diagrams presented in the following pages were constructed by a CDC 6400 computer (from a program written by T. H. Brown) using a peripheral Calcomp plotter. Because of this, the labels for the abscissa and ordinate of each diagram are printed in computer notation; e.g., $\log (a_{Mg^{++}}/a_{H^+}^2)$ is represented in the computer output by LOG A(MG++)/A(H+)2. The fugacities of S_2, O_2, and CO_2 appear in the diagrams as FS2, FO2, and FCO2, respectively. The data used in the calculations were taken from Helgeson (1969), Robie and Waldbaum (1968), Kitahara (1960) [for amorphous silica], and Hostetler (1960) [for hydromagnesite].

Activity diagrams are only as accurate as the thermodynamic data used in their construction. Because many of the data currently available are subject to large uncertainties, the diagrams presented in the following pages should be regarded as provisional representations of chemical equilibrium in geologic systems. An effort was made to be comprehensive in selecting diagrams for inclusion in this compilation, but it was not possible in most cases to make adequate critical comparisons of the phase relations predicted from thermodynamic data with those documented by experimental studies of phase equilibria and/or field observations reported in the literature.

Part of our purpose in presenting the theoretical diagrams is to provoke such critical comparisons and to stimulate experimental appraisal of the accuracy of the predicted phase relations. Because graphic depiction of phase equilibria makes inconsistencies in thermodynamic data more readily apparent, the diagrams are useful guides for such investigations.

Thermodynamic data are not yet available for all minerals. Consequently, only certain minerals (listed below) in any given system could be considered in the calculations. As a result less common counterparts of some important minerals appear in place of these minerals in the diagrams. For example, zoisite ($Ca_2Al_3Si_3O_{12}OH$) could not be included, which permits the appearance of leonhardite ($Ca_2Al_4Si_8O_{24} \cdot 7H_2O$) in its place. Other minerals such as diaspore, boehmite, andalusite, and pyrophyllite were omitted from consideration because important contradictions in the thermodynamic data reported for the minerals could not be resolved. As more and better thermodynamic data become available, revised diagrams will take into account all minerals, and the geometry of the stability fields shown in the diagrams will change accordingly.

Several apparent incompatibilities among phases in the diagrams are open to question. For example, the theoretical calculations suggest that quartz and pure magnesian chlorite are incompatible in the presence of an aqueous phase. This relation may be caused by uncertainties in the thermodynamic data used in the calculations, or it may indicate that iron is important in establishing the compatibility of chlorite and quartz in nature. Similarly, the predicted limits of talc stability preclude several compatibilities observed in geologic systems, suggesting errors in the data or supersaturation with respect to talc in nature. Other contradictions apparent in the diagrams almost certainly can be attributed to inconsistencies in the reported thermodynamic properties of minerals; e.g., those in the systems Fe—O and MgO—CaO—CO_2. Sufficient data are not yet available to resolve such uncertainties.

Although the aqueous phase is not shown explicitly in all diagrams, *it is a coexisting phase in all stability fields;* i.e., the aqueous phase is saturated with the minerals shown in the diagrams. The dashed lines in the diagrams represent saturation in the aqueous phase of various minerals that do not occupy stability fields in all diagrams for a given system. Both stable and metastable equilibria are thus represented. The diagrams are based on a standard state of unit activity of the solids, and unit activity of the aqueous species in an hypothetical one molal liquid solution at *one atmosphere and the temperature of interest*. H_2O refers to liquid water. Because increasing pressure has a slight effect on equilibrium constants at temperatures below 300°C, the diagrams can be used for pressures up to ~500 bars. HCl is shown as a component in all systems, which allowed the activity of H_2O to be specified as a constant with a value of unity in all of the calculations. The activity of H_2O departs insignificantly from one in most natural aqueous solutions at the temperatures and pressures considered here (Helgeson, 1967, 1969).

A few of the diagrams presented in the following pages are for systems in which the activity of H_2S in the aqueous phase is a specified constant. In such cases, we have selected alternate values of 10^{-2}, 10^{-3}, and 10^{-4} as typical activities of H_2S in natural aqueous solutions. Similarly, we have shown carbonate mineral saturation at fugacities of CO_2 commonly found in nature. A number of diagrams depict equilibrium phase relations in the presence of quartz or amorphous silica. Oxidation-reduction equilibria are represented in terms of the activity ratios of ions in the aqueous phase or the fugacities of S_2 and O_2 in the system. Where ion ratios are used to describe phase relations in systems involving copper, iron, and sulfur, the reversible reactions corresponding to stability field boundaries were written with H_2SO_4 as the balancing component. No provision in the calculations has been made for the effects of solid solution on the equilibrium relations portrayed; all stability field boundaries are thus shown as straight lines. Although many of the minerals considered in the calculations exhibit variable compositions in nature, where the extent of solid solution is of the order of five or ten per cent (or less), solid solubility has a slight effect on most of the phase relations depicted in the diagrams. Possible exceptions to this generalization are iron-magnesian phyllosilicates such as chlorite and biotite. Montmorillonite clay minerals are represented by end-member beidellites. Where a boundary is shown between stability fields for two beidellites, it should be regarded as an hypothetical boundary; in nature the beidellites exhibit continuous solid solution.

It should be emphasized that analytical data for natural aqueous solutions can be used to calculate the activities of species in these solutions. The requisite dissociation constants and activity coefficients can be computed from theoretical equations (Helgeson and James, 1968; Helgeson, 1969) for dilute as well as concentrated solutions up to 300°C. Such calculations make it possible to compare equilibrium activity ratios taken from activity diagrams with those in natural solutions, and to interpret directly phase relations observed in geologic systems in terms of their idealized counterparts in activity diagrams. This approach affords an insight into the chemical environment in which geochemical processes occur and facilitates prediction of mass transfer in geologic systems (c.f. Helgeson, 1968; Helgeson, Garrels, and Mackenzie, 1969; Helgeson, Brown, Nigrini, and Jones, 1970).

References

Garrels, R. M., and Christ, C. L. (1965) *Solutions, Minerals, and Equilibria:* Harper and Row, N. Y., 450 pp.

Helgeson, H. C. (1967) Solution chemistry and metamorphism: *in* Researches in Geochemistry (P. H. Abelson, ed.), Vol. II, Wiley, N. Y., 362–404.

Helgeson, H. C. (1968) Evaluation of irreversible reactions in geochemical processes involving minerals and aqueous solutions—I. Thermodynamic relations: *Geochim. et Cosmochim. Acta*, 32, 853–877.

Helgeson, H. C. (1969) Thermodynamics of hydrothermal systems at elevated temperatures and pressures: *Amer. J. Sci.*, 267, 729–804.

Helgeson, H. C. (1970) Description and interpretation of phase relations in geochemical processes involving aqueous electrolyte solutions: *Geochim. et Cosmochim. Acta* (in press).

Helgeson, H. C., Brown, T. H., Nigrini, A., and Jones, T. A. (1970) Calculation of mass transfer in geochemical processes involving aqueous solutions: *Geochim. et Cosmochim. Acta* (in press).

Helgeson, H. C., Garrels, R. M., and Mackenzie, F. T. (1969) Evaluation of irreversible reactions in geochemical processes involving minerals and aqueous solutions—II. Applications: *Geochim. et Cosmochim. Acta*, 33, 455–481.

Helgeson, H. C., and James, W. R. (1968) Activity coefficients in concentrated electrolyte solutions at elevated temperatures (an abstract): *Abstracts of Papers, 155th Natl. Meeting, Amer. Chem. Soc.*, April, 1968, San Francisco, California, S-130.

Hostetler, P. B. (1960) Low temperature relations in the system MgO—SiO_2—CO_2—H_2O: Ph.D. thesis, Department of Geological Sciences, Harvard University, Cambridge, Massachusetts.

Kitahara, S. (1960) The polymerization of silicic acid obtained by the hydrothermal treatment of quartz and the solubility of amorphous silica: *Rev. Phys. Chem. Japan*, 30 (2), 131–137.

Levin, E. M., Robbins, C. R., and McMurdie, H. F. (1964) Phase Diagrams for Ceramists: American Ceramic Society, Columbus, Ohio, 601 pp.

Robie, R. A., and Waldbaum, D. R. (1968) Thermodynamic properties of minerals and related substances at 298.15°K (25.0°C) and one atmosphere (1.013 Bars) pressure and at higher temperatures: *U. S. Geol. Survey Bull.* 1259, 256 pp.

Schmitt, H. H., ed. (1962) Equilibrium diagrams for minerals at low temperature and pressure: *Geological Club of Harvard, Special Publication*, 199 pp.

List of Minerals Considered in the Calculations

Mineral	Composition
Akermanite	$Ca_2MgSi_2O_7$
Albite (high and low)	$NaAlSi_3O_8$
Amorphous silica	SiO_2
Analcime	$NaAlSi_2O_6 \cdot H_2O$
Annite	$KFe_3AlSi_3O_{10}(OH)_2$
Anorthite	$CaAl_2Si_2O_8$
Argentite	Ag_2S
Bornite	Cu_5FeS_4
Brucite	$Mg(OH)_2$
Calcite	$CaCO_3$
Calcium olivine	γCa_2SiO_4
Cerrusite	$PbCO_3$
Chalcocite	Cu_2S
Chalcopyrite	$CuFeS_2$
Chlorite (Mg)	$Mg_5Al_2Si_3O_{10}(OH)_8$
Chrysotile	$Mg_3Si_2O_5(OH)_4$
Cinnabar	HgS
Clinoenstatite	$MgSiO_3$
Copper	Cu
Corundum	Al_2O_3
Covellite	CuS
Cuprite	Cu_2O
Dickite	$Al_2Si_2O_5(OH)_4$
Diopside	$CaMg(SiO_3)_2$
Dolomite	$CaMg(CO_3)_2$
Ferrous oxide	FeO
Forsterite	Mg_2SiO_4
Galena	PbS
Gehlenite	$Ca_2Al_2SiO_7$
Gibbsite	$Al(OH)_3$
Grossular	$Ca_3Al_2Si_3O_{12}$

List of Minerals

Mineral	Composition
Halloysite	$Al_2Si_2O_5(OH)_4$
Hematite	Fe_2O_3
Huntite (25° only)	$Mg_3Ca(CO_3)_4$
Hydromagnesite (25° only)	$Mg_4(CO_3)_3(OH)_2$
Illite	$K_{0.6}Mg_{0.25}Al_{2.3}Si_{3.5}O_{10}(OH)_2$
Iron	Fe
Jadeite	$NaAl(SiO_3)_2$
Kaliophilite	$KAlSiO_4$
Kaolinite	$Al_2Si_2O_5(OH)_4$
Larnite	$\beta\text{-}Ca_2SiO_4$
Lawsonite	$CaAl_2Si_2O_7(OH)_2 \cdot H_2O$
Leonhardite	$Ca_2Al_4Si_8O_{24} \cdot 7H_2O$
Leucite	$KAlSi_2O_6$
Lime	CaO
Magnesite	$MgCO_3$
Magnetite	Fe_3O_4
Metacinnabar	HgS
Merwinite	$Ca_3Mg(SiO_4)_2$
Microcline	$KAlSi_3O_8$
Monticellite	$CaMgSiO_4$
Montmorillonite (Ca)	$Ca_{0.167}Al_{2.33}Si_{3.67}O_{10}(OH)_2$
Montmorillonite (K)	$K_{0.33}Al_{2.33}Si_{3.67}O_{10}(OH)_2$
Montmorillonite (Mg)	$Mg_{0.167}Al_{2.33}Si_{3.67}O_{10}(OH)_2$
Montmorillonite (Na)	$Na_{0.33}Al_{2.33}Si_{3.67}O_{10}(OH)_2$
Muscovite	$KAl_3Si_3O_{10}(OH)_2$
Nepheline	$NaAlSiO_4$
Periclase	MgO
Portlandite	$Ca(OH)_2$
Pyrite	FeS_2
Pyroxene (Ca—Al)	$CaAl_2Si_2O_6$
Pyrrhotite	FeS
Quartz	SiO_2
Sepiolite	$Mg_2Si_3O_8 \cdot 2H_2O$
Siderite	$FeCO_3$
Sphalerite	ZnS
Talc	$Mg_3Si_4O_{10}(OH)_2$
Tenorite	CuO
Tremolite	$Ca_2Mg_5Si_8O_{22}(OH)_2$
Wollastonite	$CaSiO_3$
Wurtzite	ZnS

ACTIVITY DIAGRAMS

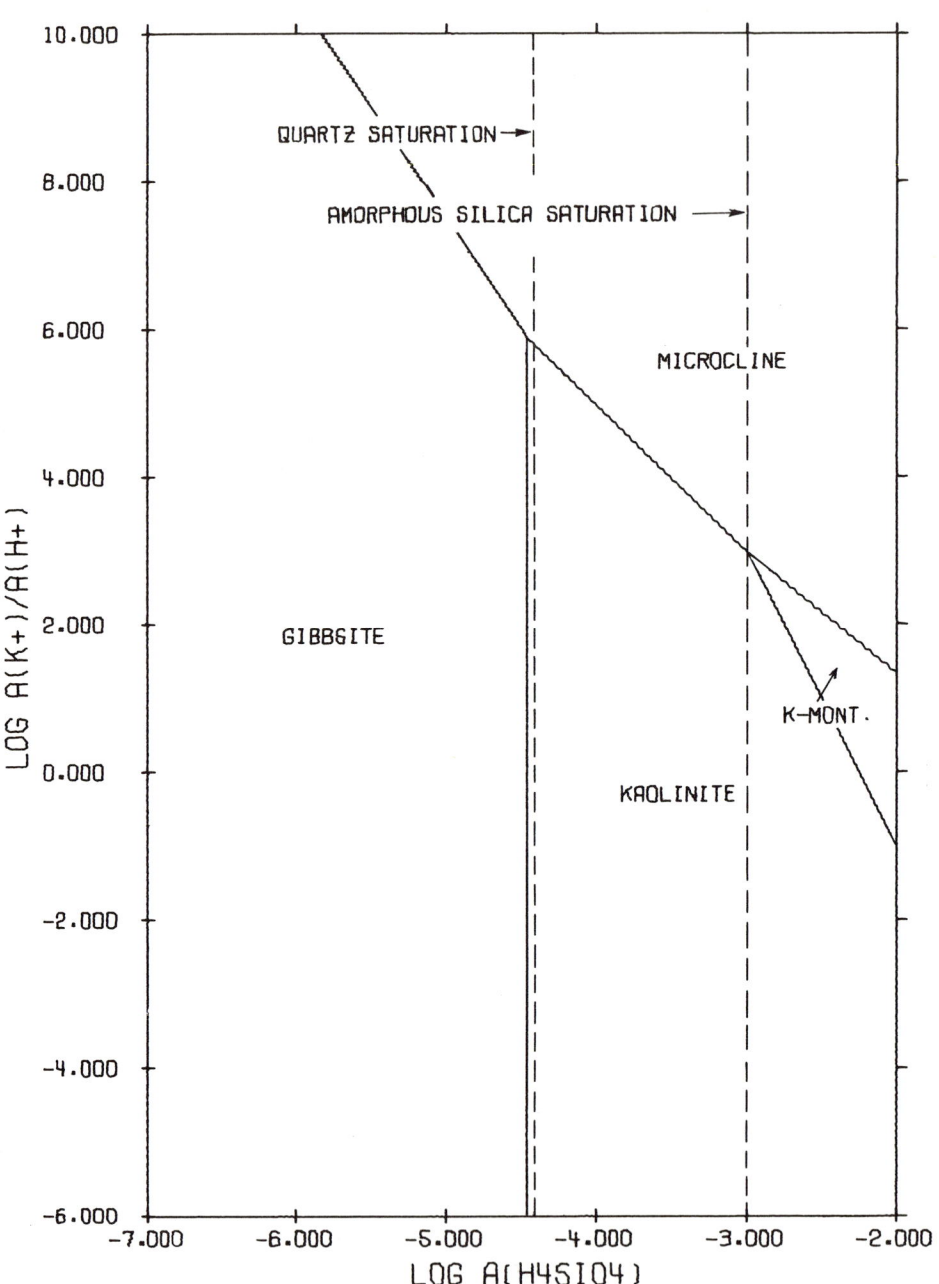

The System HCl—H₂O—Al₂O₃—K₂O—SiO₂ at 0°C.

Activity Diagrams

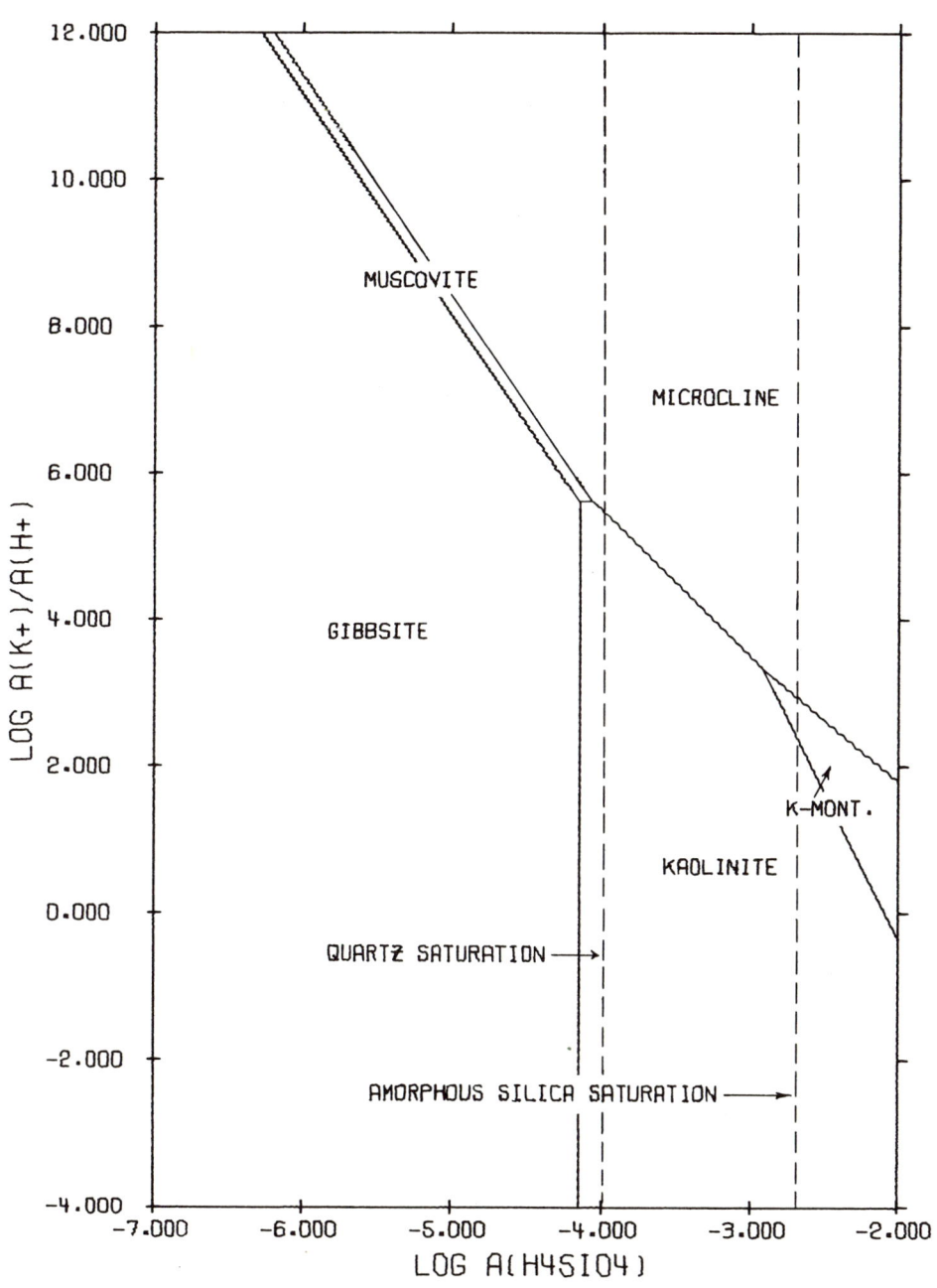

The System HCl—H$_2$O—Al$_2$O$_3$—K$_2$O—SiO$_2$ at 25°C.

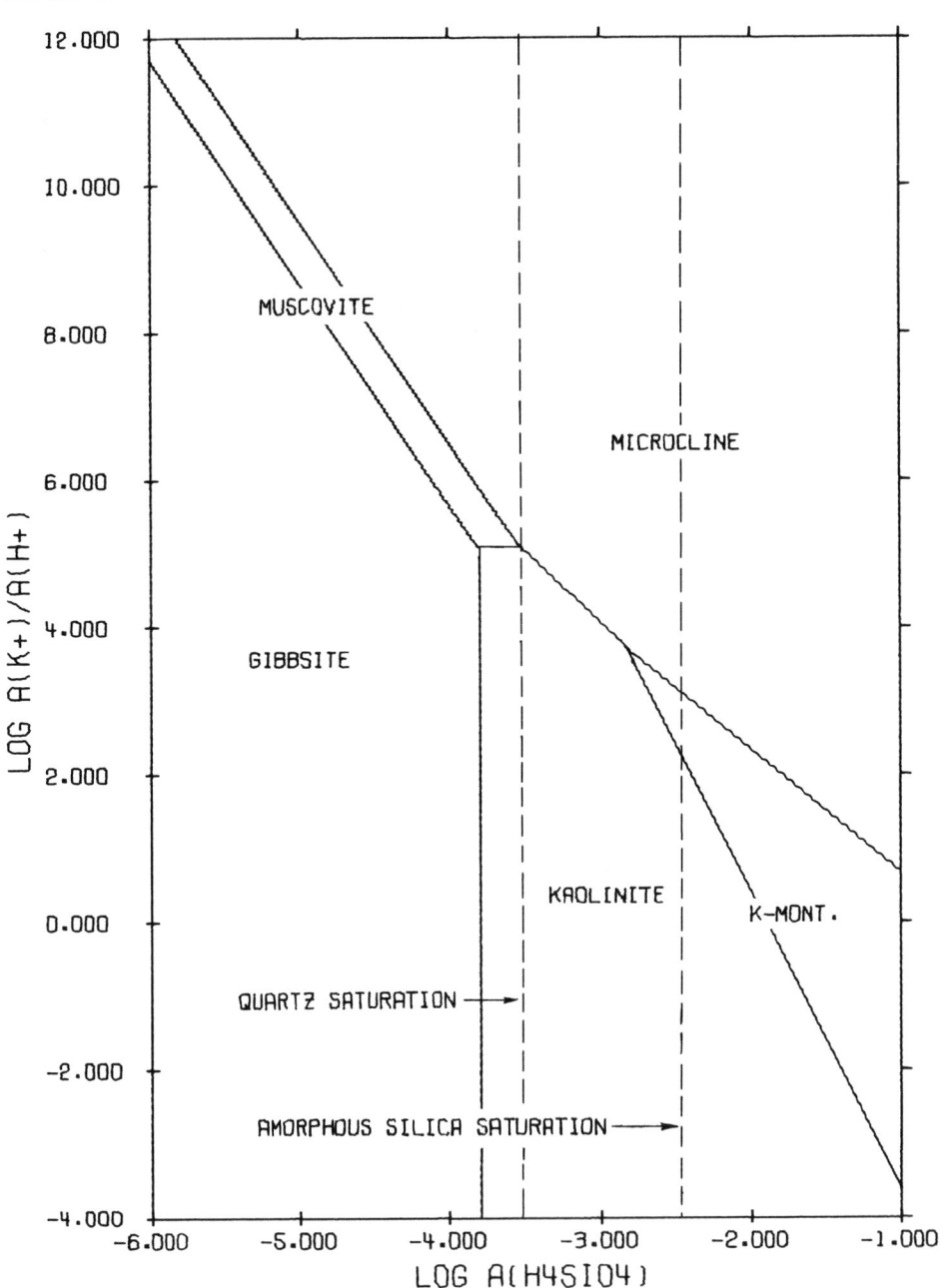

The System HCl—H_2O—Al_2O_3—K_2O—SiO_2 at 60°C.

Activity Diagrams

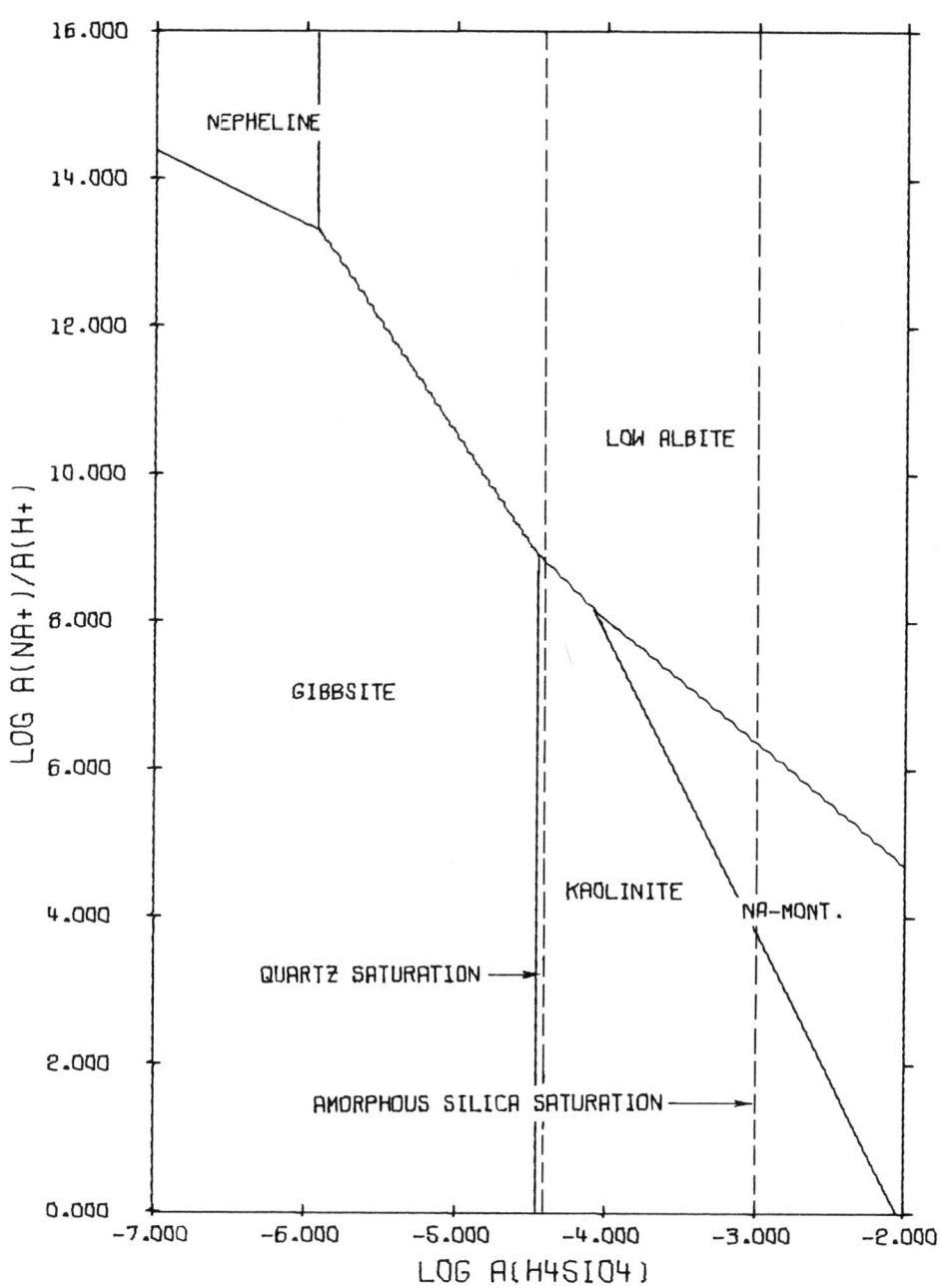

The System HCl—H₂O—Al₂O₃—Na₂O—SiO₂ at 0°C.

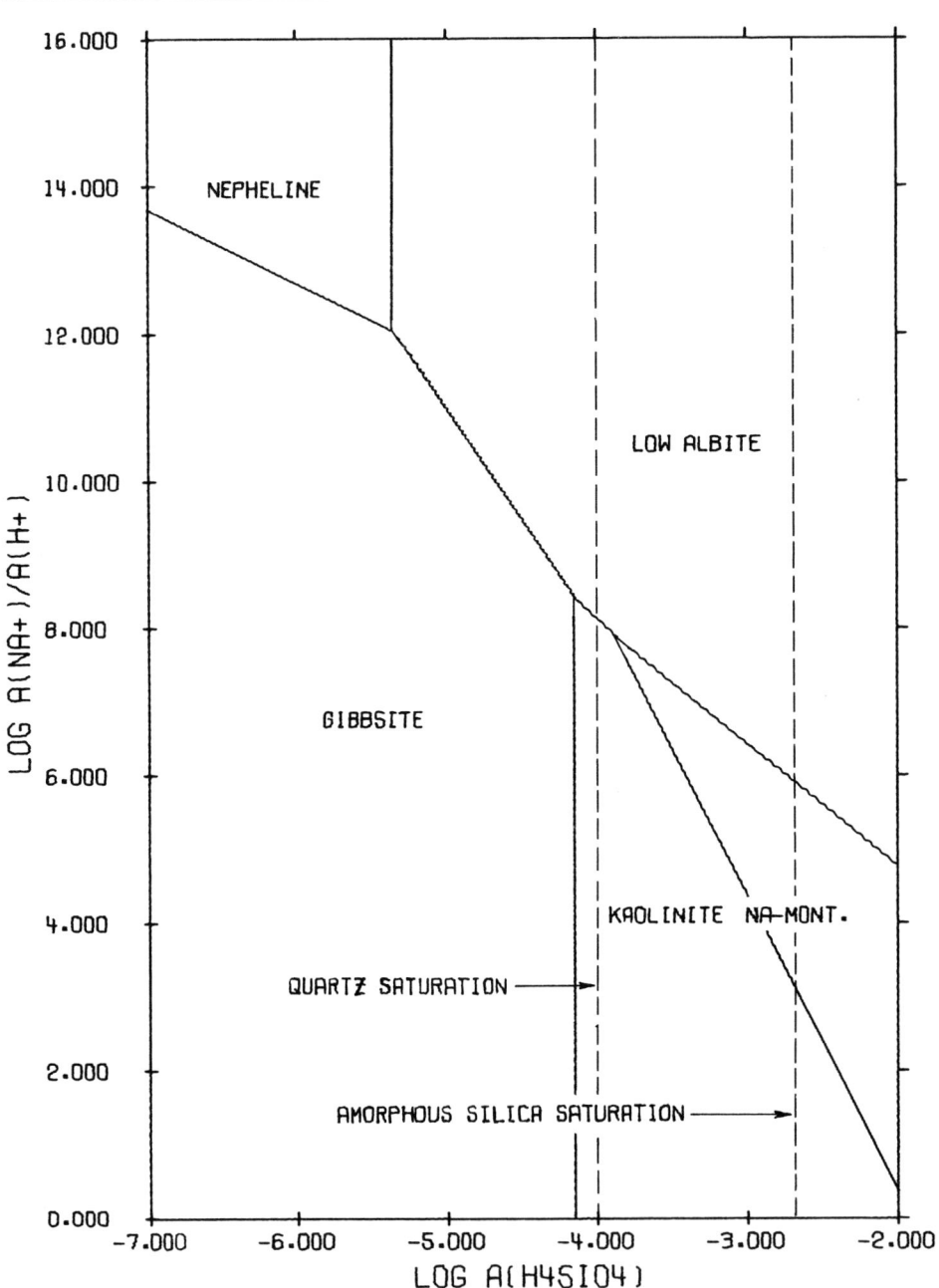

The System HCl—H$_2$O—Al$_2$O$_3$—Na$_2$O—SiO$_2$ at 25°C.

Activity Diagrams

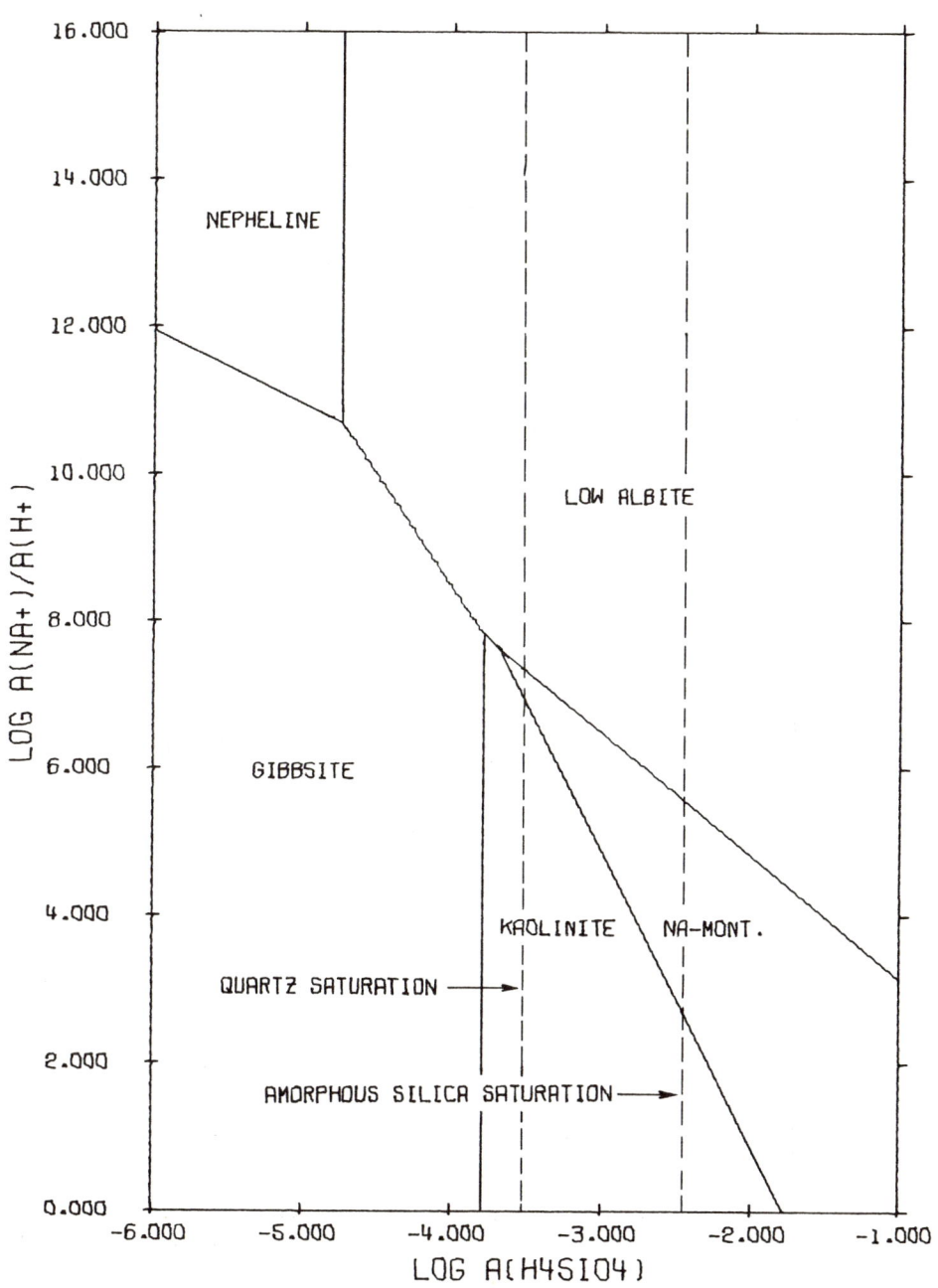

The System HCl—H$_2$O—Al$_2$O$_3$—Na$_2$O—SiO$_2$ at 60°C.

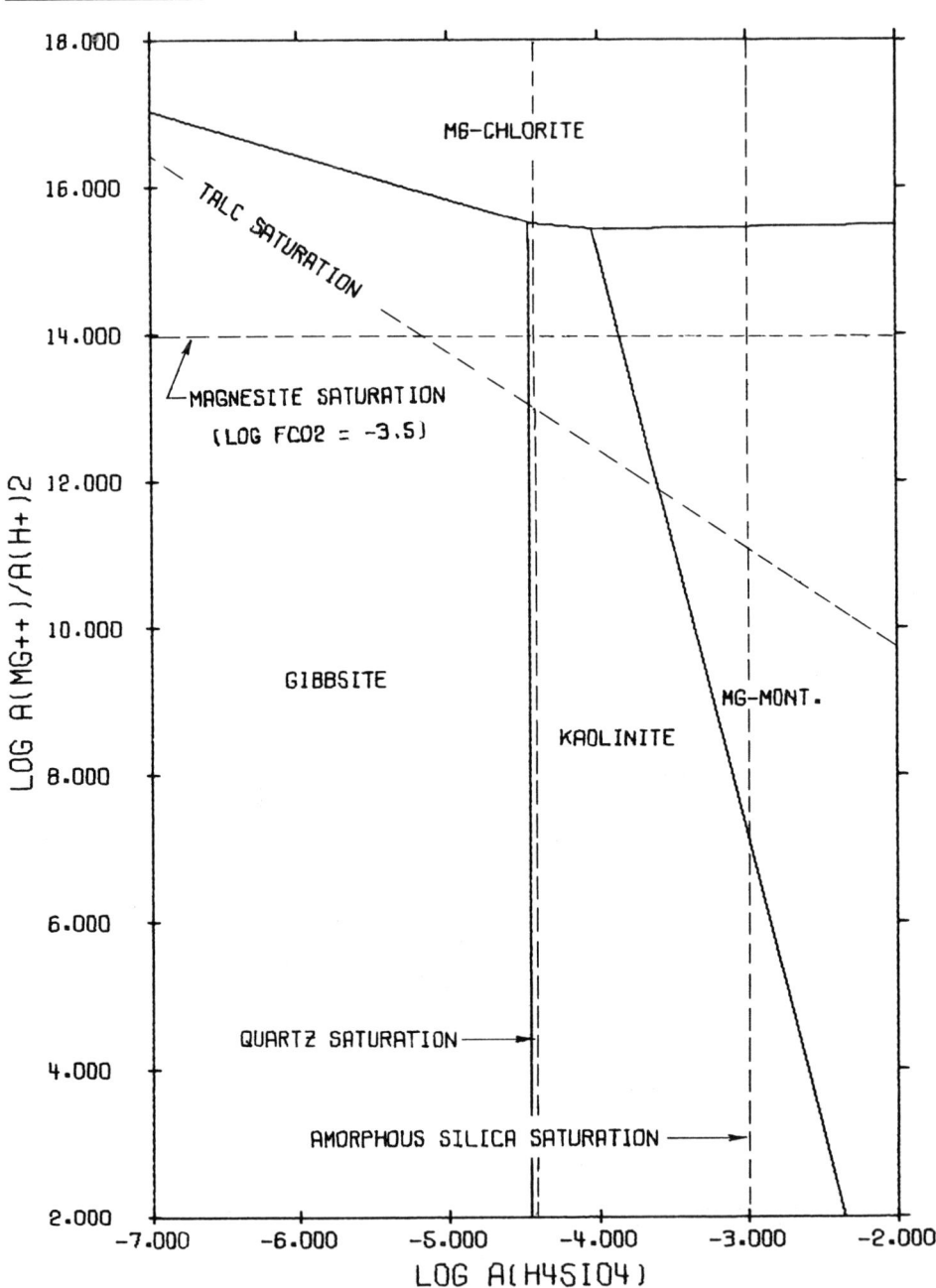

The System HCl—H$_2$O—Al$_2$O$_3$—CO$_2$—MgO—SiO$_2$ at 0°C.

Activity Diagrams

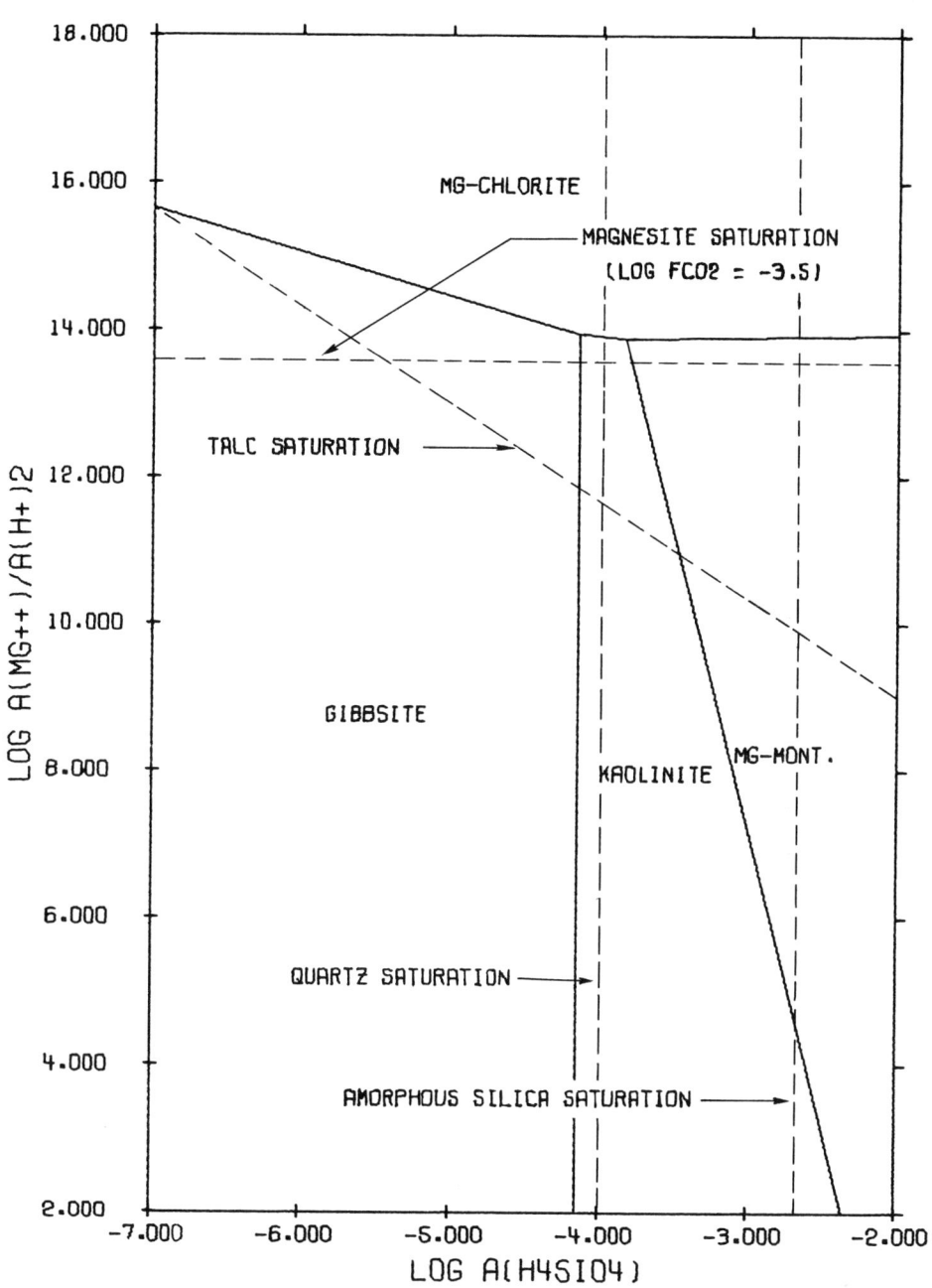

The System HCl—H$_2$O—Al$_2$O$_3$—CO$_2$—MgO—SiO$_2$ at 25°C.

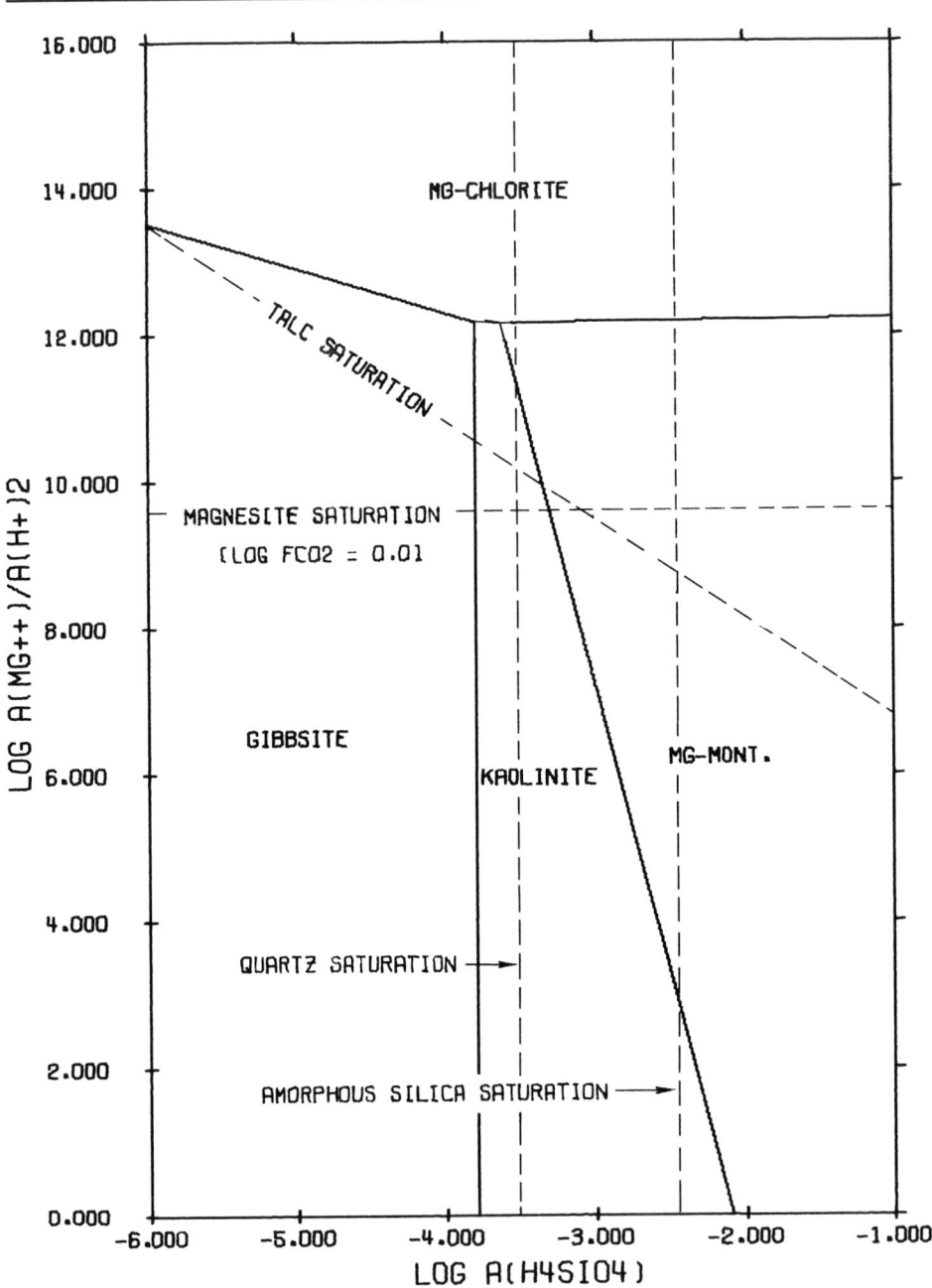

The System HCl—H$_2$O—Al$_2$O$_3$—CO$_2$—MgO—SiO$_2$ at 60°C.

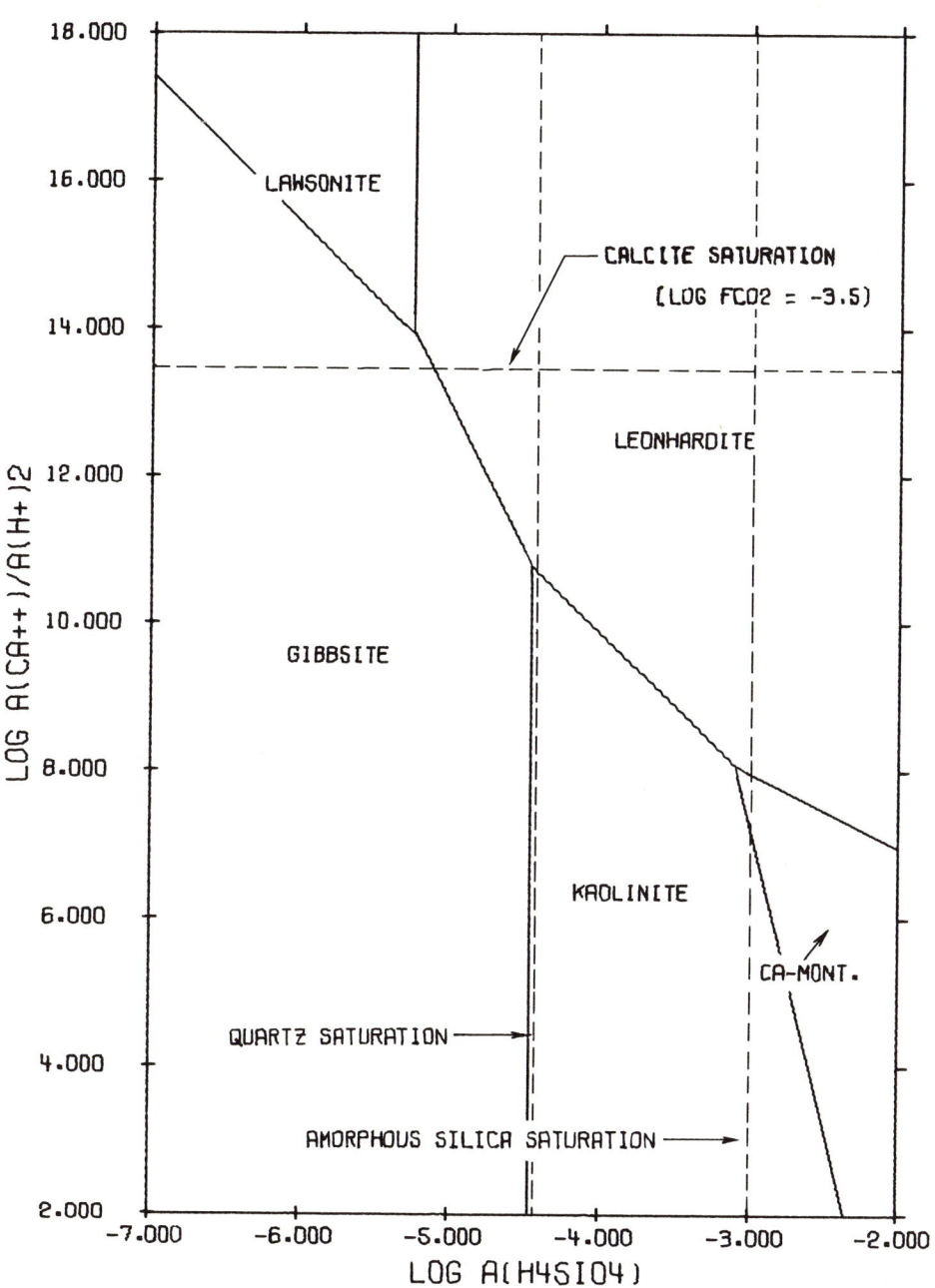

The System HCl—H$_2$O—Al$_2$O$_3$—CaO—CO$_2$—SiO$_2$ at 0°C.

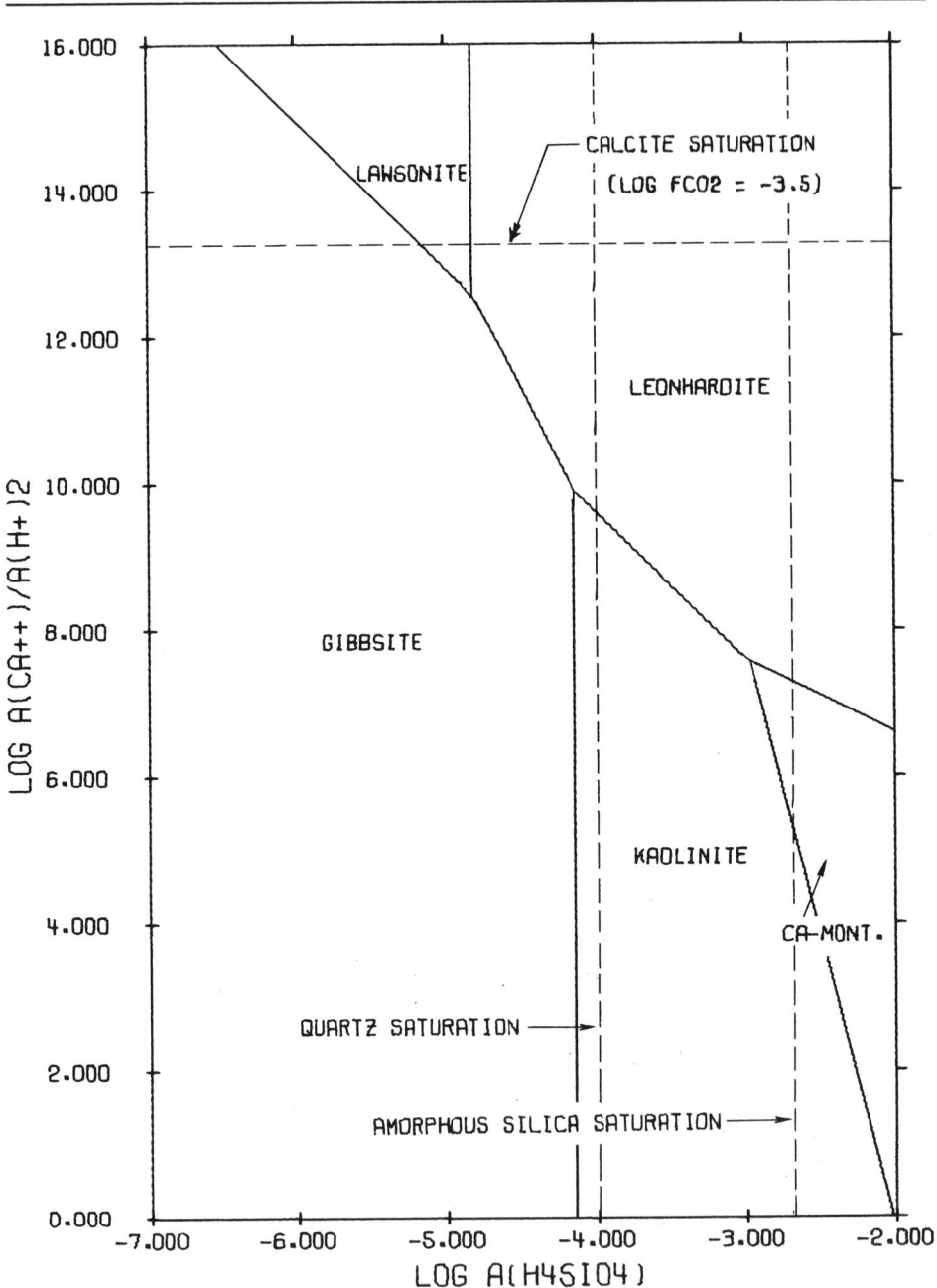

The System HCl—H$_2$O—Al$_2$O$_3$—CaO—CO$_2$—SiO$_2$ at 25°C.

Activity Diagrams

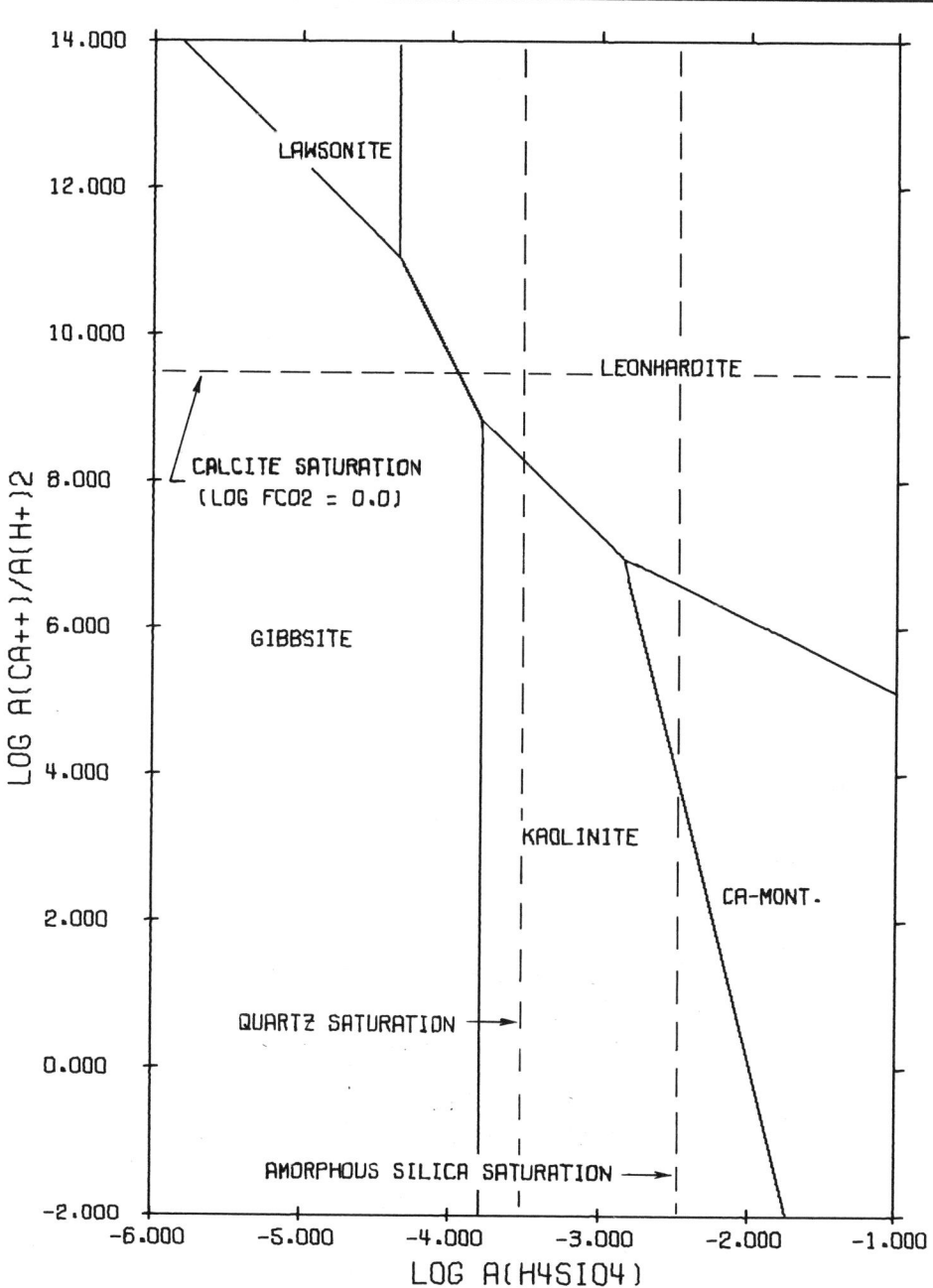

The System HCl—H$_2$O—Al$_2$O$_3$—CaO—CO$_2$—SiO$_2$ at 60°C.

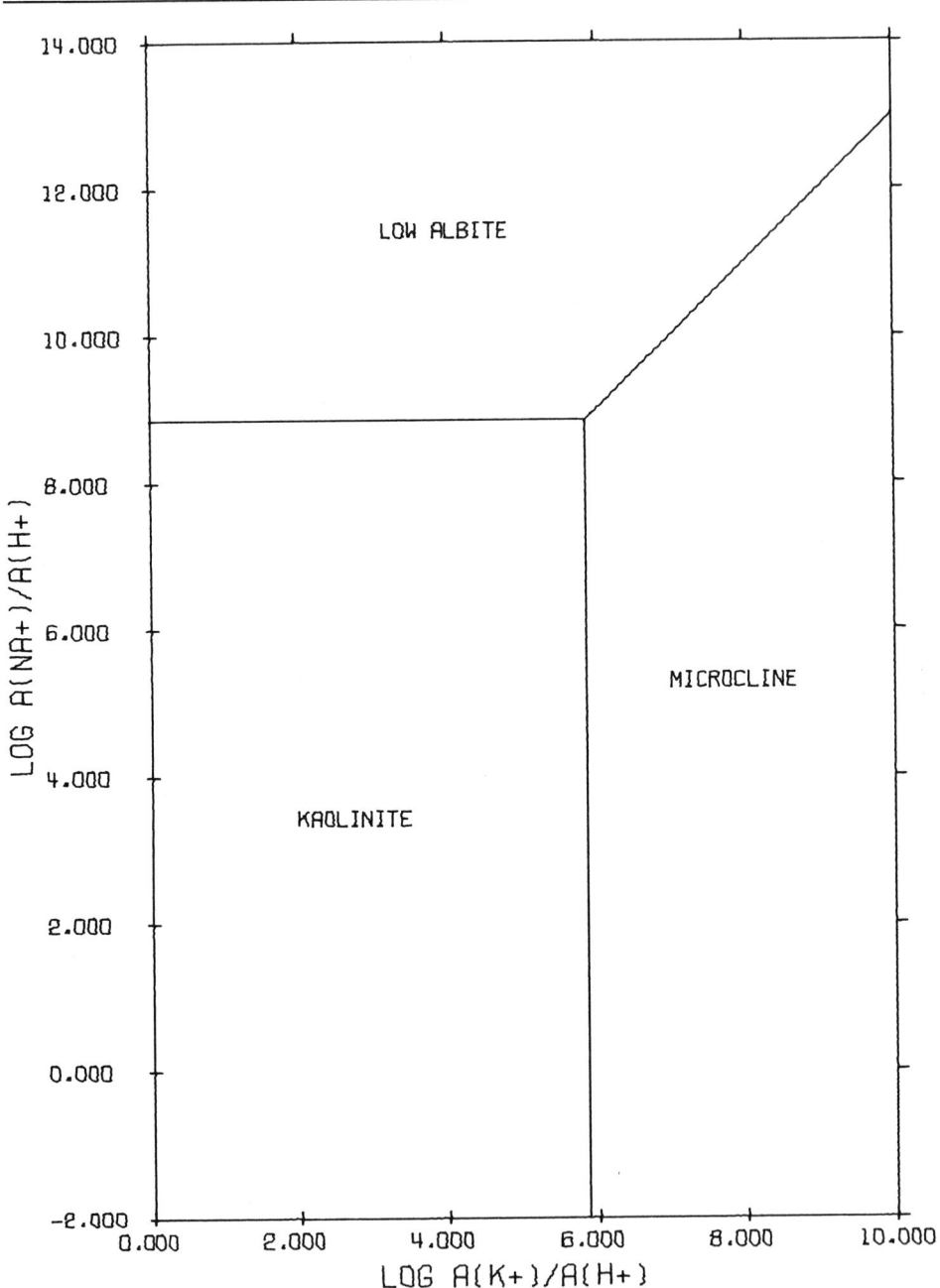

The System HCl—H_2O—Al_2O_3—K_2O—Na_2O—SiO_2 at 0°C; $\log a_{H_4SiO_4} = -4.44$ = quartz saturation.

Activity Diagrams

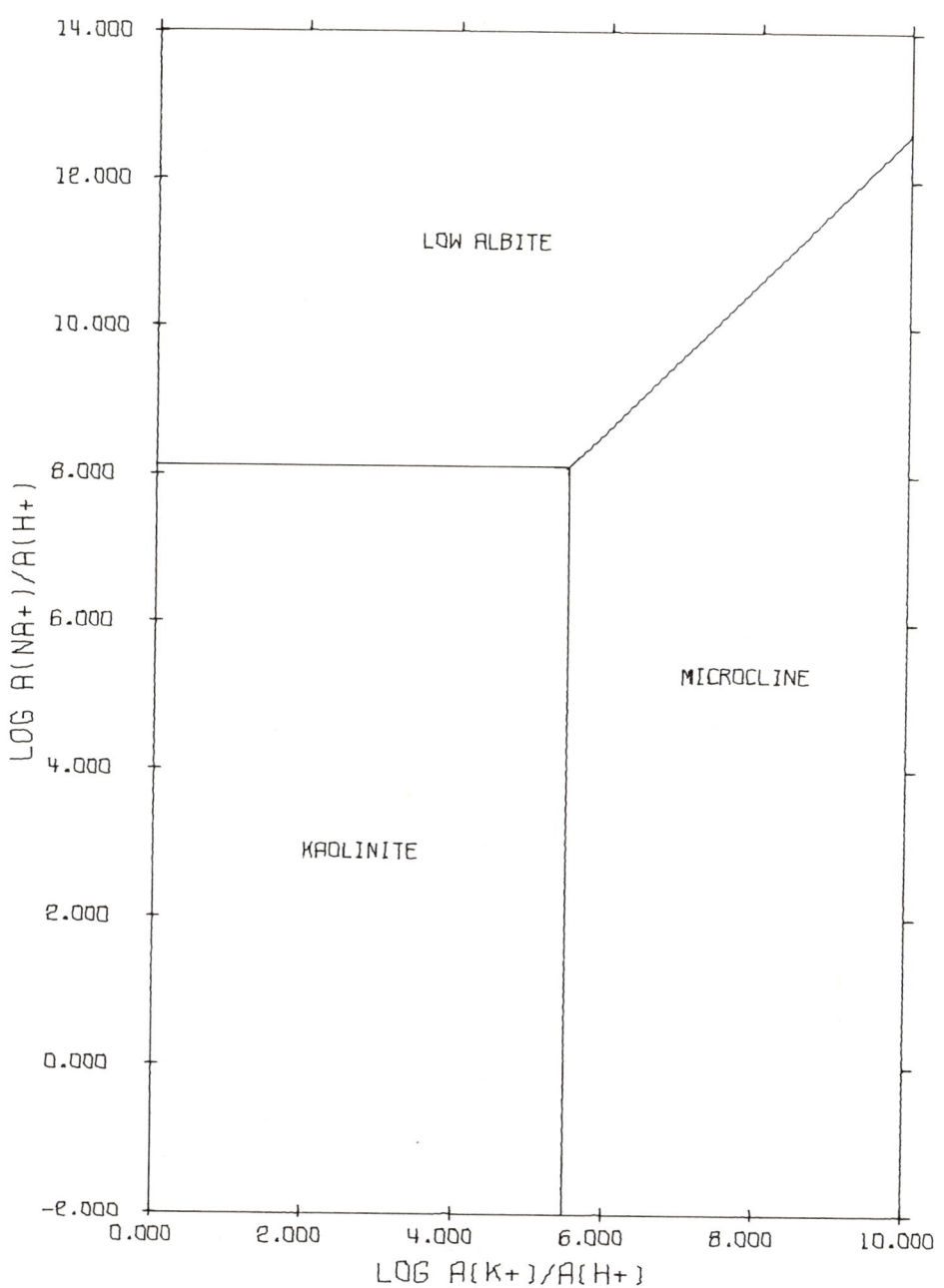

The System HCl—H$_2$O—Al$_2$O$_3$—K$_2$O—Na$_2$O—SiO$_2$ at 25°C; log $a_{H_4SiO_4}$ = −4.00 = quartz saturation.

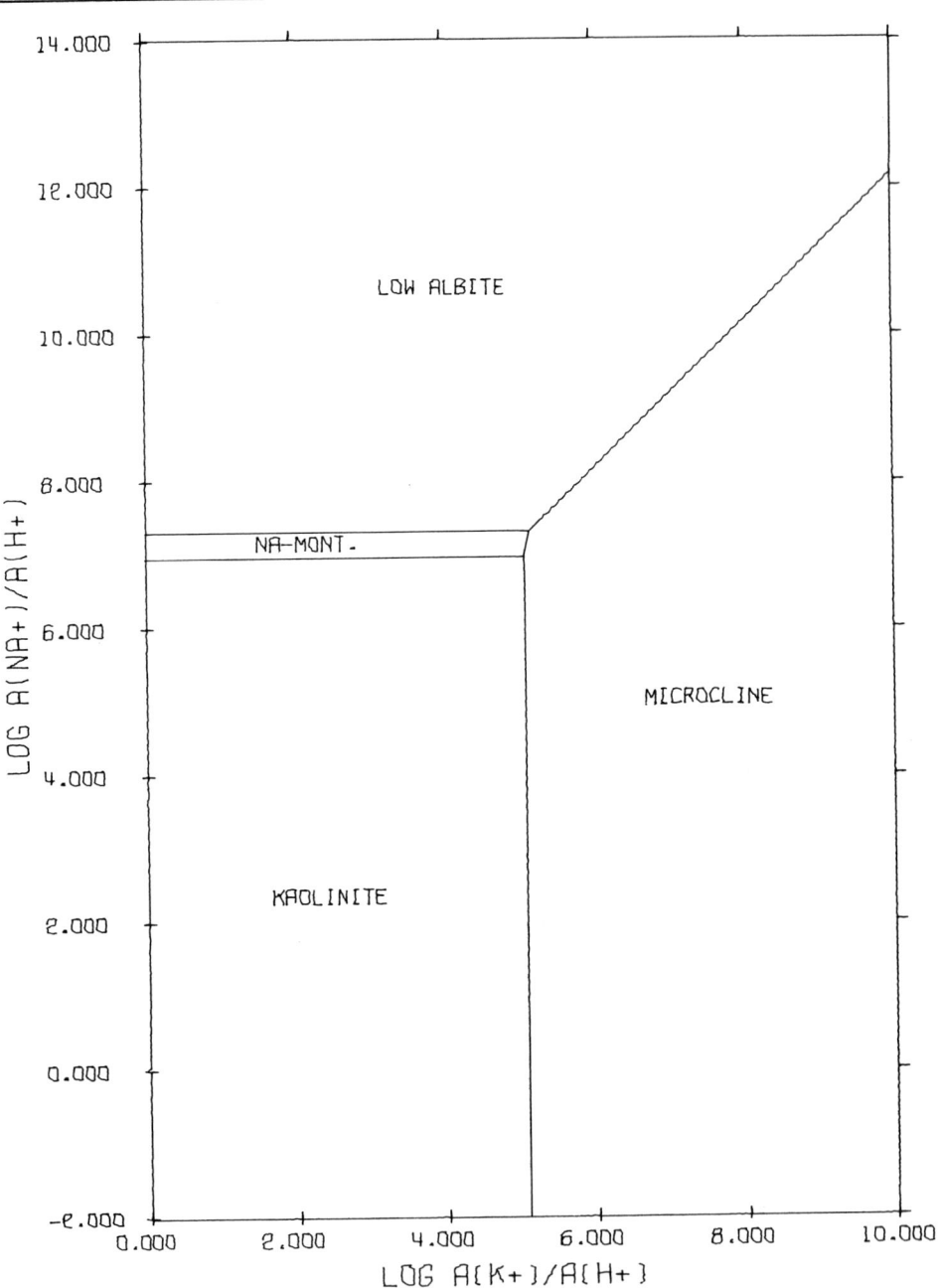

The System HCl—H_2O—Al_2O_3—K_2O—Na_2O—SiO_2 at 60°C; log $a_{H_4SiO_4}$ = −3.52 = quartz saturation.

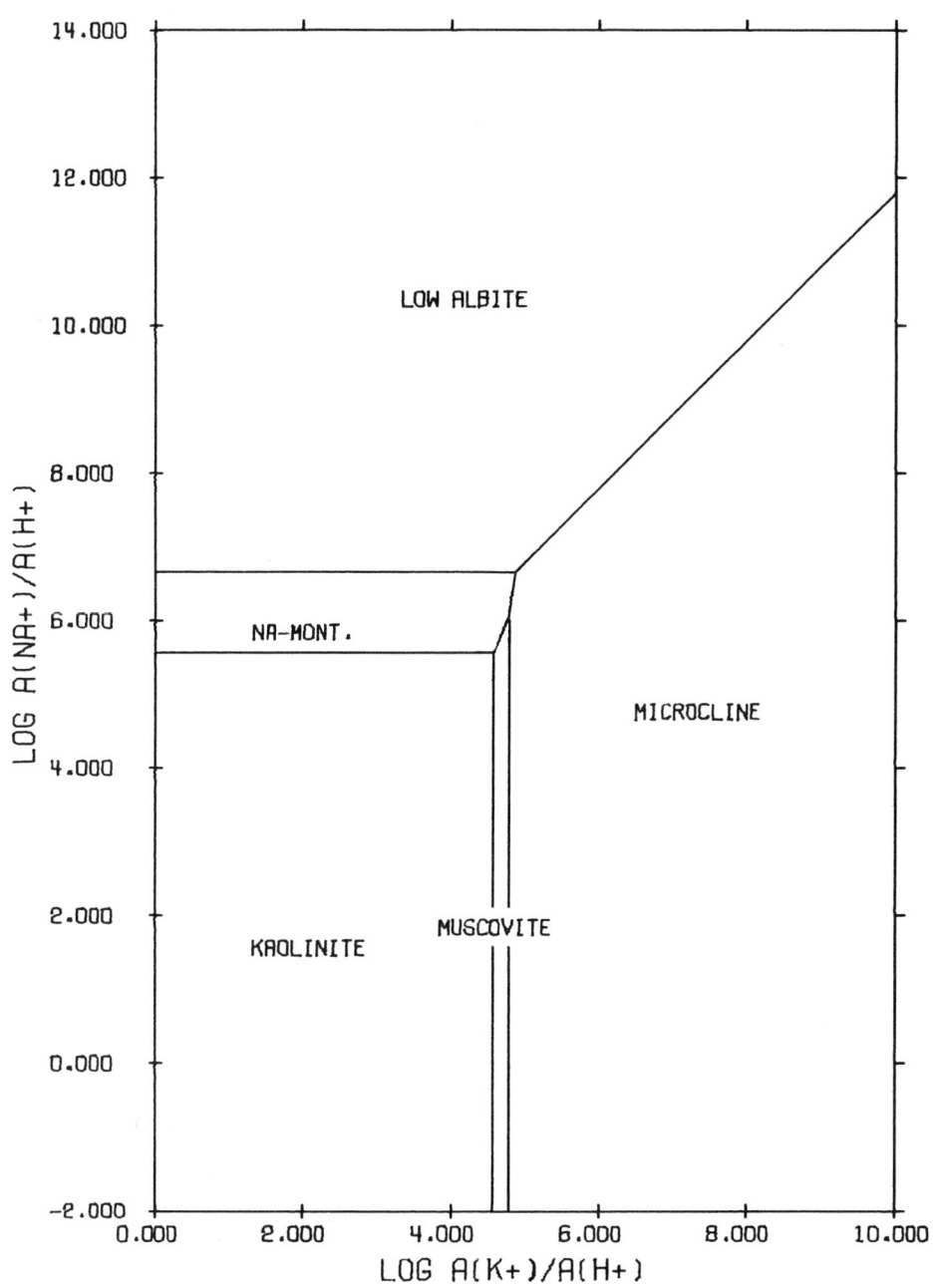

The System HCl—H$_2$O—Al$_2$O$_3$—K$_2$O—Na$_2$O—SiO$_2$ at 100°C; log $a_{\text{H}_4\text{SiO}_4}$ = −3.08 = quartz saturation.

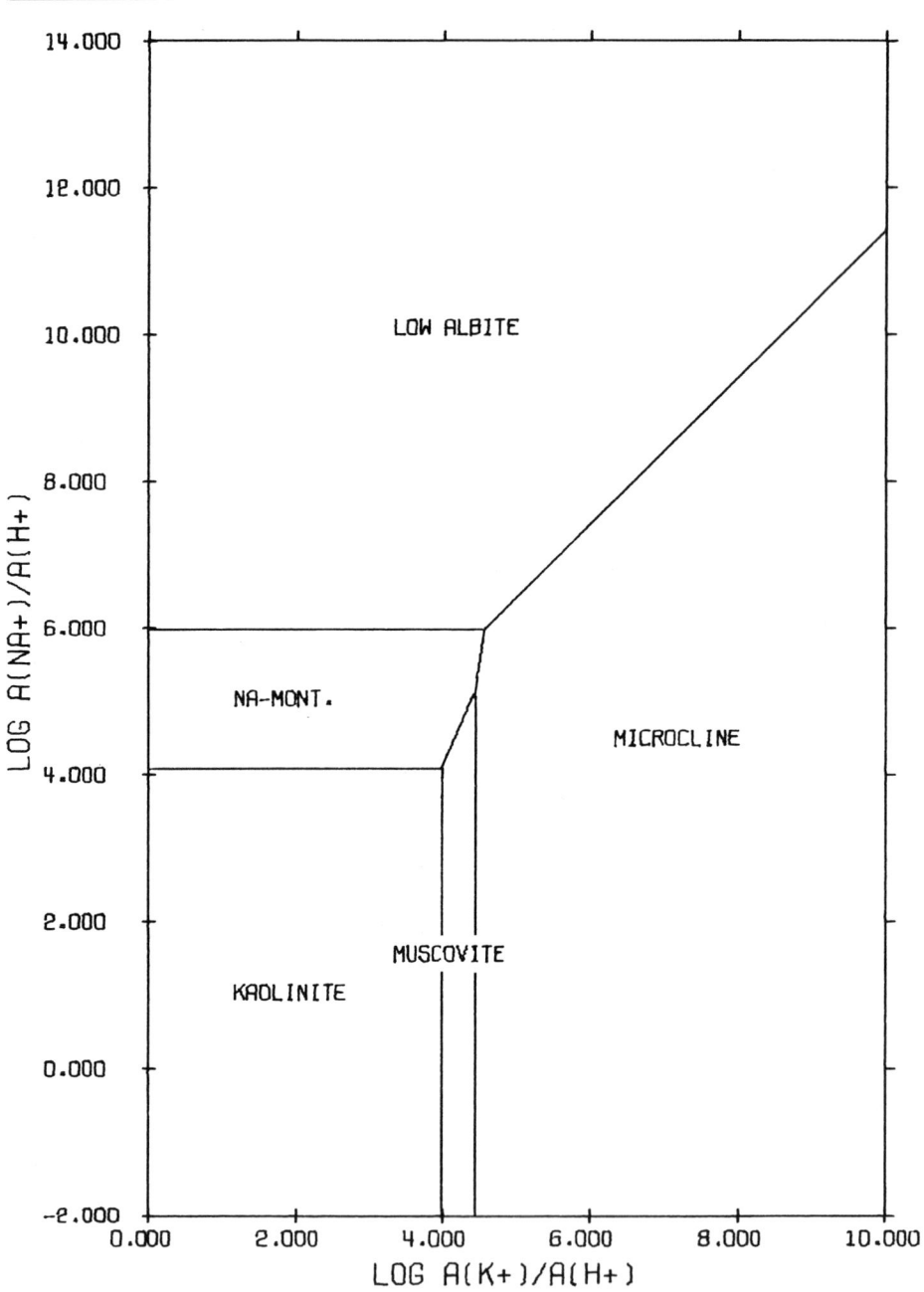

The System HCl—H$_2$O—Al$_2$O$_3$—K$_2$O—Na$_2$O—SiO$_2$ at 150°C; log $a_{H_4SiO_4}$ = −2.67 = quartz saturation.

Activity Diagrams

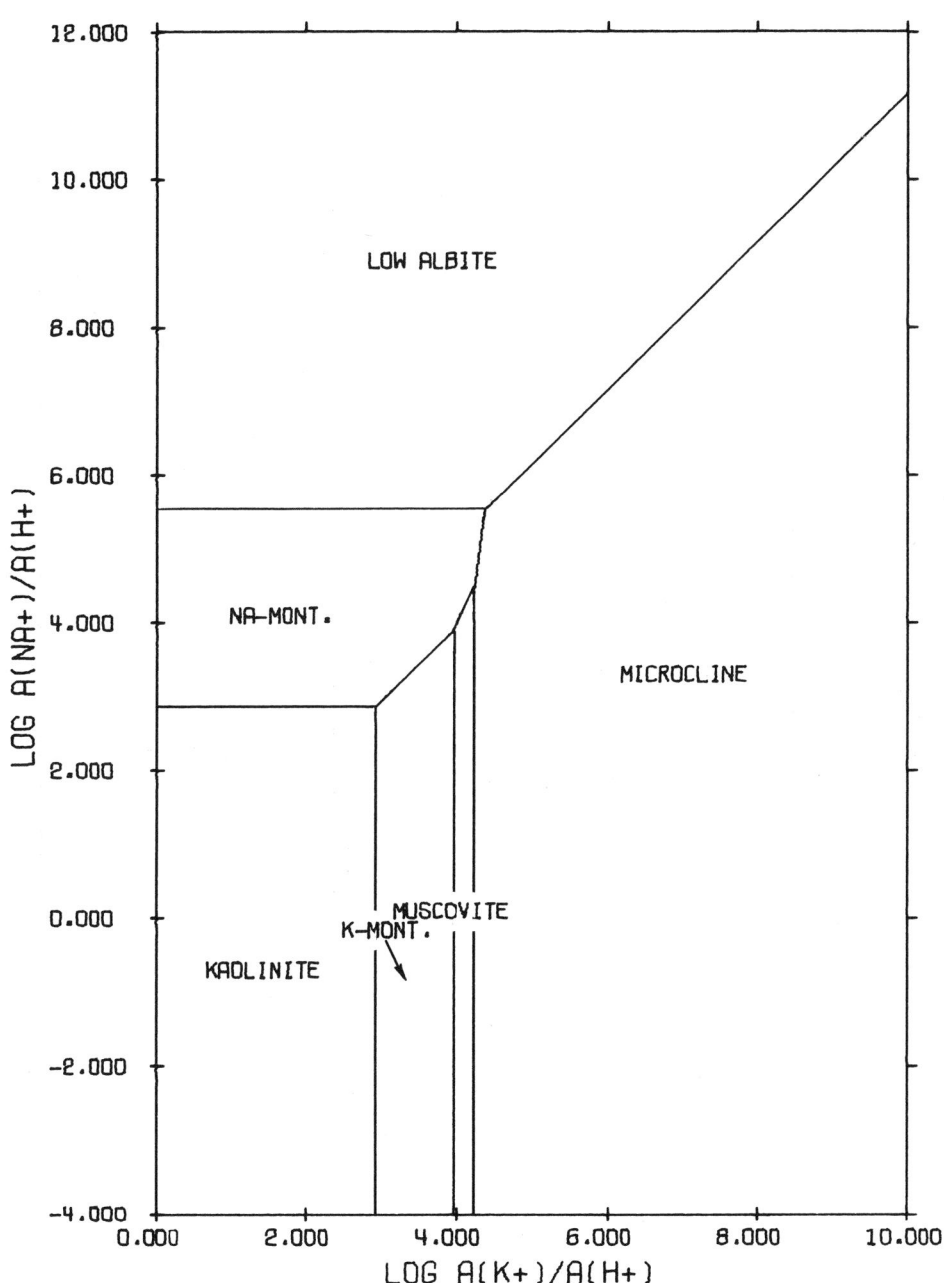

The System HCl—H_2O—Al_2O_3—K_2O—Na_2O—SiO_2 at 200°C; $\log a_{H_4SiO_4} = -2.35$ = quartz saturation.

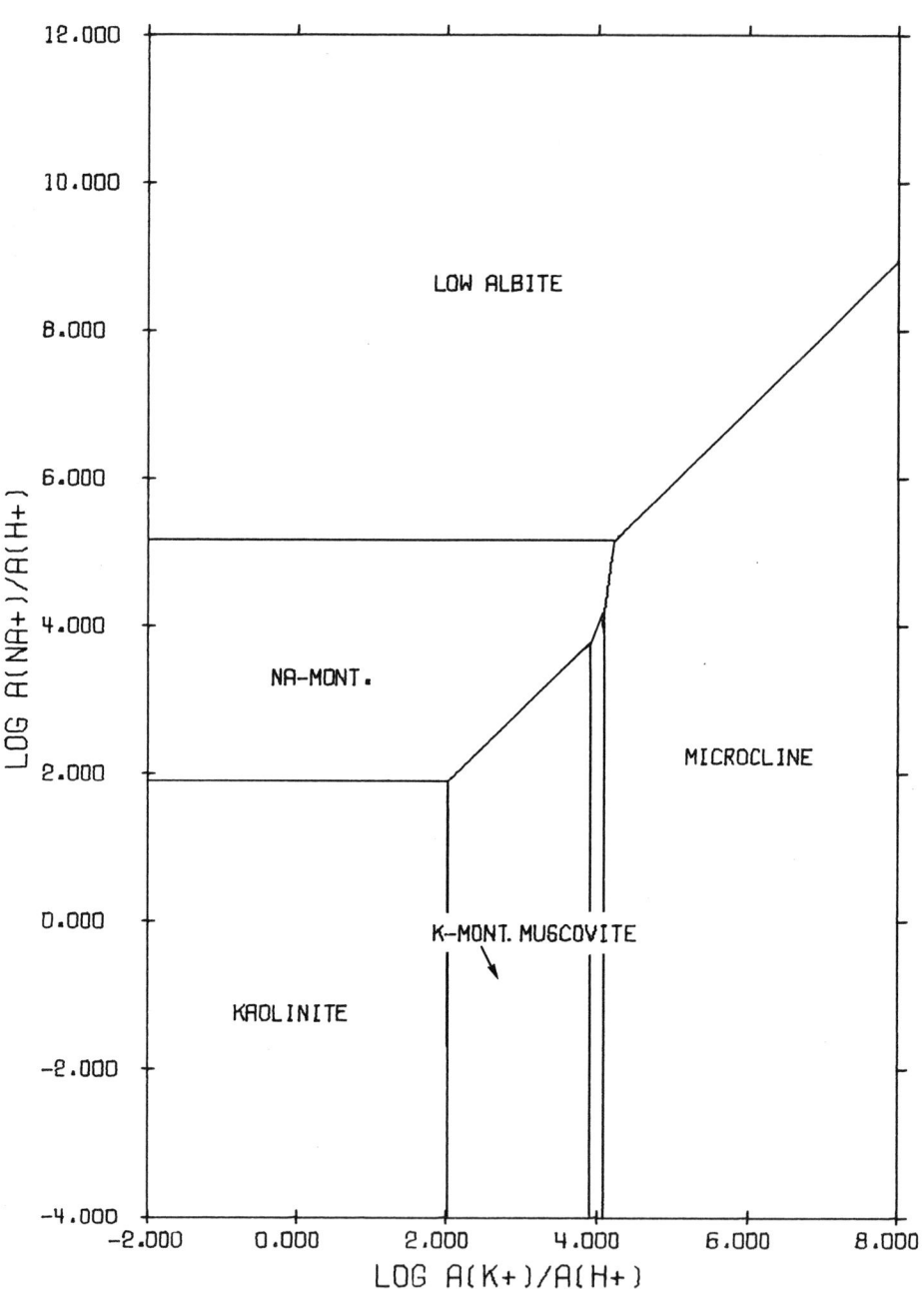

The System HCl—H$_2$O—Al$_2$O$_3$—K$_2$O—Na$_2$O—SiO$_2$ at 250°C; log $a_{H_4SiO_4}$ = -2.11 = quartz saturation.

Activity Diagrams

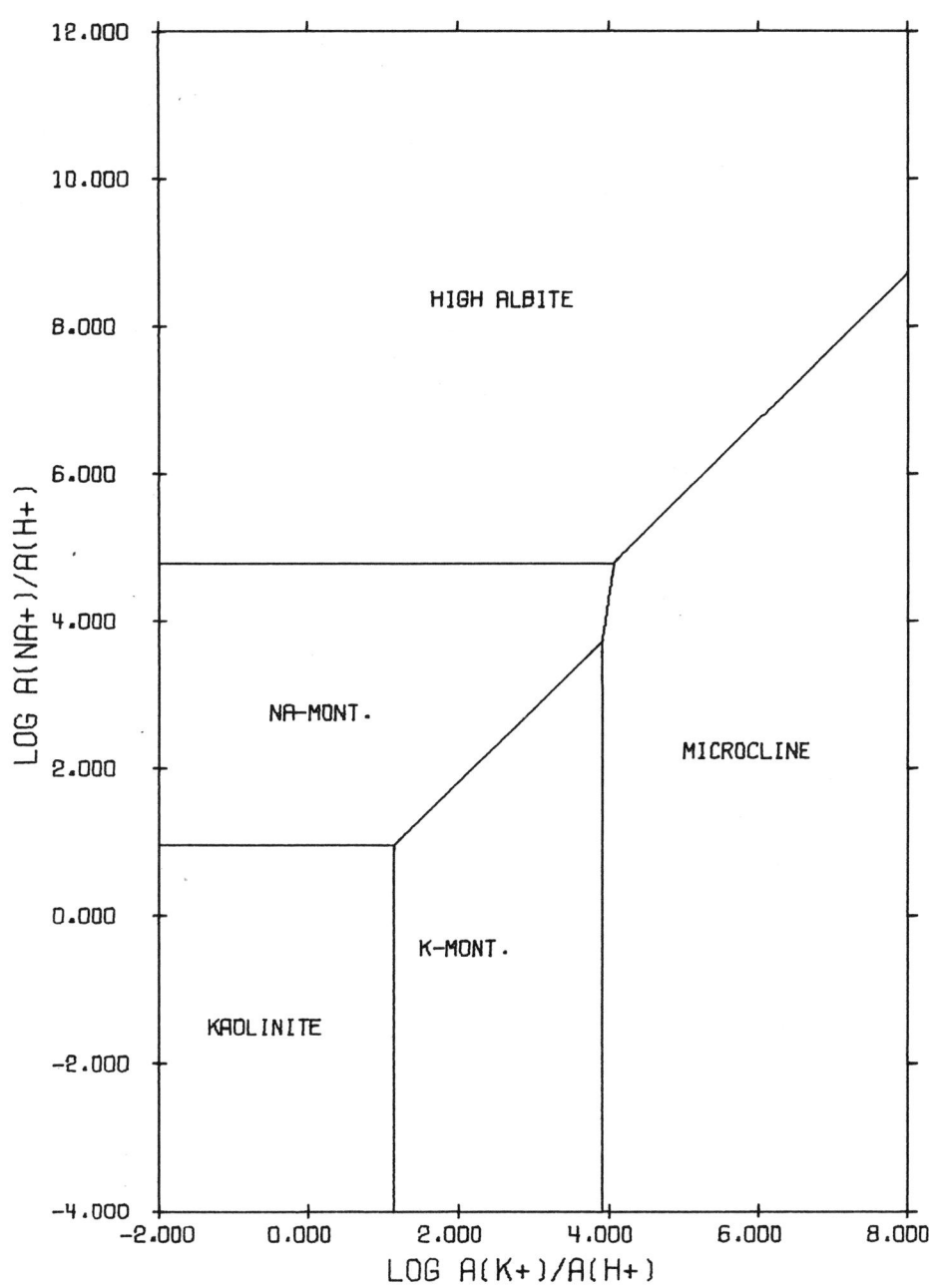

The System HCl—H$_2$O—Al$_2$O$_3$—K$_2$O—Na$_2$O—SiO$_2$ at 300°C; log $a_{H_4SiO_4}$ = −1.94 = quartz saturation.

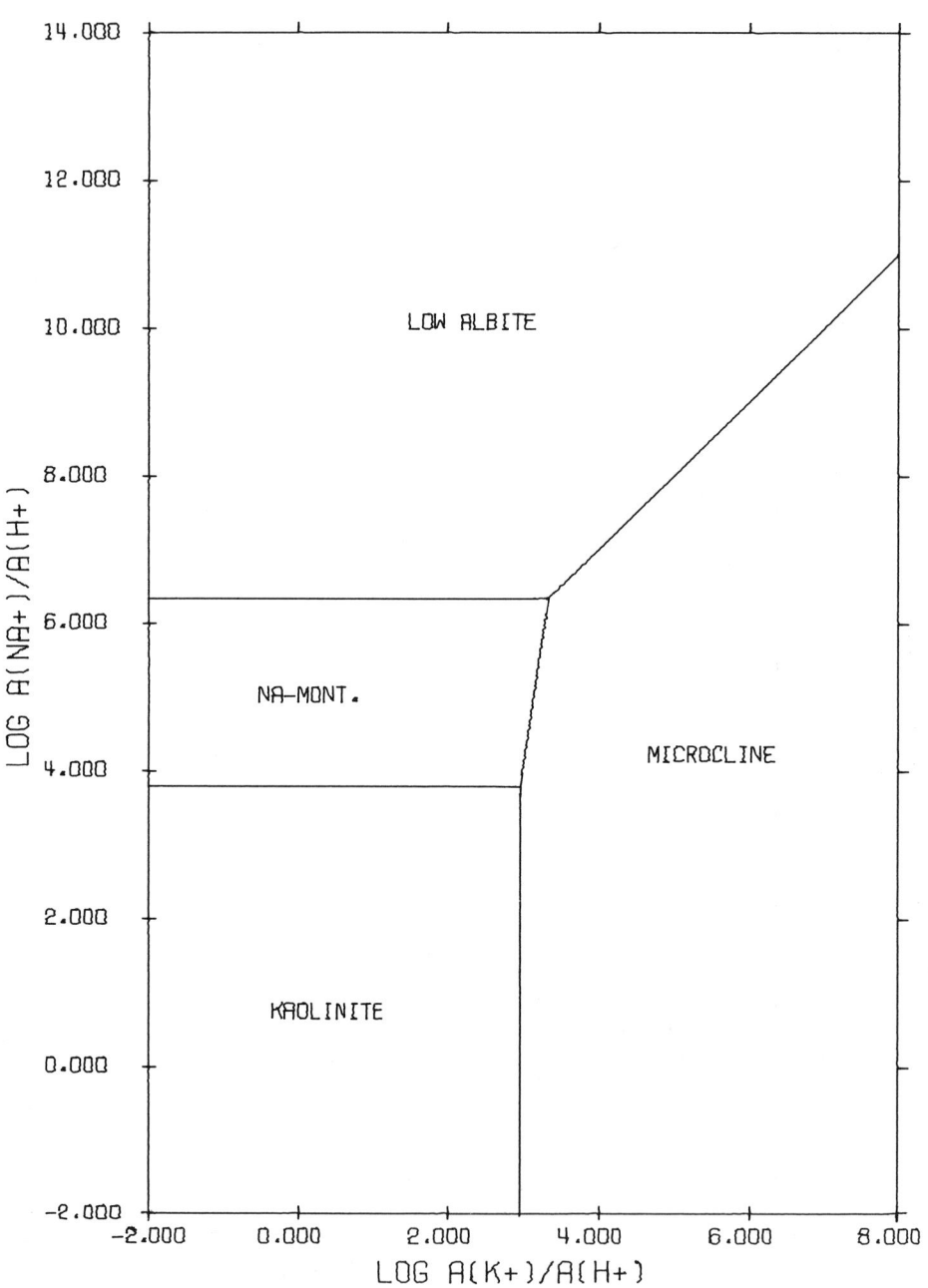

The System HCl—H$_2$O—Al$_2$O$_3$—K$_2$O—Na$_2$O—SiO$_2$ at 0°C; log $a_{H_4SiO_4}$ = −3.00 = amorphous silica saturation.

Activity Diagrams

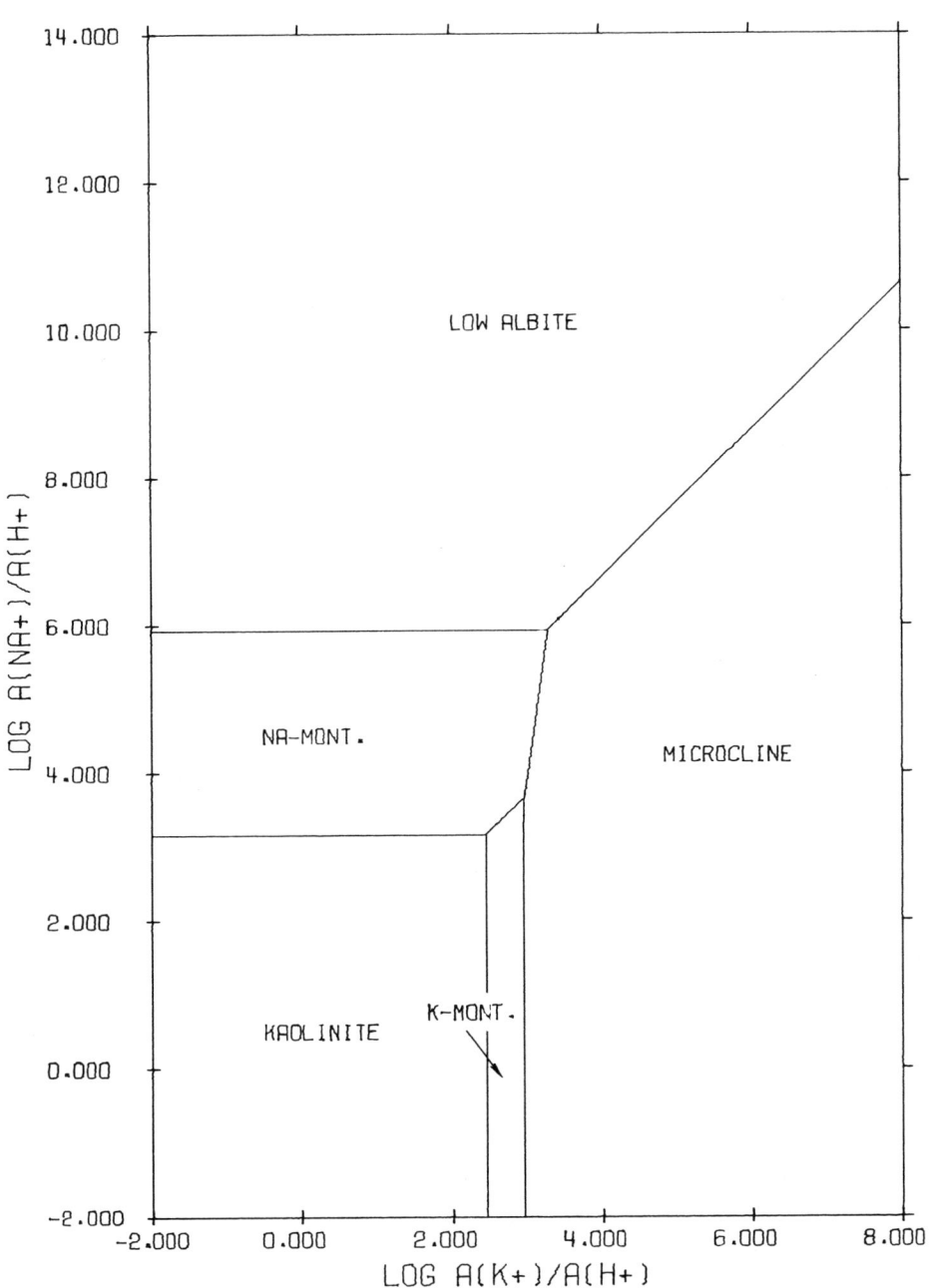

The System HCl—H$_2$O—Al$_2$O$_3$—K$_2$O—Na$_2$O—SiO$_2$ at 25°C; log $a_{H_4SiO_4}$ = −2.70 = amorphous silica saturation.

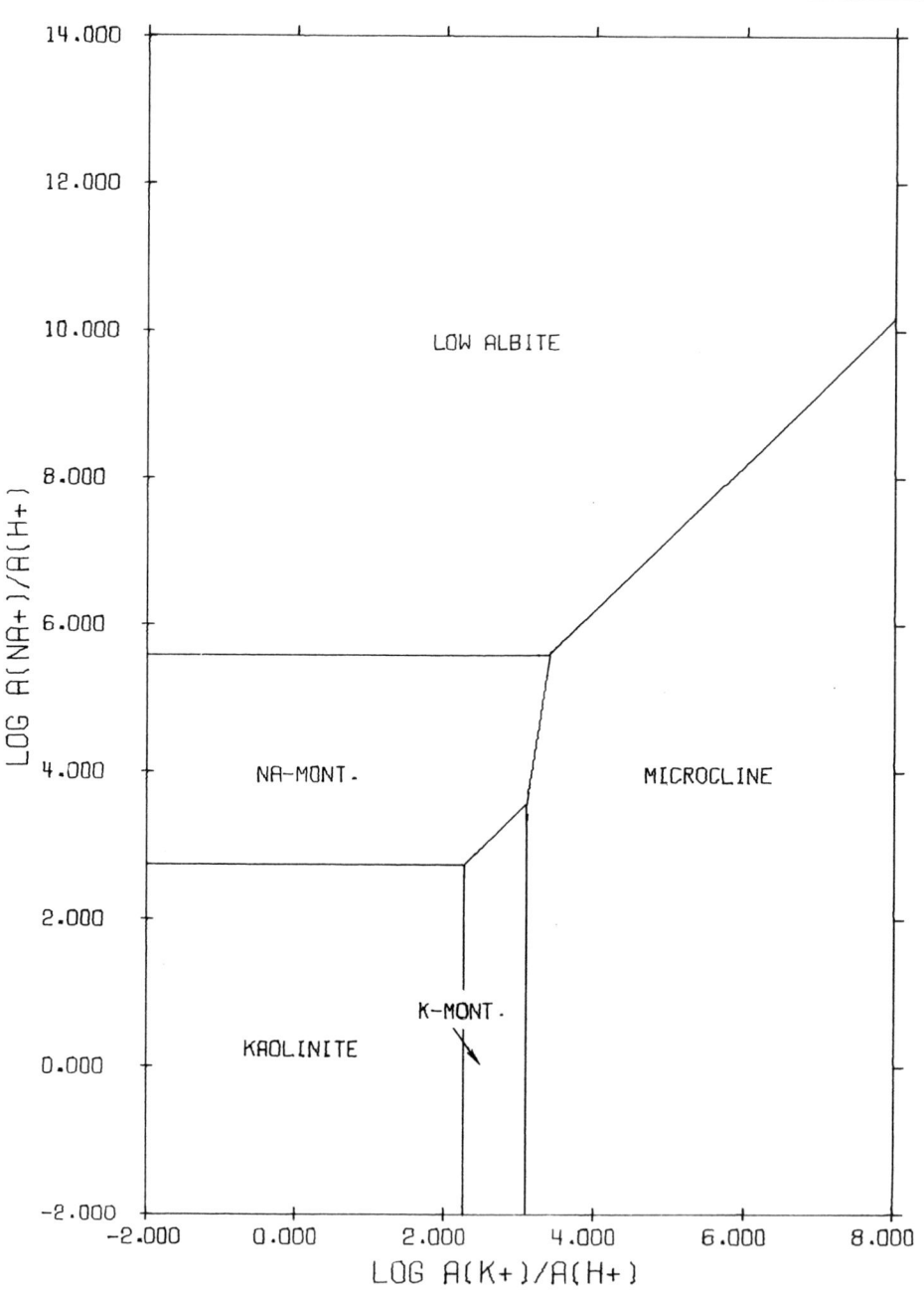

The System HCl—H$_2$O—Al$_2$O$_3$—K$_2$O—Na$_2$O—SiO$_2$ at 60°C; log $a_{H_4SiO_4}$ = −2.47 = amorphous silica saturation.

Activity Diagrams

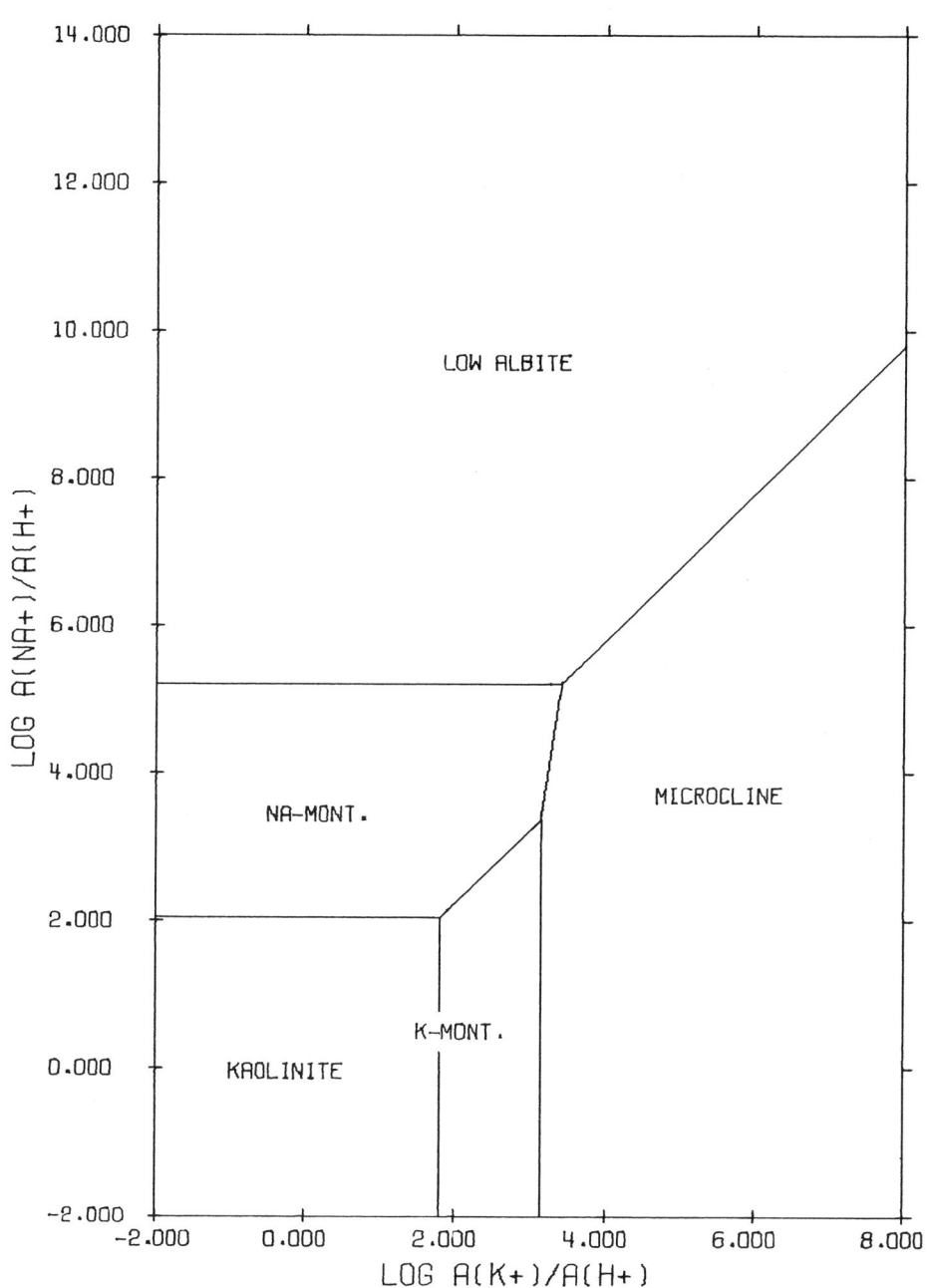

The System HCl—H$_2$O—Al$_2$O$_3$—K$_2$O—Na$_2$O—SiO$_2$ at 100°C; log $a_{H_4SiO_4}$ = -2.20 = amorphous silica saturation.

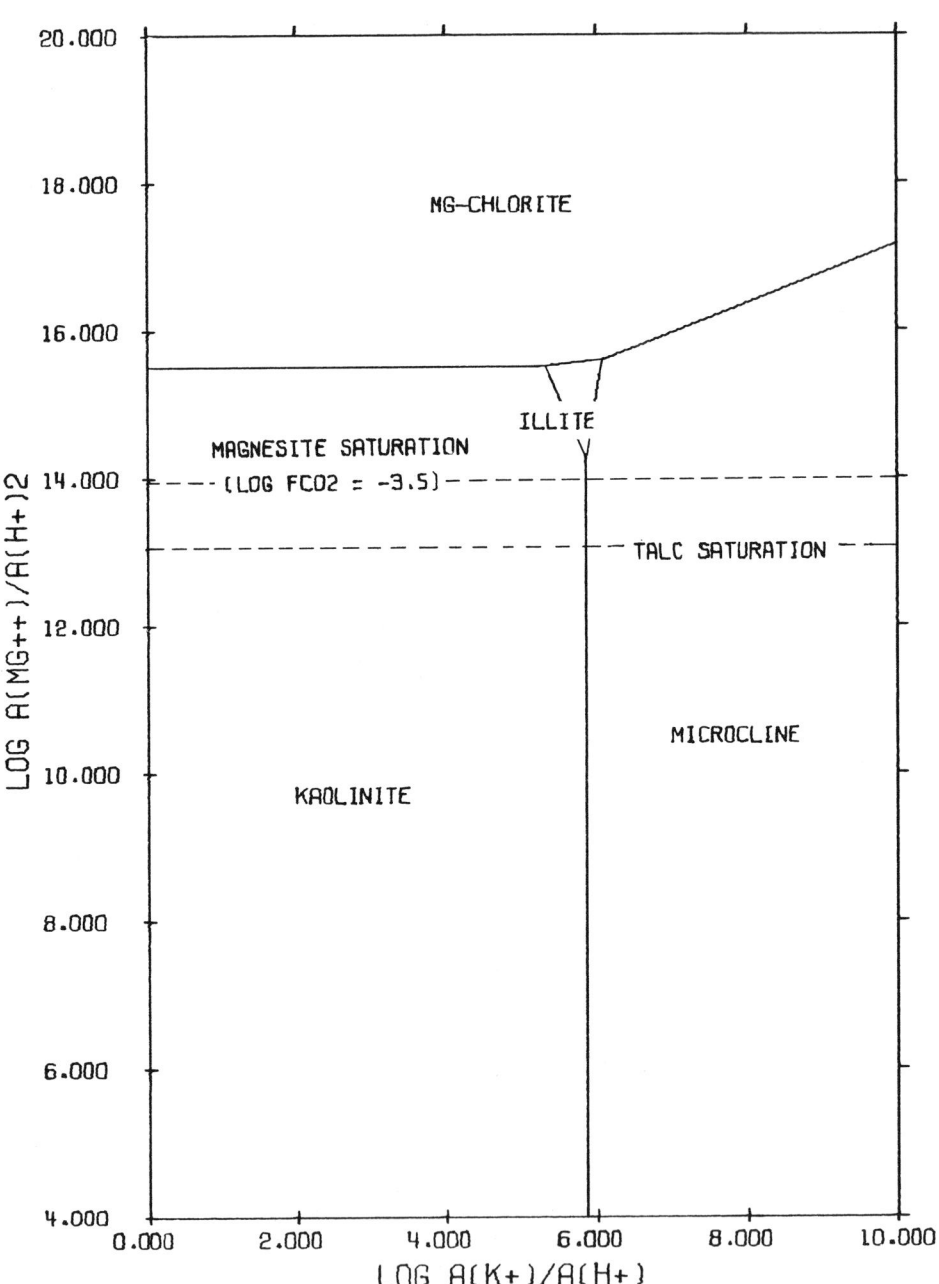

The System HCl—H$_2$O—Al$_2$O$_3$—CO$_2$—K$_2$O—MgO—SiO$_2$ at 0°C; log $a_{\text{H}_4\text{SiO}_4}$ = −4.44 = quartz saturation.

Activity Diagrams

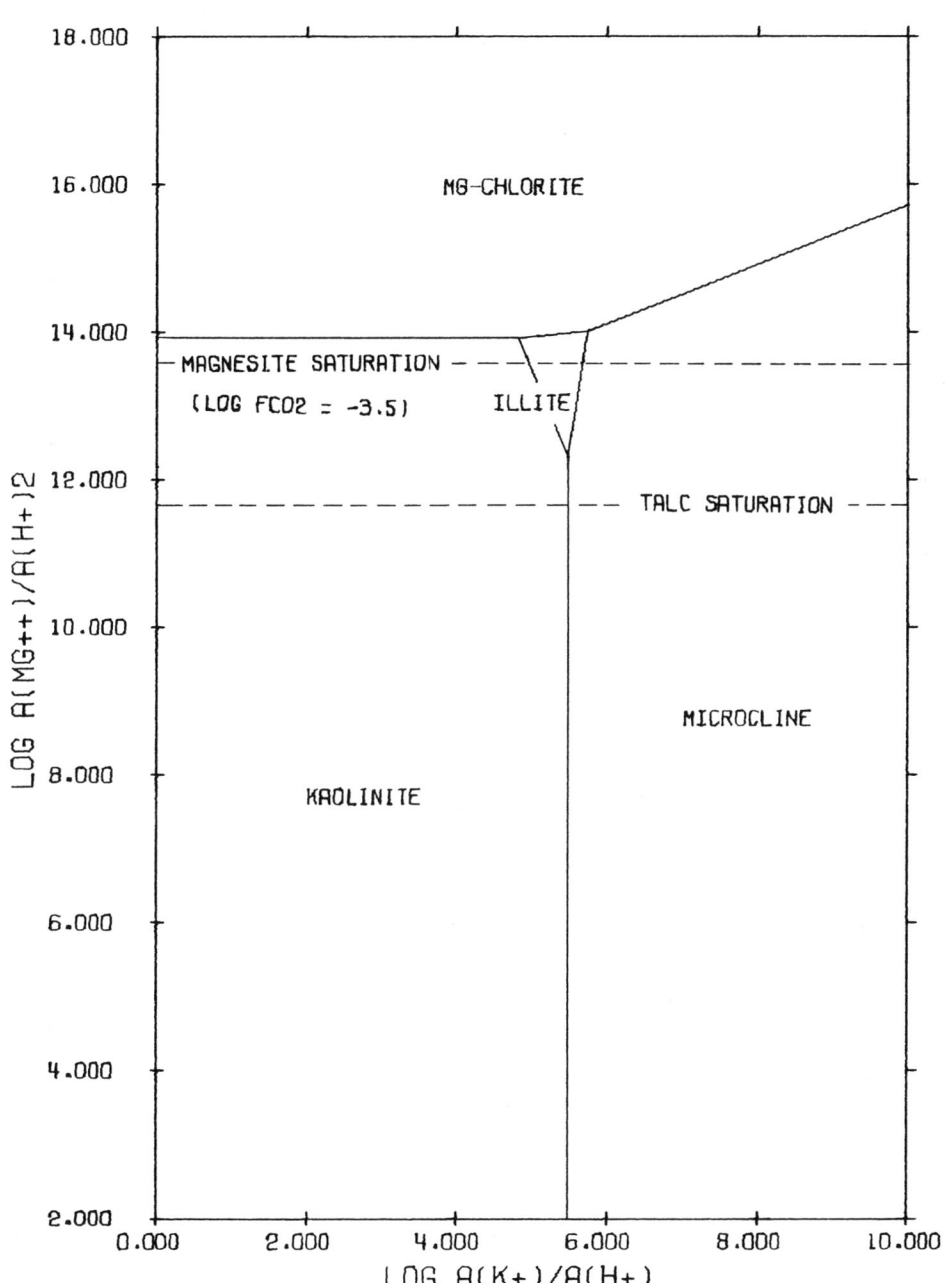

The System HCl—H$_2$O—Al$_2$O$_3$—CO$_2$—K$_2$O—MgO—SiO$_2$ at 25°C; log $a_{H_4SiO_4}$ = −4.00 = quartz saturation.

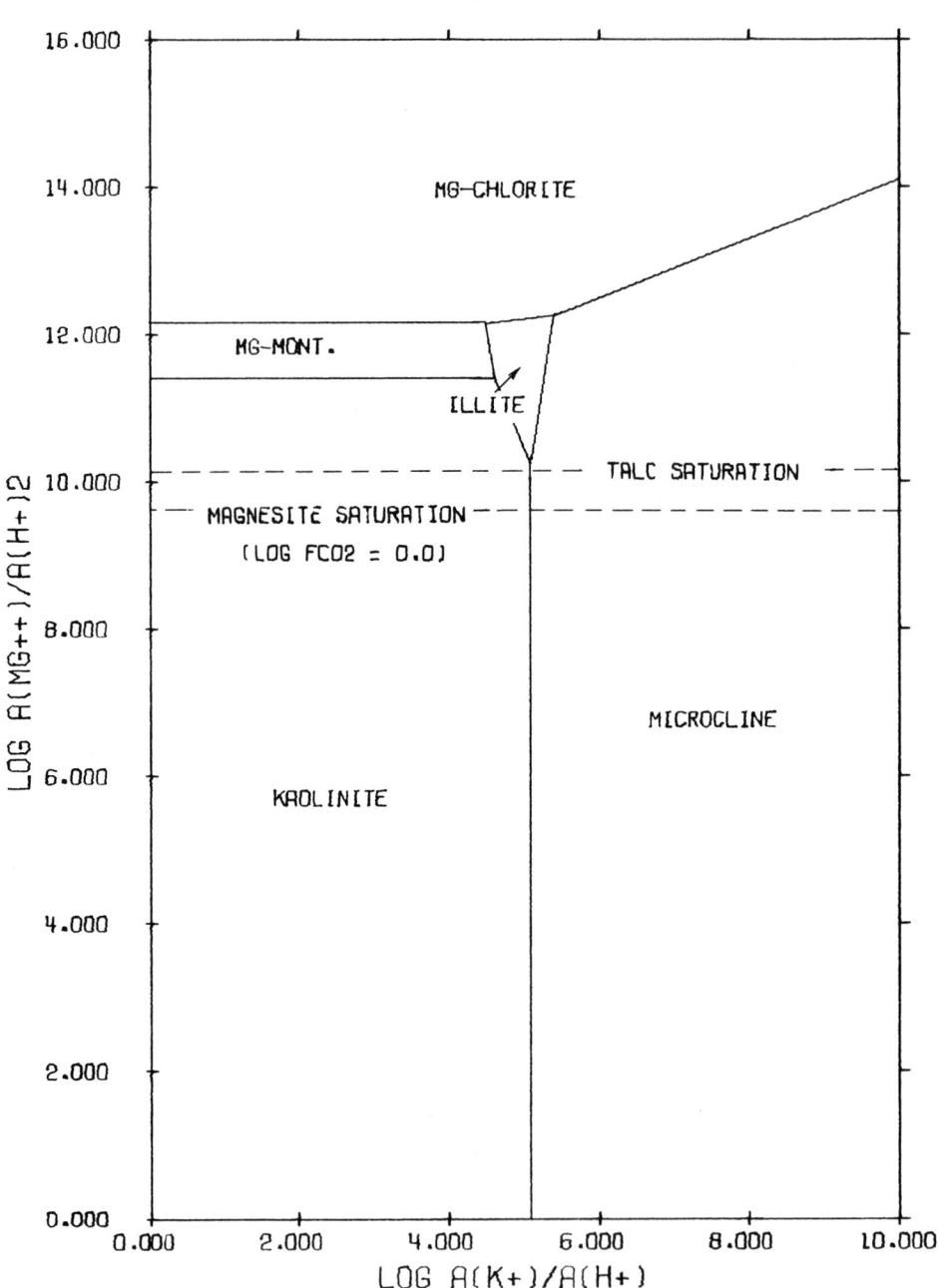

The System HCl—H$_2$O—Al$_2$O$_3$—CO$_2$—K$_2$O—MgO—SiO$_2$ at 60°C; log $a_{H_4SiO_4}$ = −3.52 = quartz saturation.

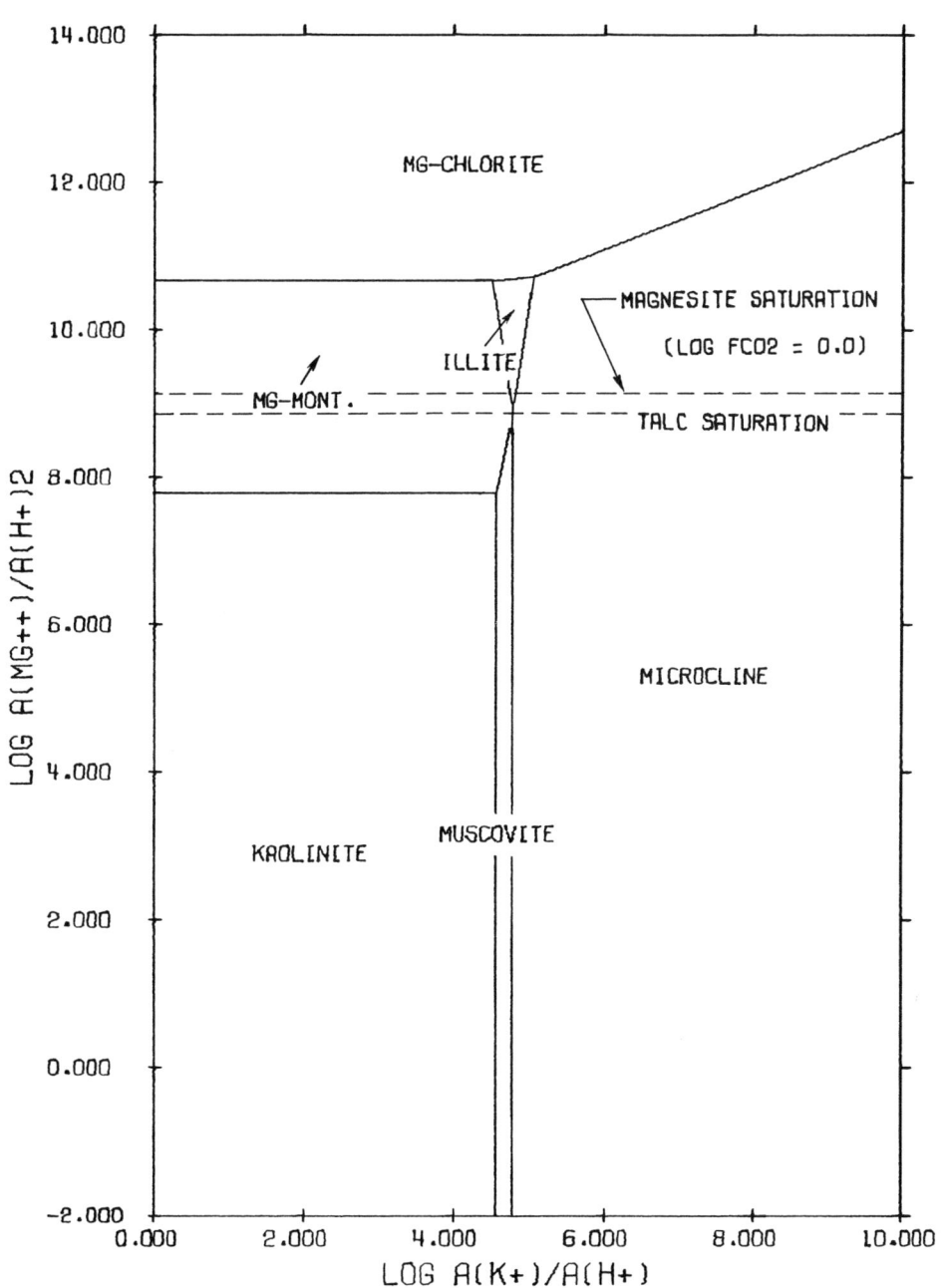

The System HCl—H_2O—Al_2O_3—CO_2—K_2O—MgO—SiO_2 at 100°C; log $a_{H_4SiO_4}$ = −3.08 = quartz saturation.

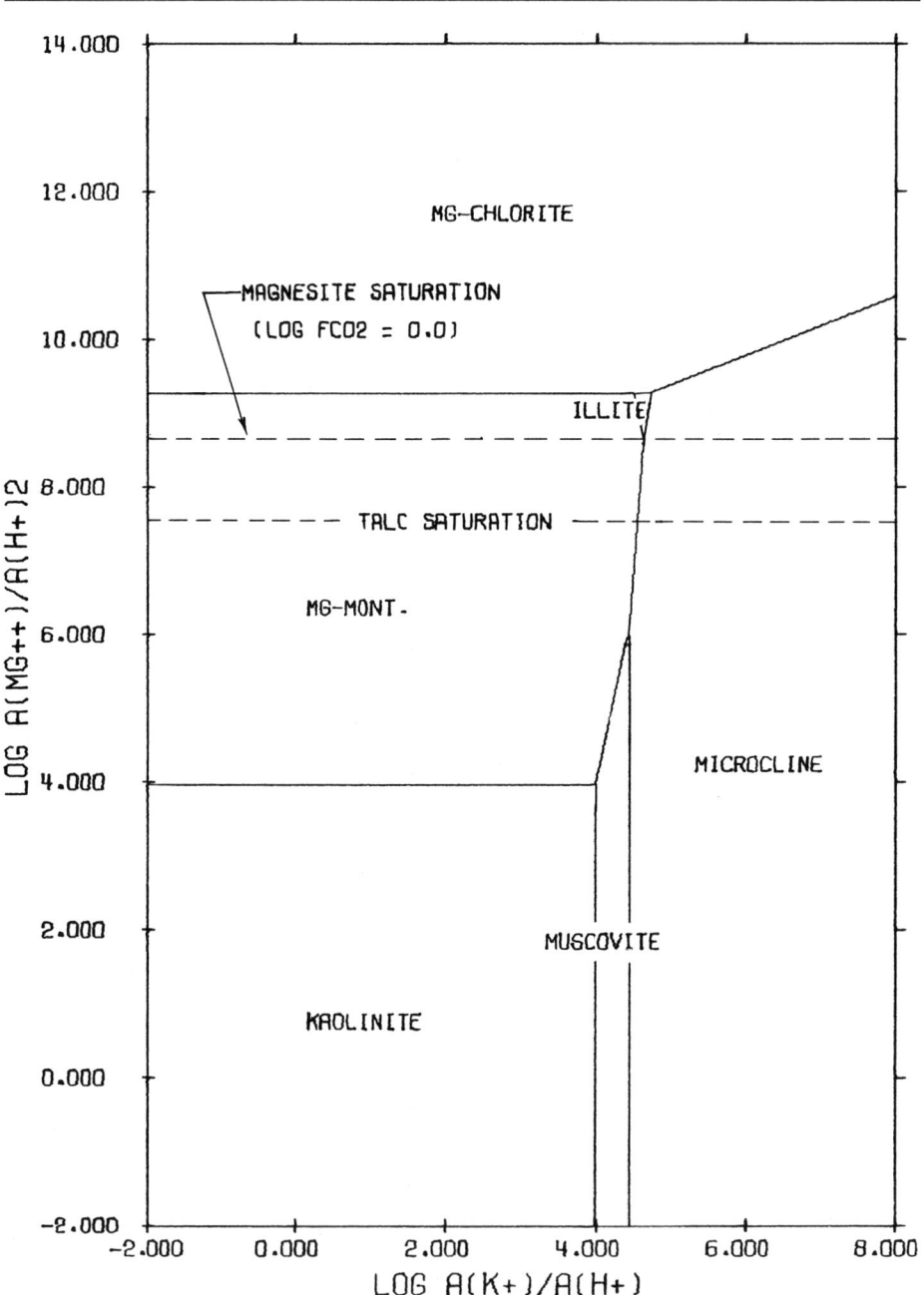

The System HCl—H_2O—Al_2O_3—CO_2—K_2O—MgO—SiO_2 at 150°C; log $a_{H_4SiO_4}$ = −2.67 = quartz saturation.

Activity Diagrams

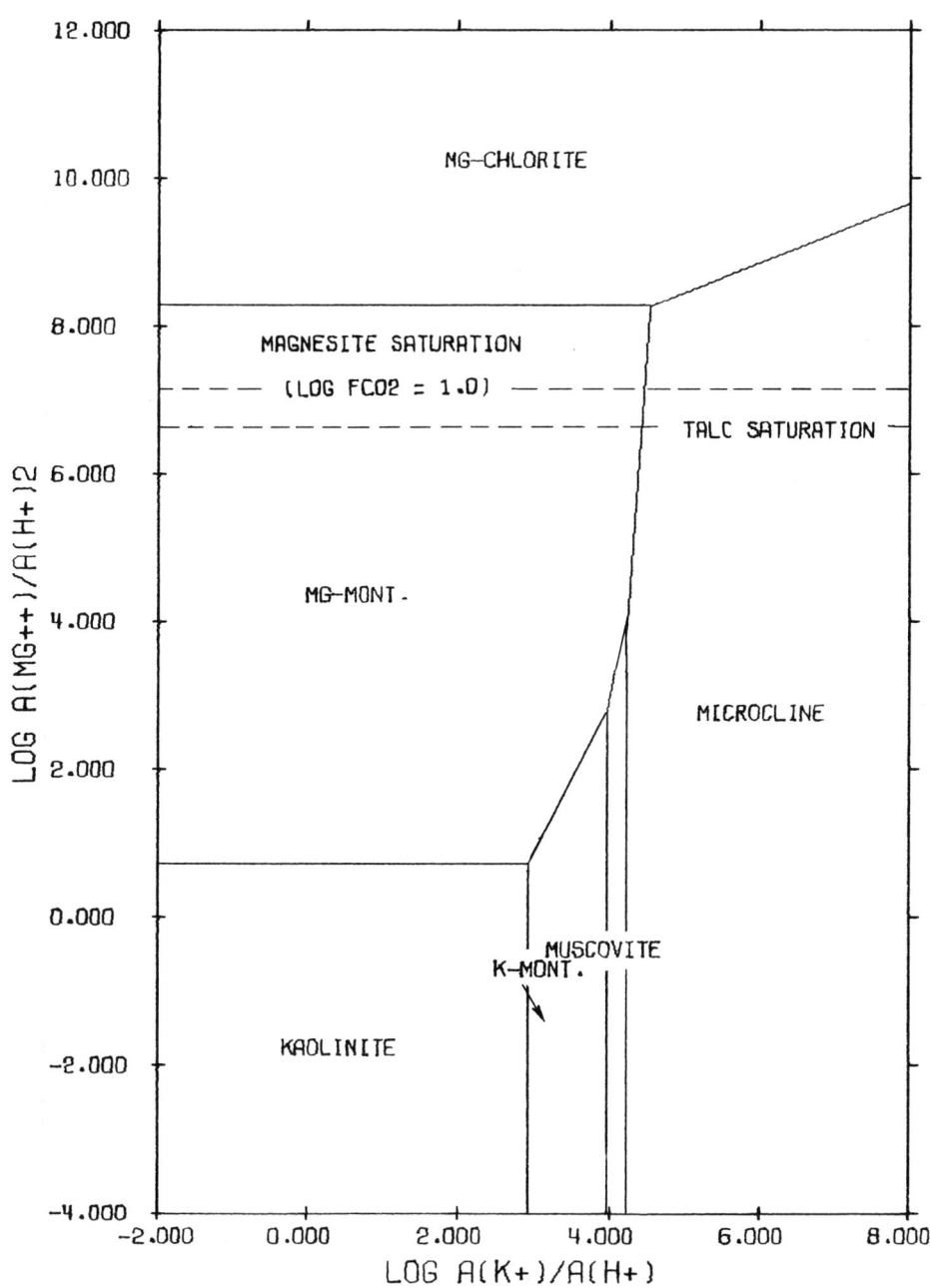

The System HCl—H$_2$O—Al$_2$O$_3$—CO$_2$—K$_2$O—MgO—SiO$_2$ at 200°C; log $a_{H_4SiO_4}$ = −2.35 = quartz saturation.

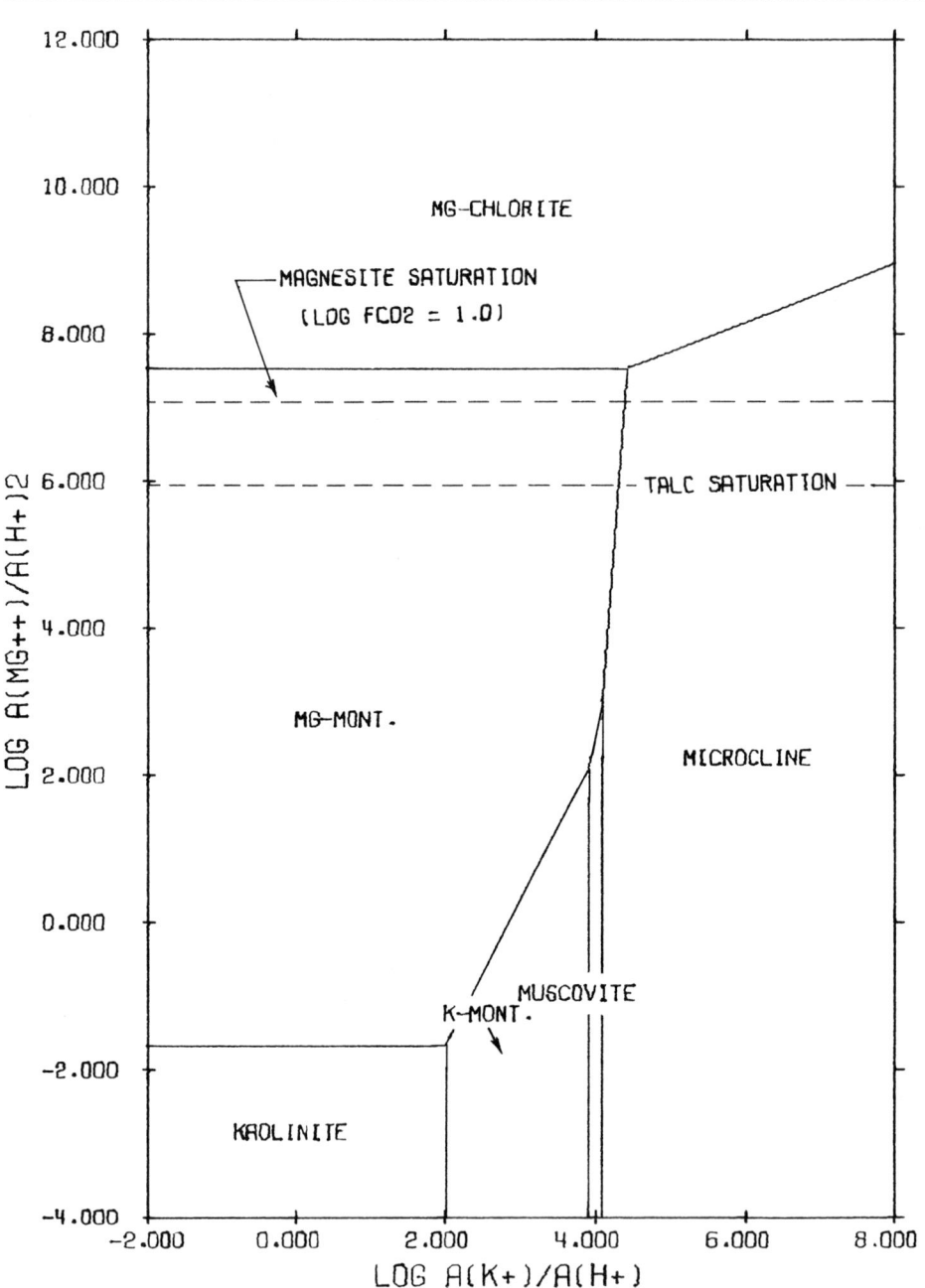

The System HCl—H_2O—Al_2O_3—CO_2—K_2O—MgO—SiO_2 at 250°C; log $a_{H_4SiO_4}$ = −2.11 = quartz saturation.

Activity Diagrams

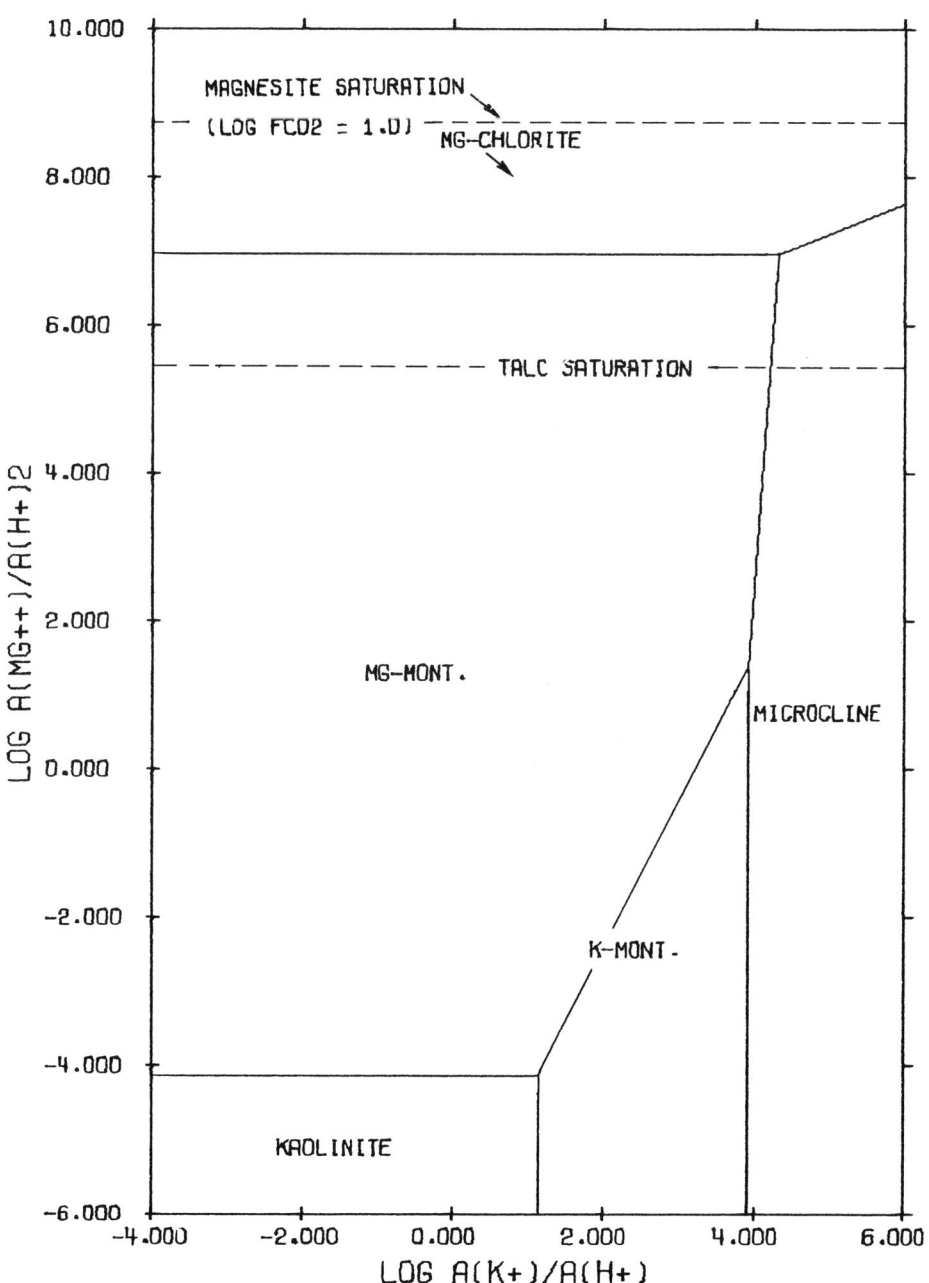

The System HCl—H_2O—Al_2O_3—CO_2—K_2O—MgO—SiO_2 at 300°C; log $a_{H_4SiO_4}$ = −1.94 = quartz saturation.

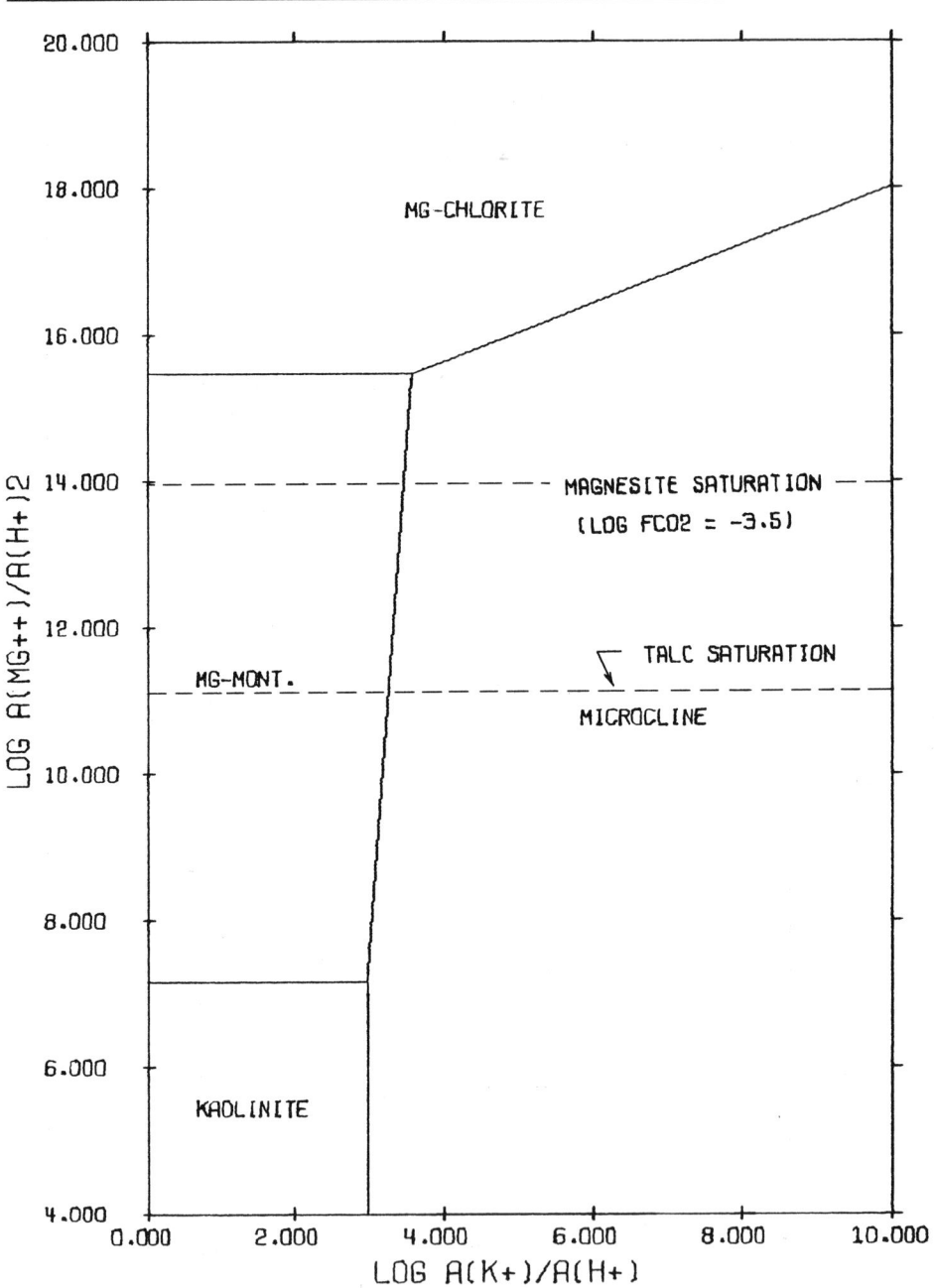

The System HCl—H$_2$O—Al$_2$O$_3$—CO$_2$—K$_2$O—MgO—SiO$_2$ at 0°C; log $a_{H_4SiO_4}$ = −3.00 = amorphous silica saturation.

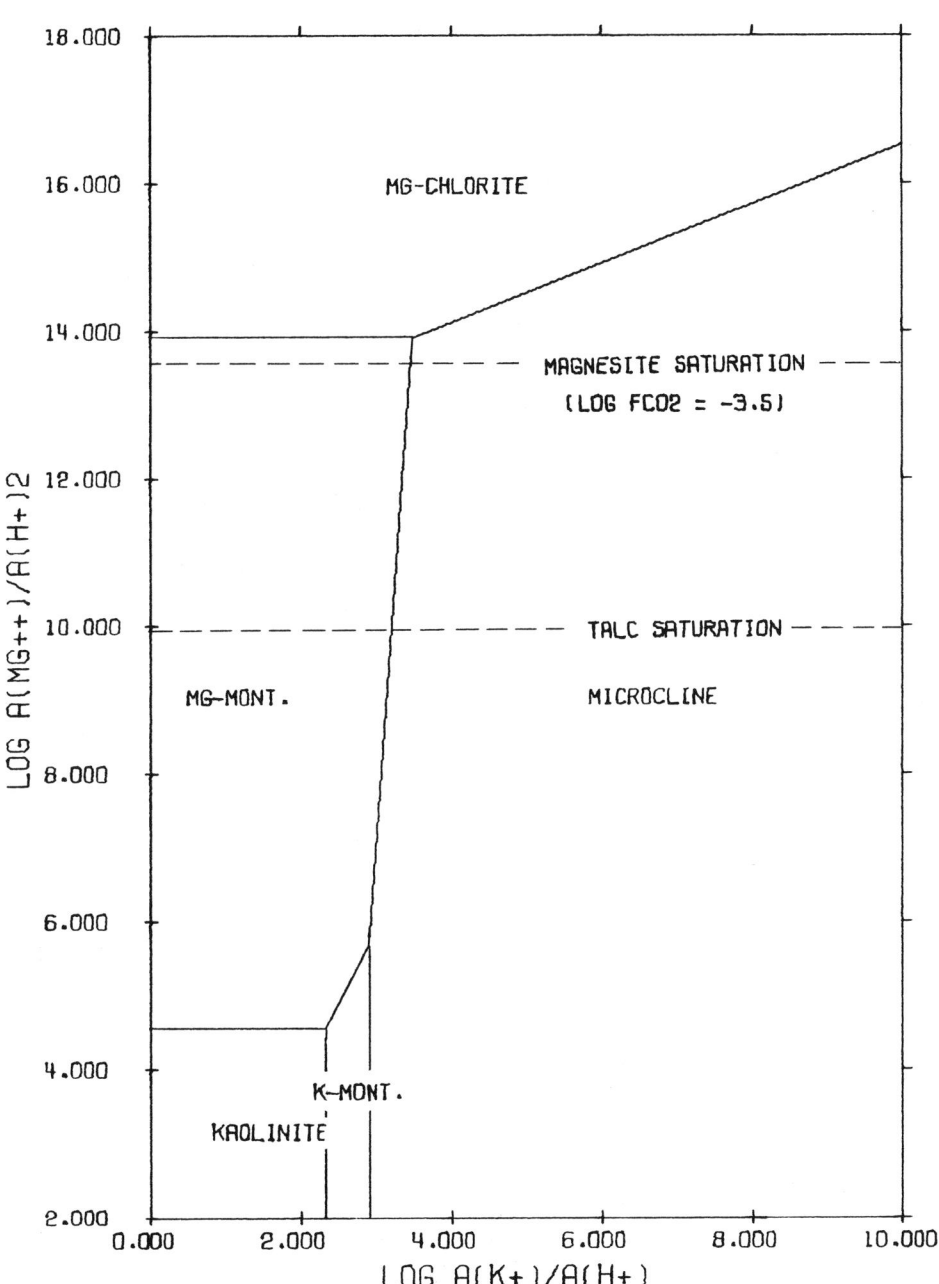

The System HCl—H_2O—Al_2O_3—CO_2—K_2O—MgO—SiO_2 at 25°C; $\log a_{H_4SiO_4} = -2.70$ = amorphous silica saturation.

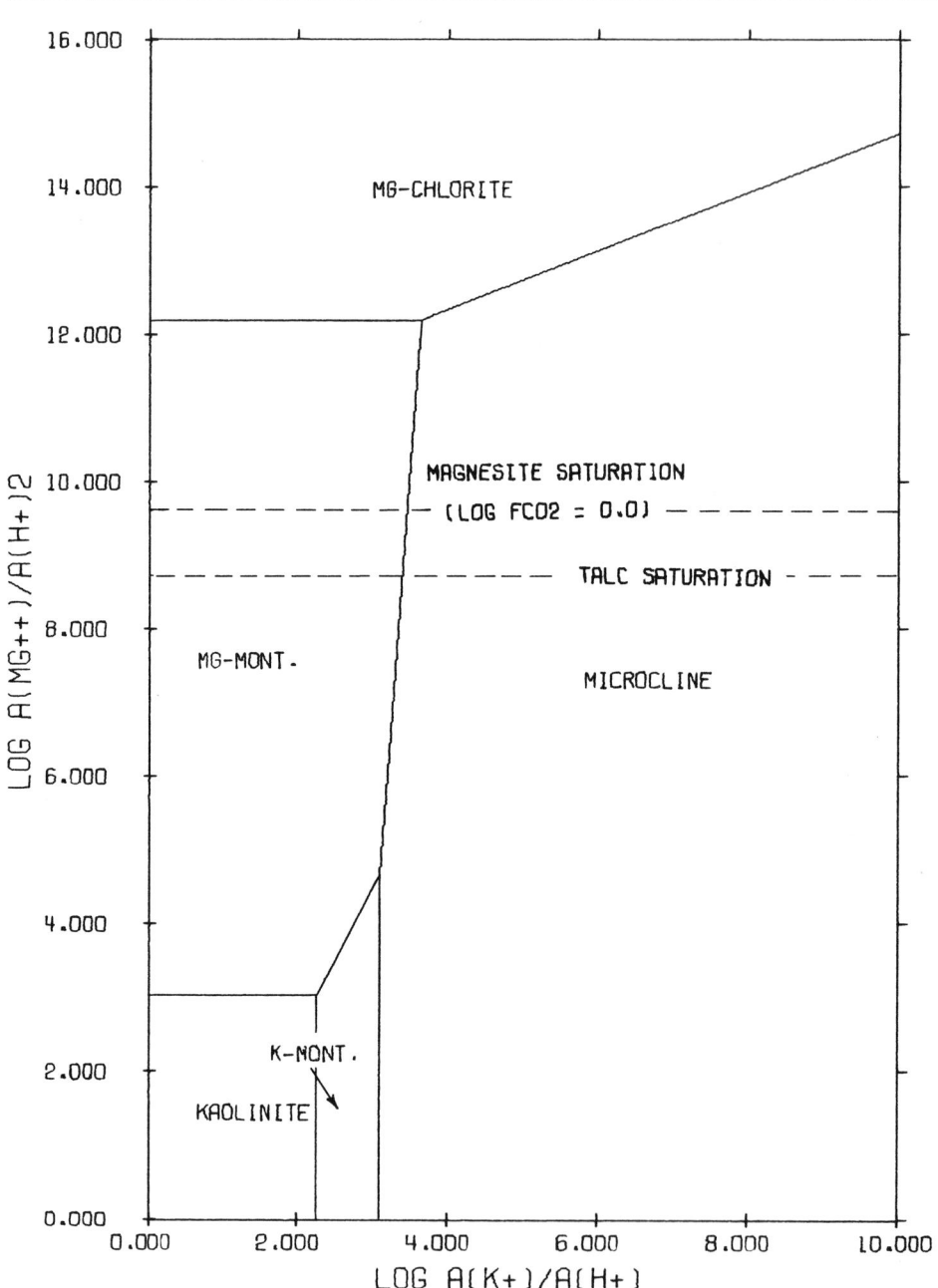

The System HCl—H_2O—Al_2O_3—CO_2—K_2O—MgO—SiO_2 at 60°C; log $a_{H_4SiO_4}$ = −2.47 = amorphous silica saturation.

Activity Diagrams

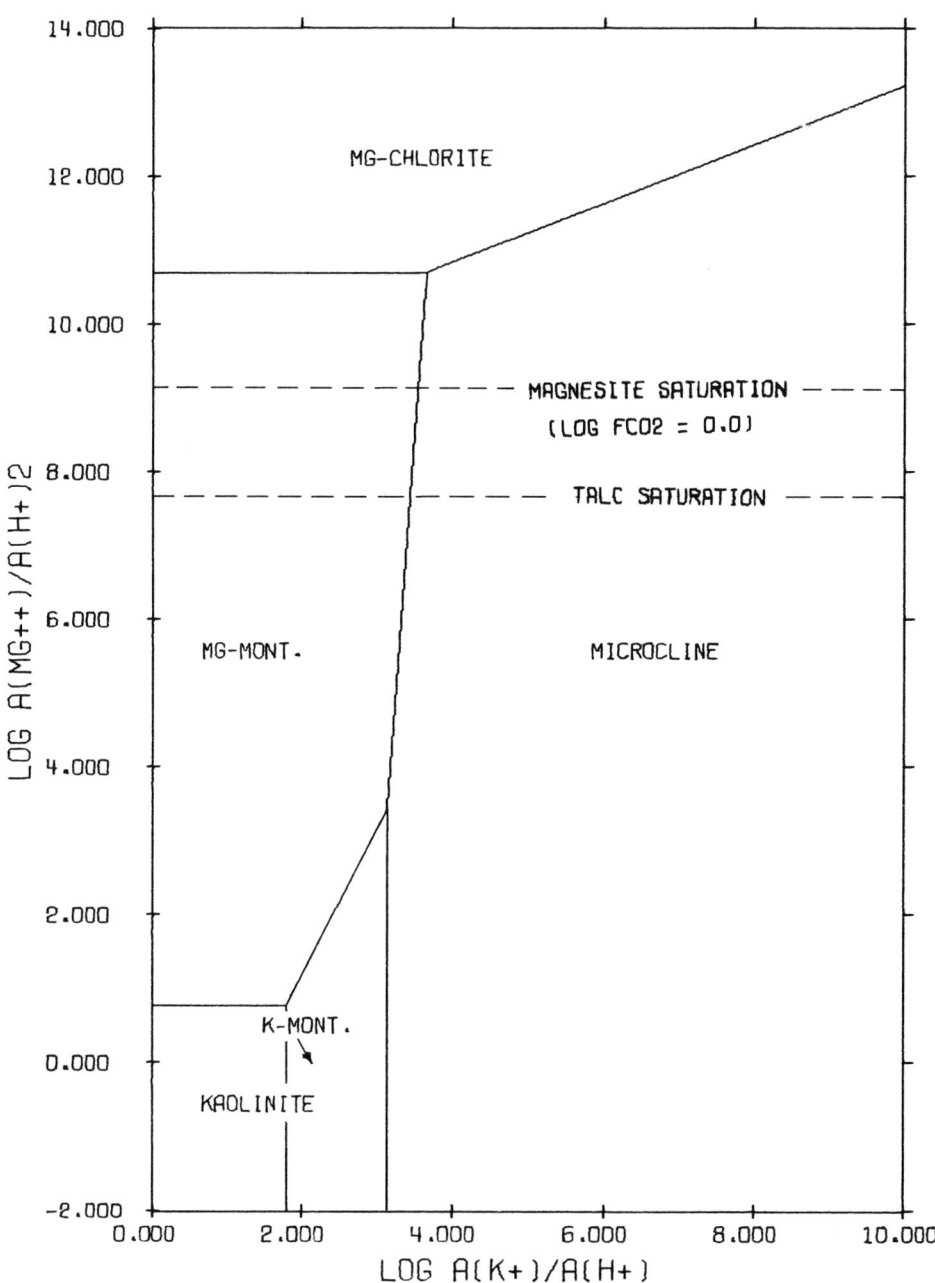

The System $HCl—H_2O—Al_2O_3—CO_2—K_2O—MgO—SiO_2$ at 100°C; log $a_{H_4SiO_4}$ = -2.20 = amorphous silica saturation.

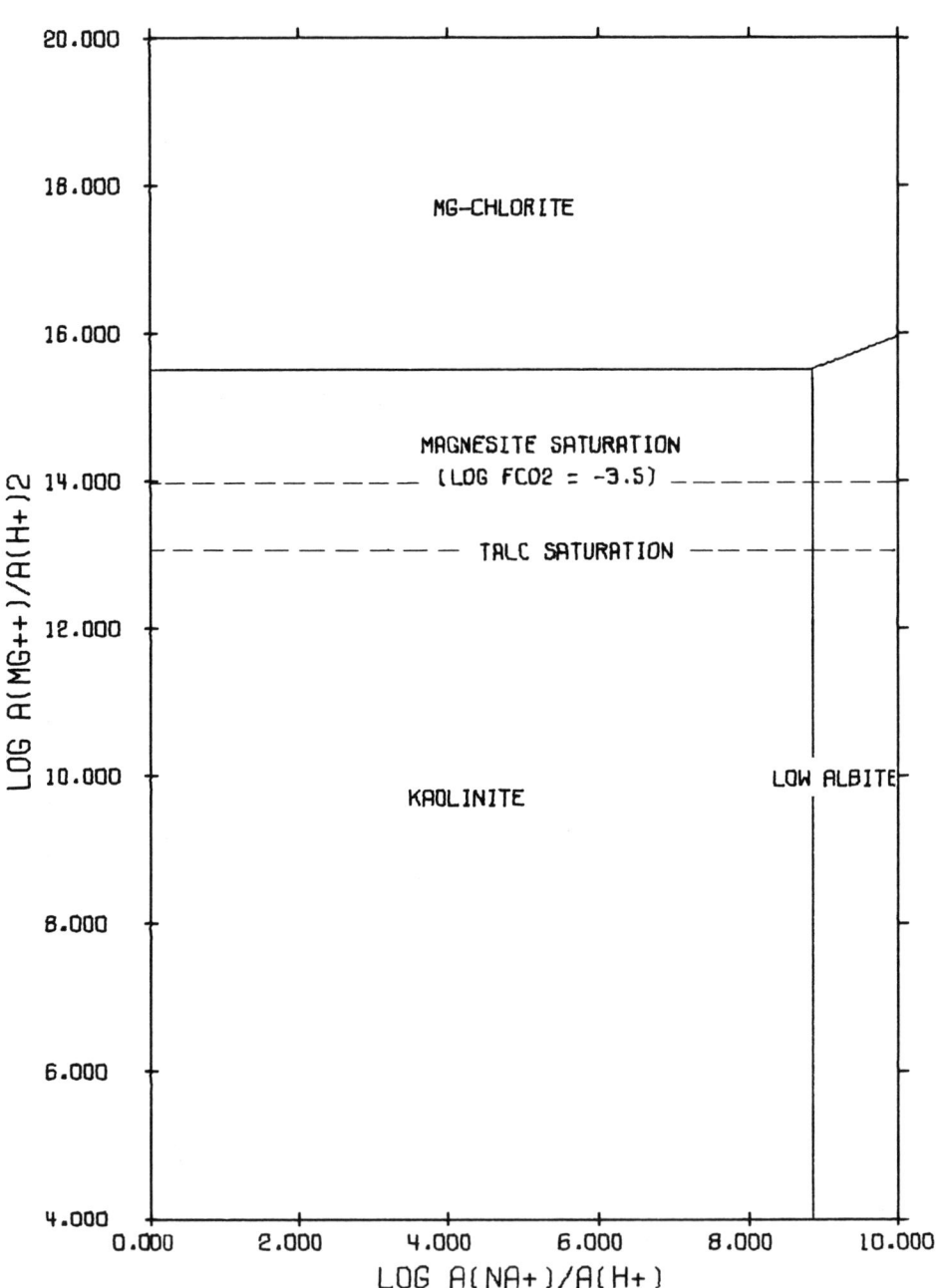

The System HCl—H_2O—Al_2O_3—CO_2—MgO—Na_2O—SiO_2 at 0°C; log $a_{H_4SiO_4}$ = −4.44 = quartz saturation.

Activity Diagrams

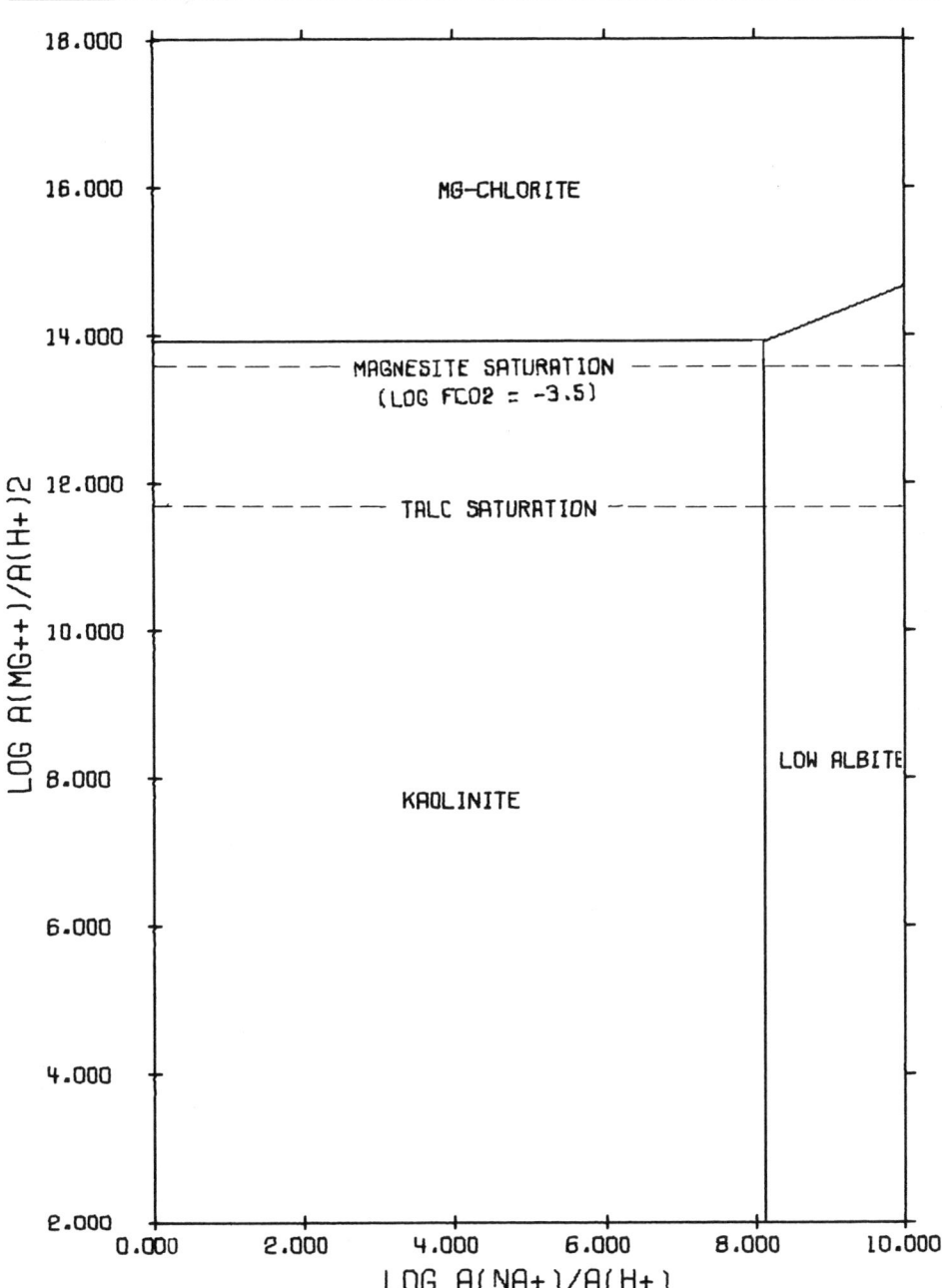

The System HCl—H_2O—Al_2O_3—CO_2—MgO—Na_2O—SiO_2 at 25°C; log $a_{H_4SiO_4}$ = −4.00 = quartz saturation.

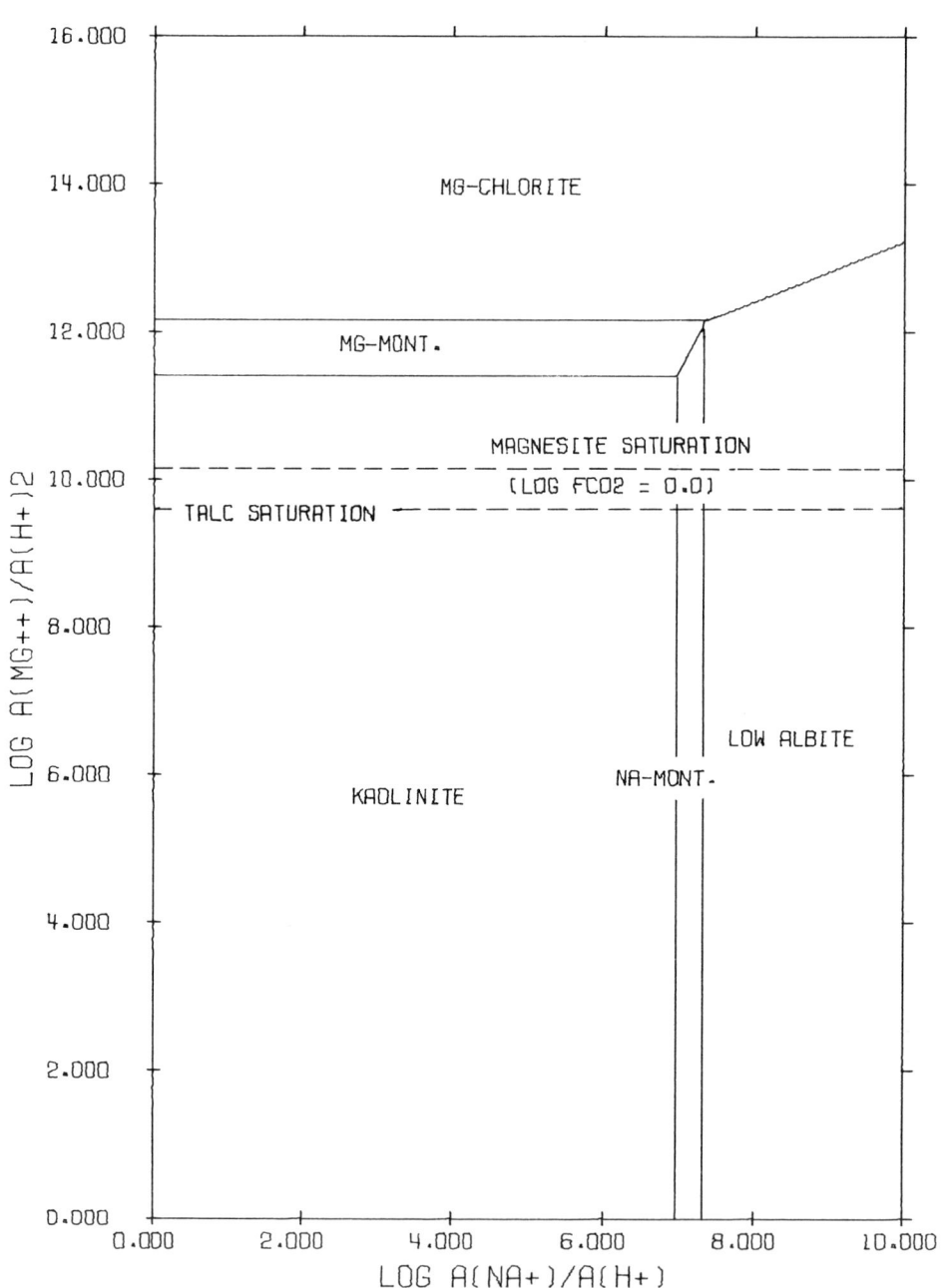

The System HCl—H₂O—Al₂O₃—CO₂—MgO—Na₂O—SiO₂ at 60°C; log $a_{H_4SiO_4}$ = −3.52 = quartz saturation.

Activity Diagrams

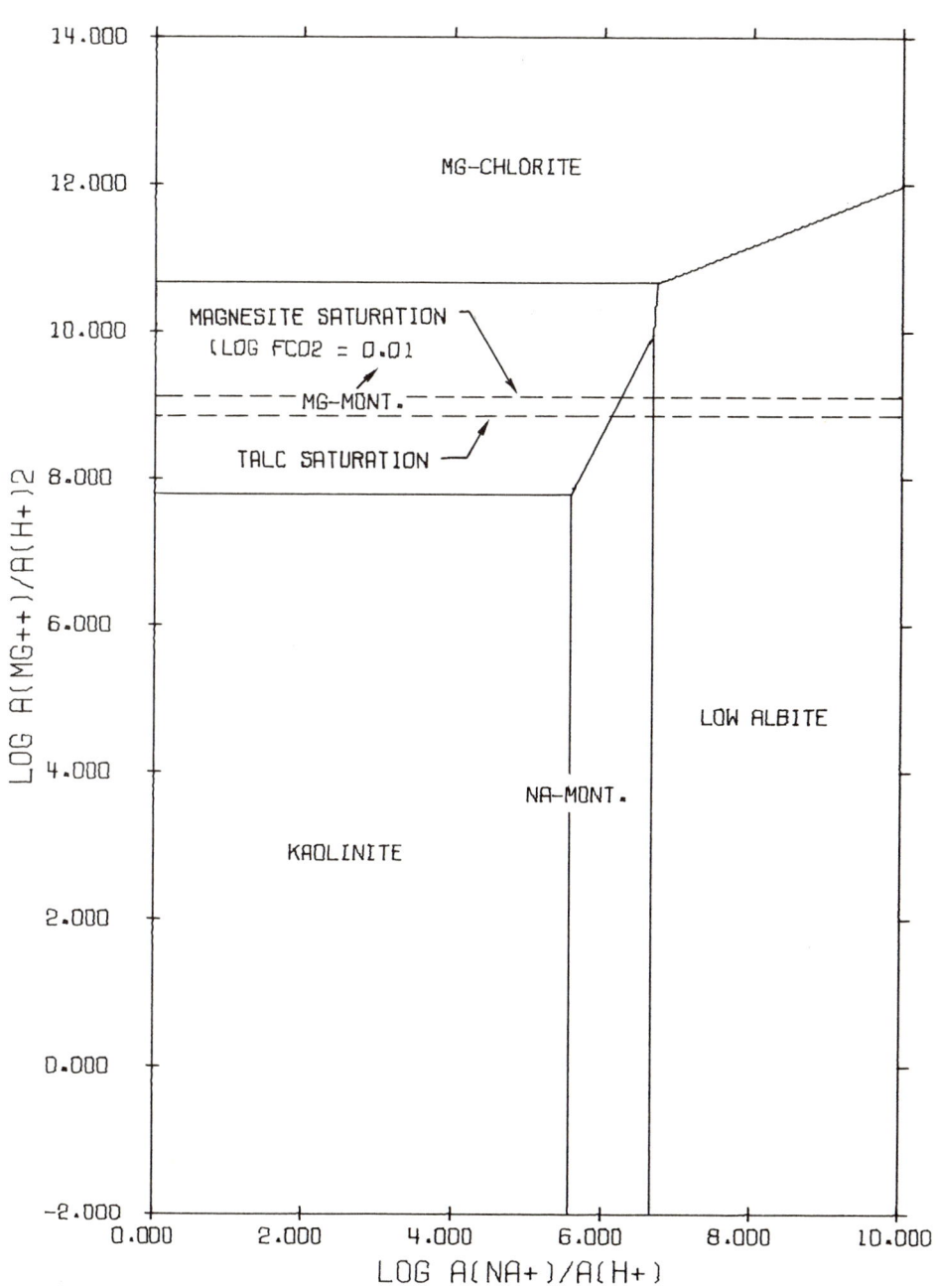

The System HCl—H_2O—Al_2O_3—CO_2—MgO—Na_2O—SiO_2 at 100°C; log $a_{H_4SiO_4}$ = −3.08 = quartz saturation.

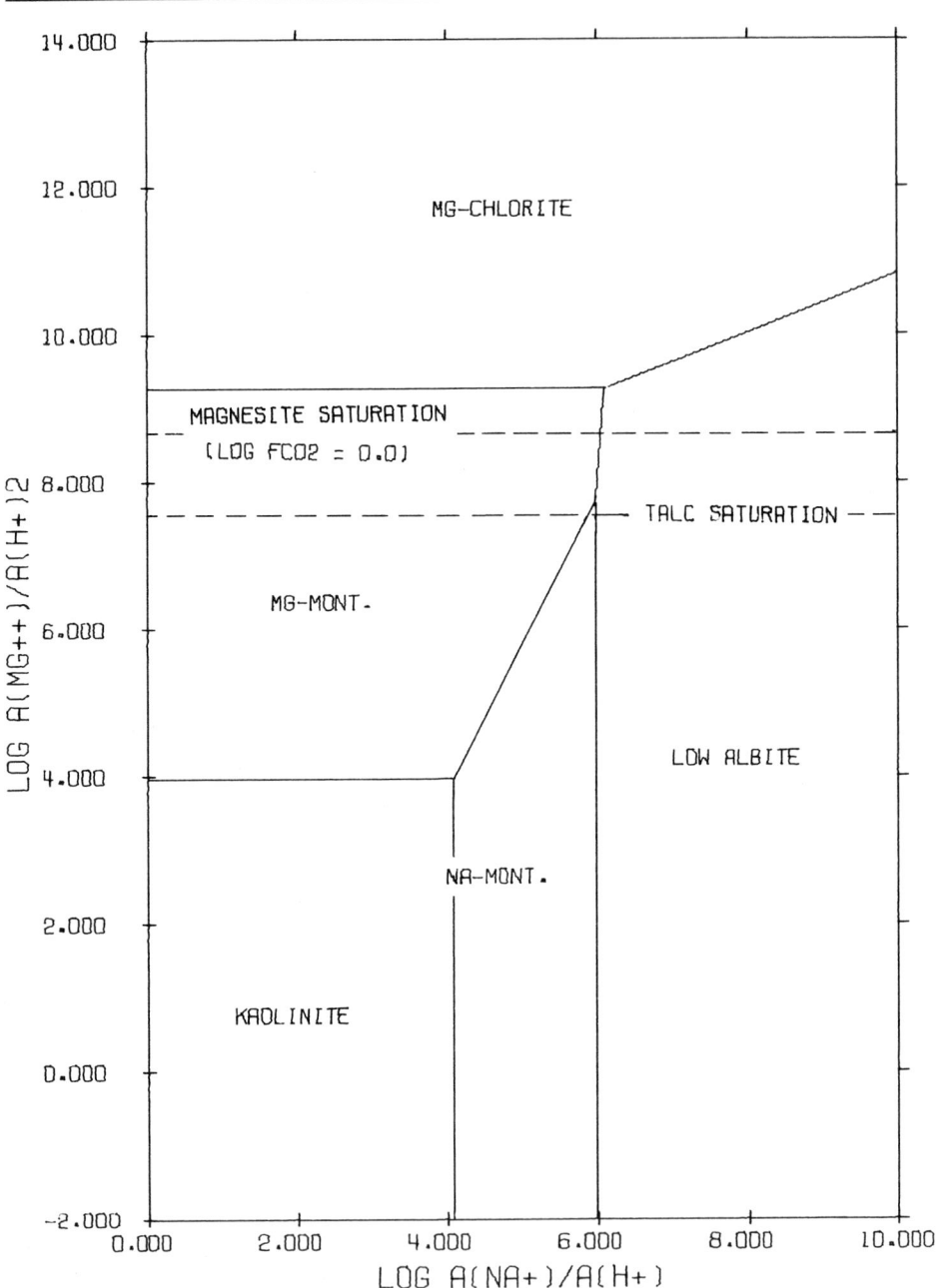

The System HCl—H_2O—Al_2O_3—CO_2—MgO—Na_2O—SiO_2 at 150°C; log $a_{H_4SiO_4}$ = −2.67 = quartz saturation.

Activity Diagrams

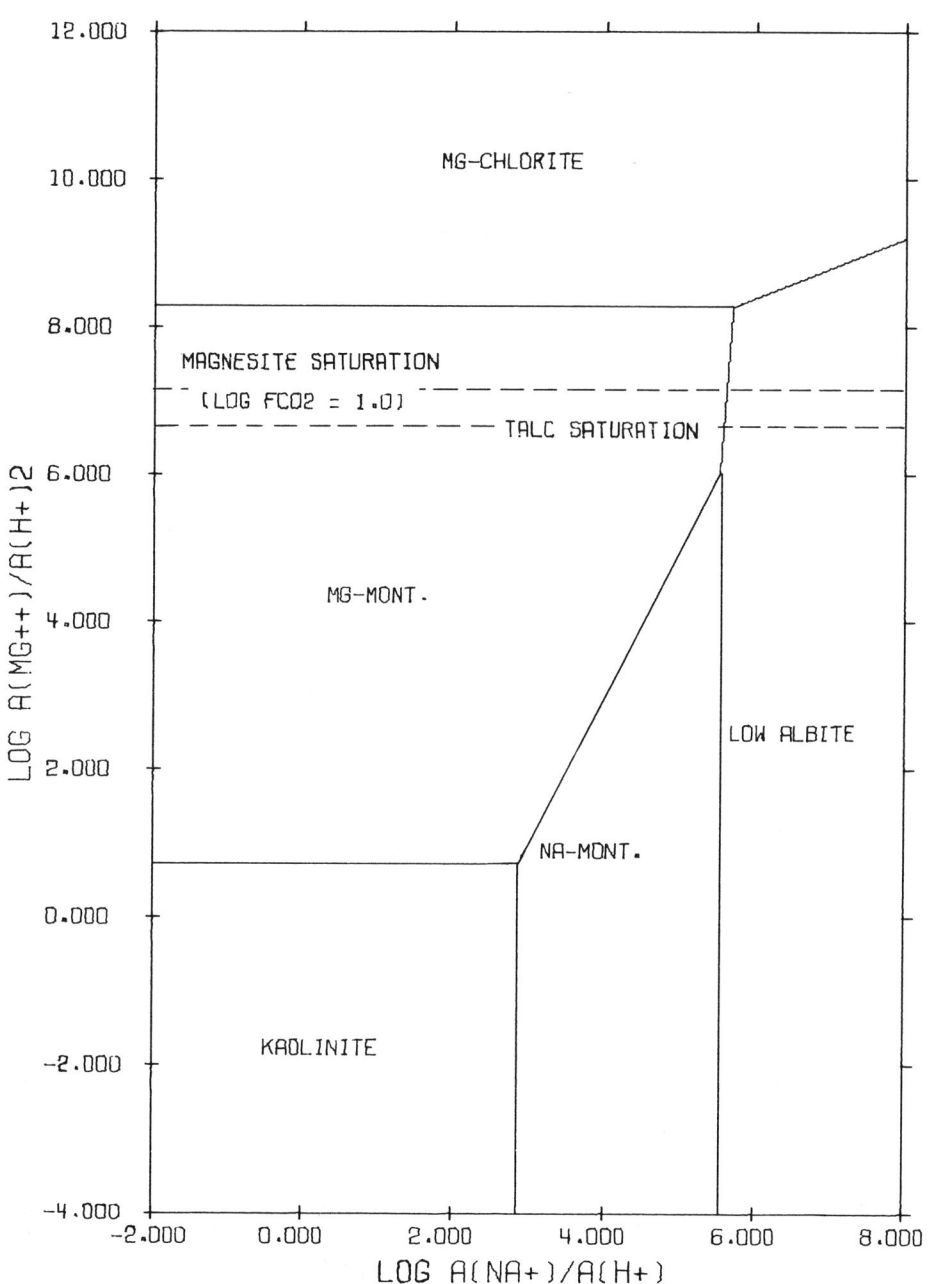

The System HCl—H$_2$O—Al$_2$O$_3$—CO$_2$—MgO—Na$_2$O—SiO$_2$ at 200°C; log $a_{H_4SiO_4}$ = -2.35 = quartz saturation.

Activity Diagrams

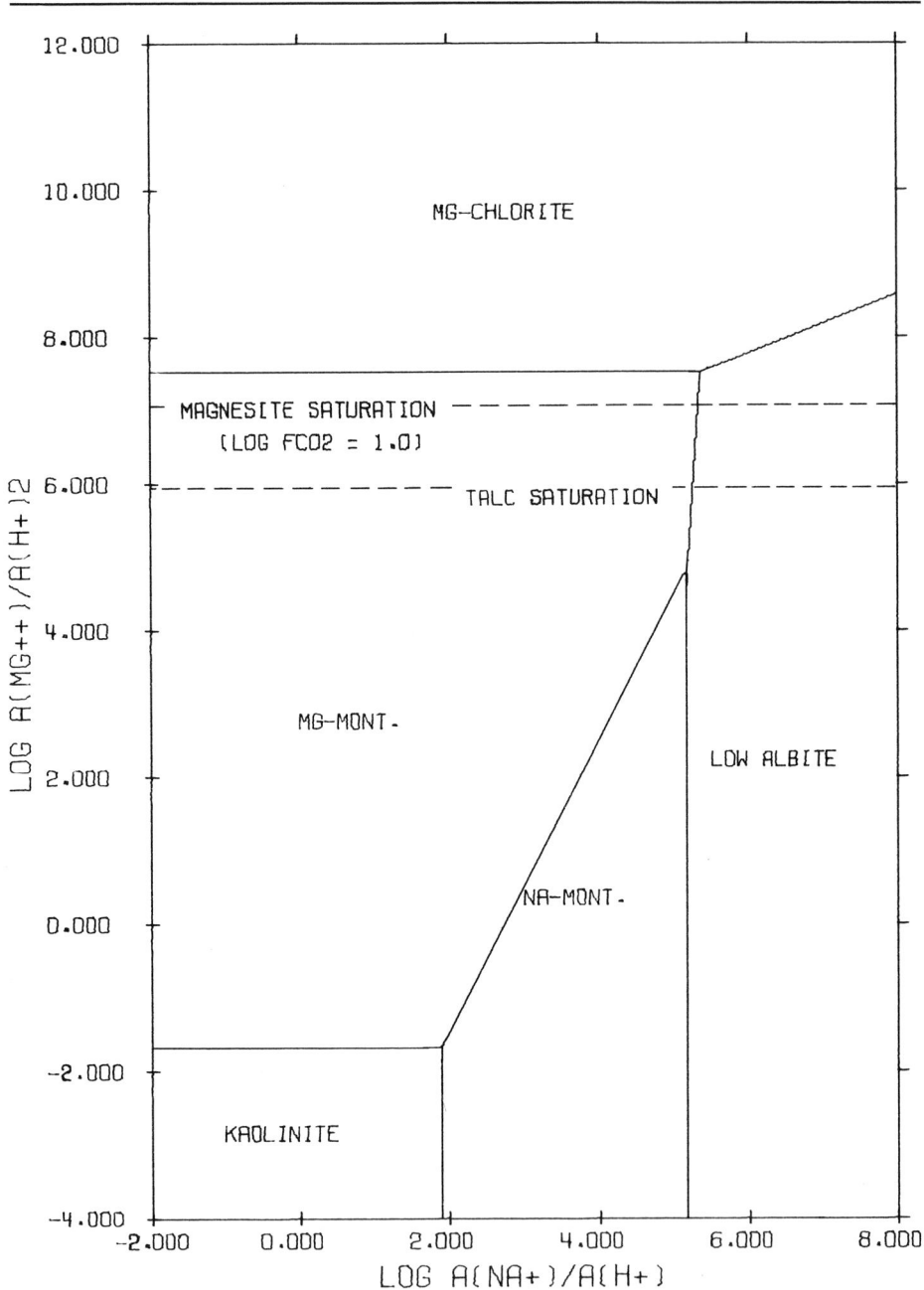

The System HCl—H_2O—Al_2O_3—CO_2—MgO—Na_2O—SiO_2 at 250°C; log $a_{H_4SiO_4}$ = −2.11 = quartz saturation.

Activity Diagrams

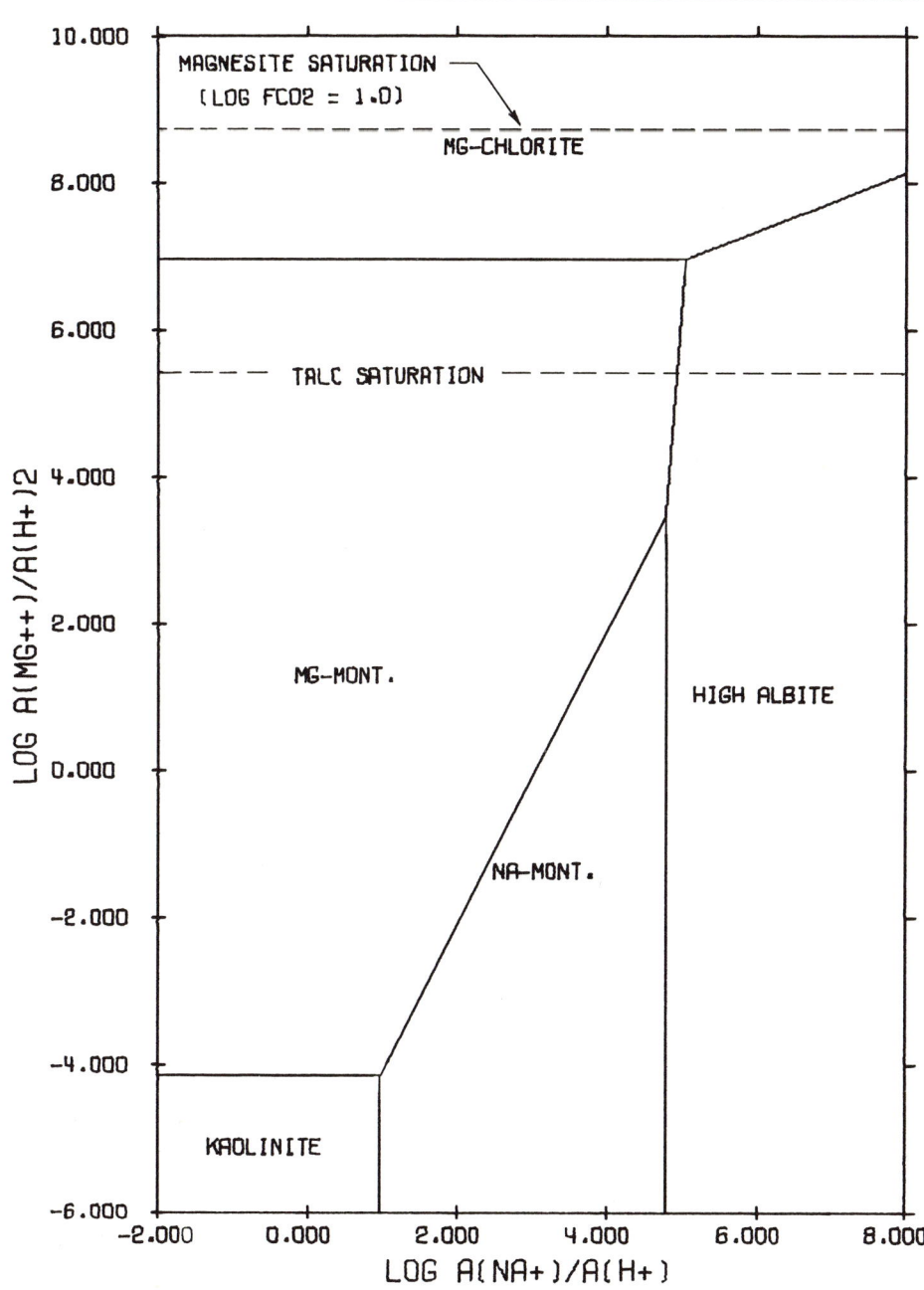

The System HCl—H_2O—Al_2O_3—CO_2—MgO—Na_2O—SiO_2 at 300°C; log $a_{H_4SiO_4}$ = −1.94 = quartz saturation.

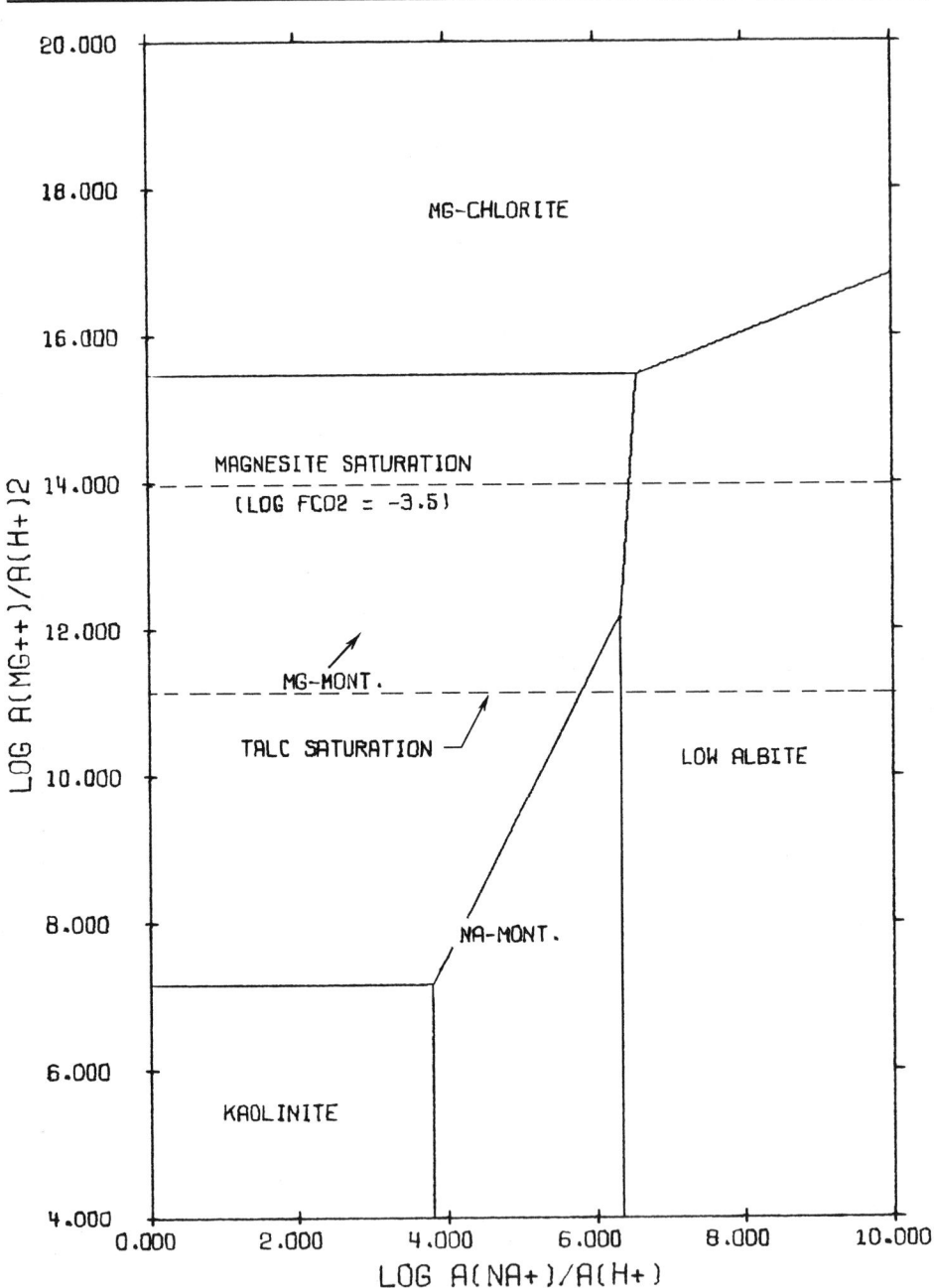

The System HCl—H$_2$O—Al$_2$O$_3$—CO$_2$—MgO—Na$_2$O—SiO$_2$ at 0°C; log $a_{H_4SiO_4}$ = −3.00 = amorphous silica saturation.

Activity Diagrams

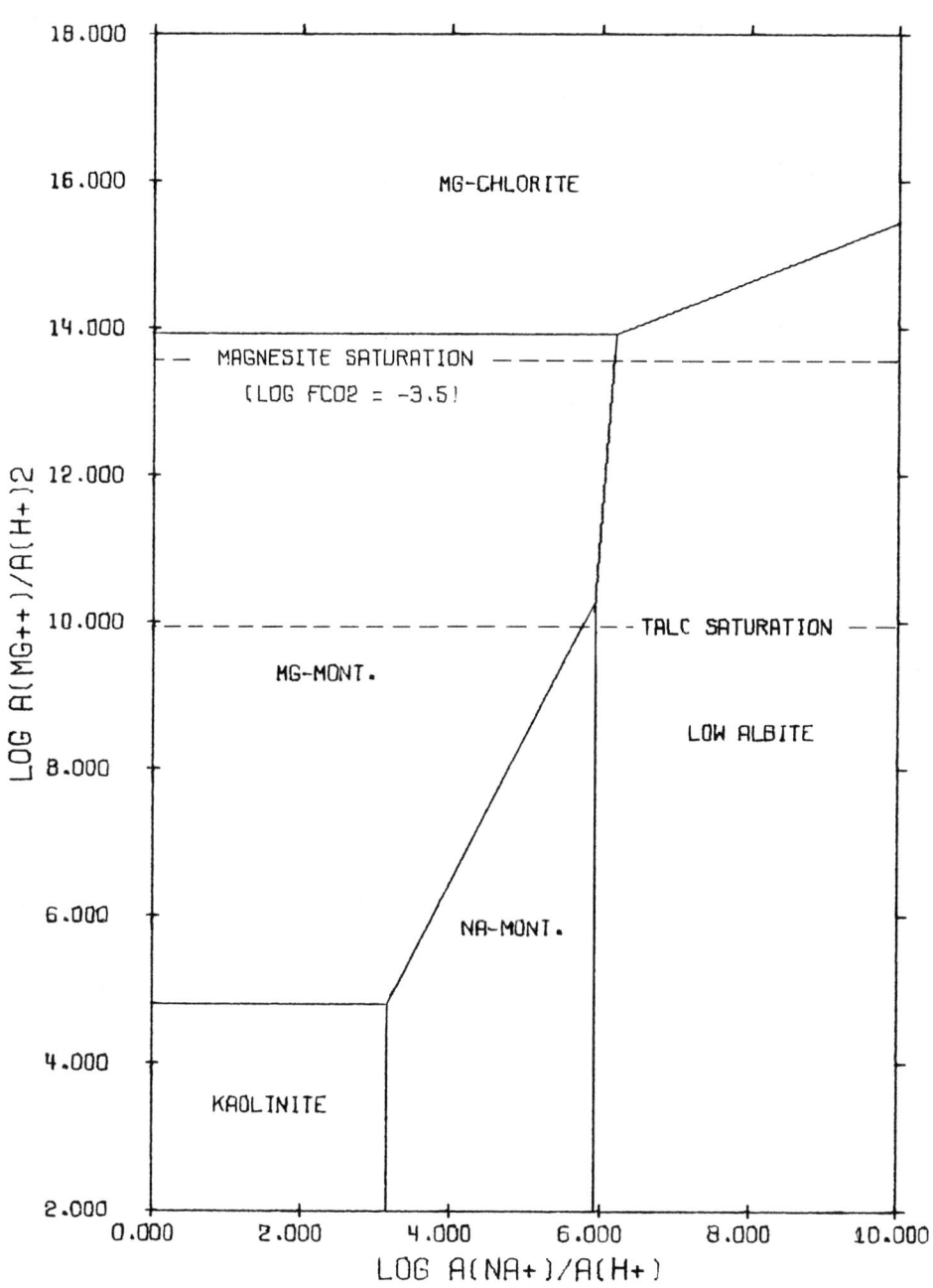

The System HCl—H_2O—Al_2O_3—CO_2—MgO—Na_2O—SiO_2 at 25°C; log $a_{H_4SiO_4}$ = −2.70 = amorphous silica saturation.

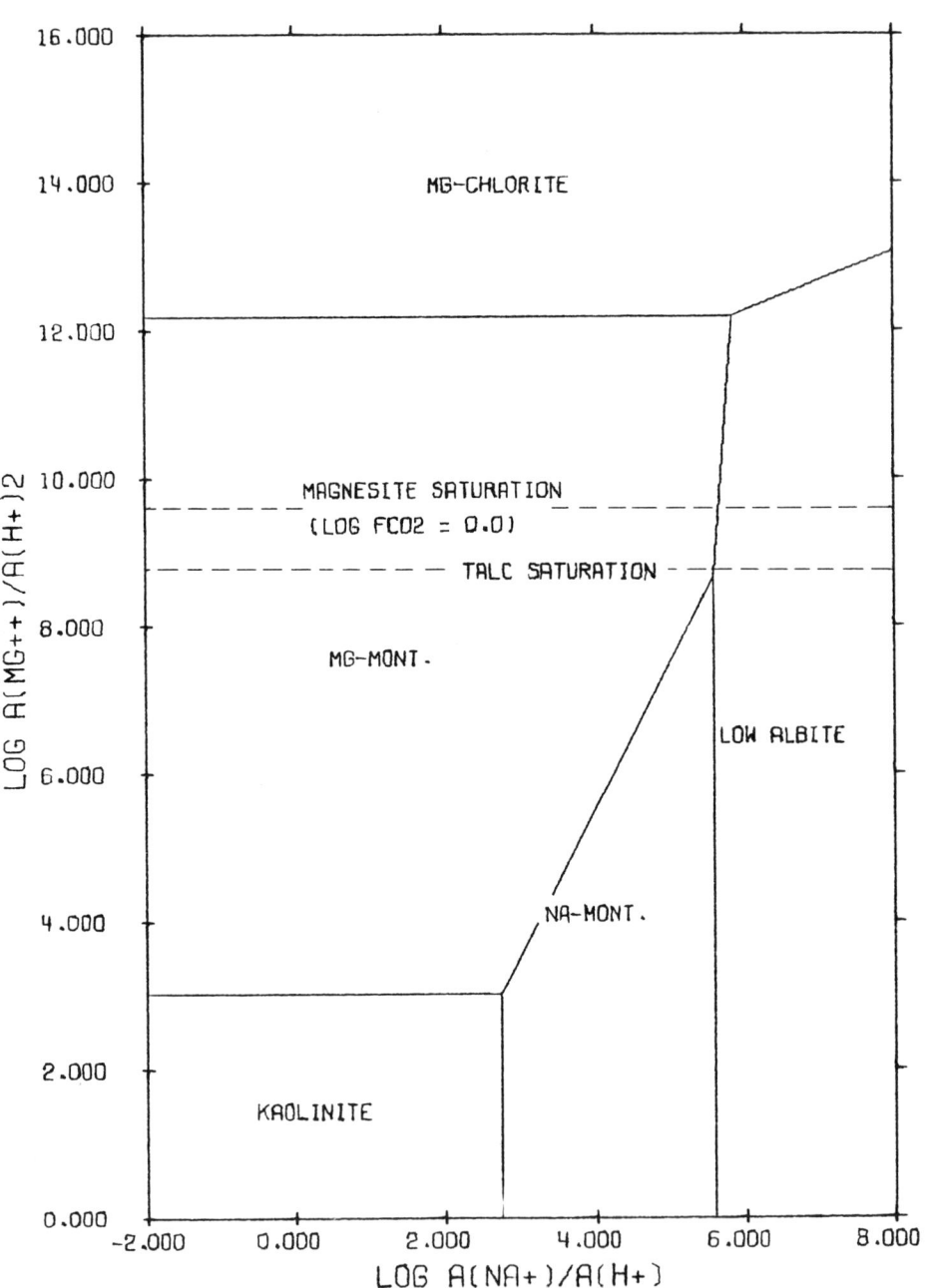

The System HCl—H$_2$O—Al$_2$O$_3$—CO$_2$—MgO—Na$_2$O—SiO$_2$ at 60°C; log $a_{\text{H}_4\text{SiO}_4}$ = −2.47 = amorphous silica saturation.

Activity Diagrams

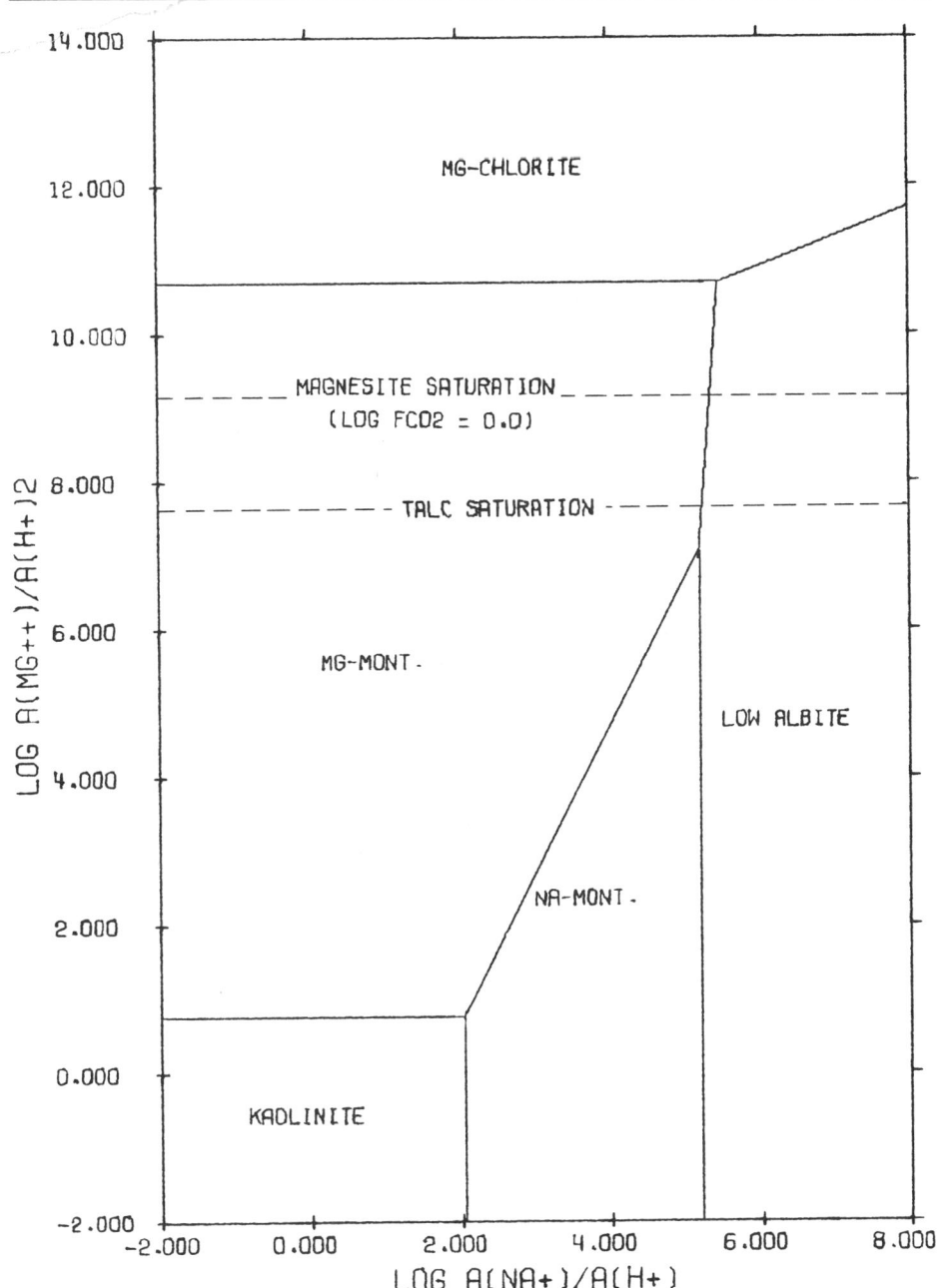

The System HCl—H_2O—Al_2O_3—CO_2—MgO—Na_2O—SiO_2 at 100°C; $\log a_{H_4SiO_4} = -2.20$ = amorphous silica saturation.

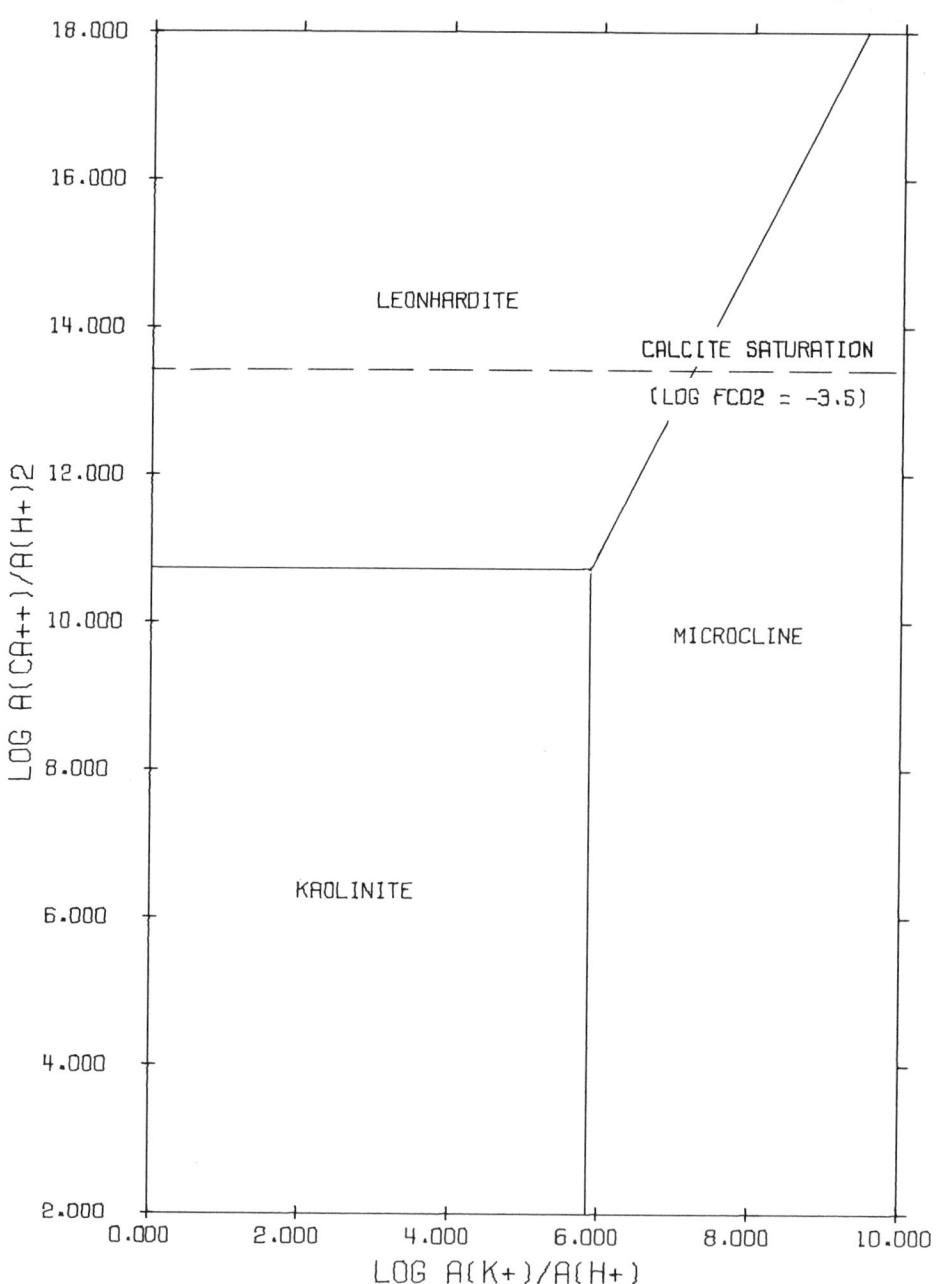

The System HCl—H_2O—Al_2O_3—CaO—CO_2—K_2O—SiO_2 at 0°C; log $a_{H_4SiO_4}$ = −4.44 = quartz saturation.

Activity Diagrams

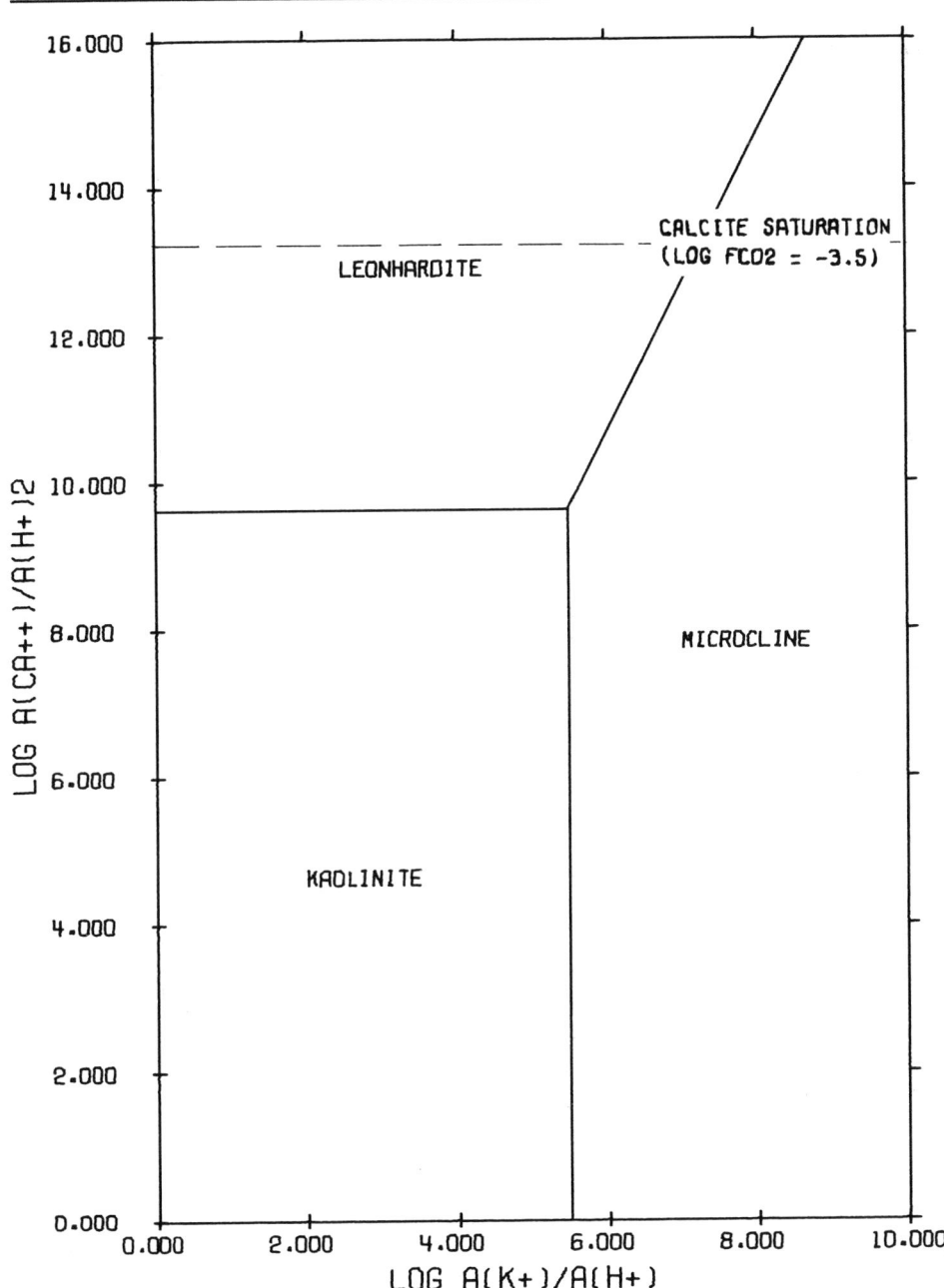

The System HCl—H_2O—Al_2O_3—CaO—CO_2—K_2O—SiO_2 at 25°C; $\log a_{H_4SiO_4} = -4.00 =$ quartz saturation.

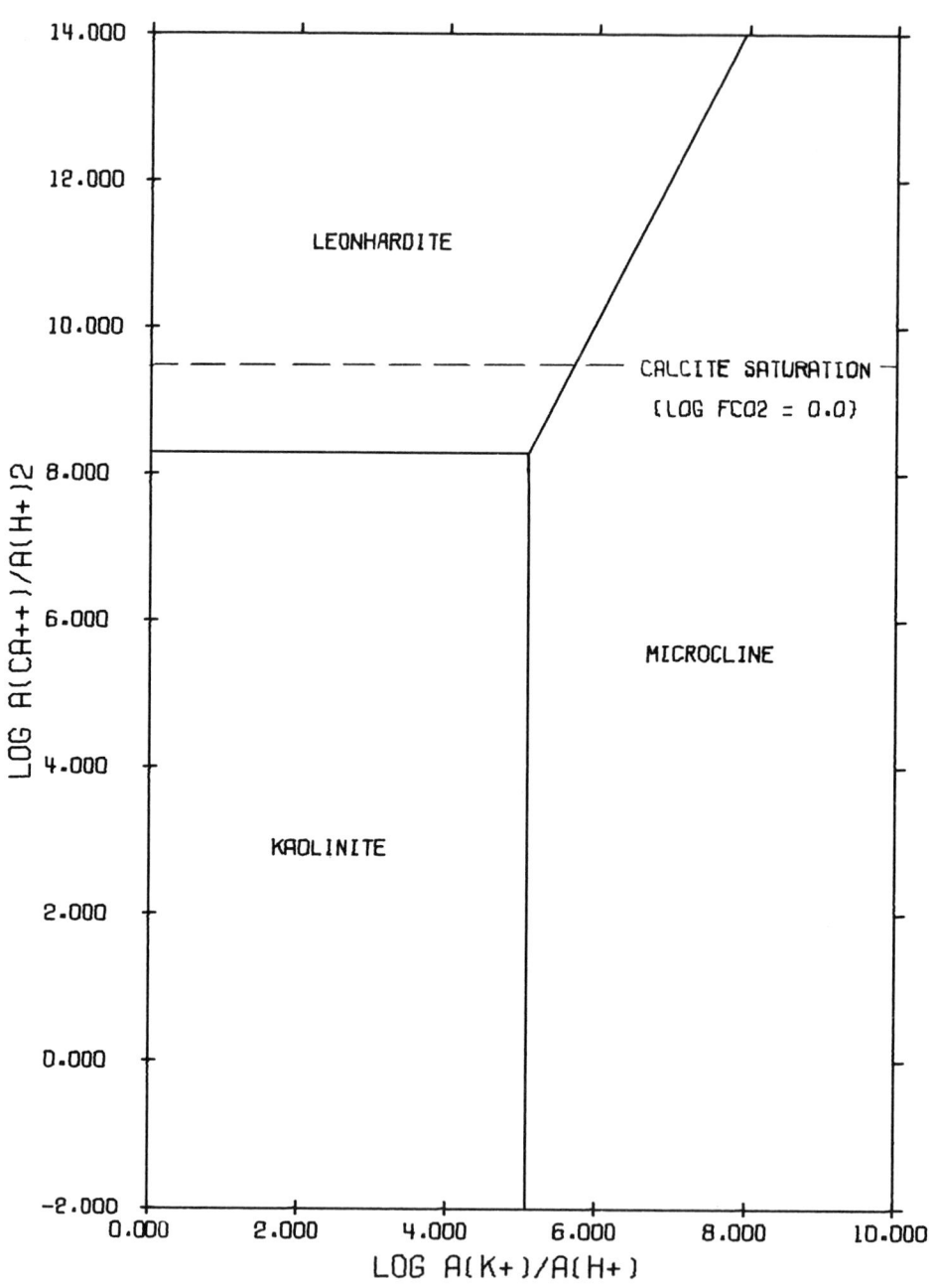

The System HCl—H$_2$O—Al$_2$O$_3$—CaO—CO$_2$—K$_2$O—SiO$_2$ at 60°C; log $a_{\text{H}_4\text{SiO}_4}$ = −3.52 = quartz saturation.

Activity Diagrams

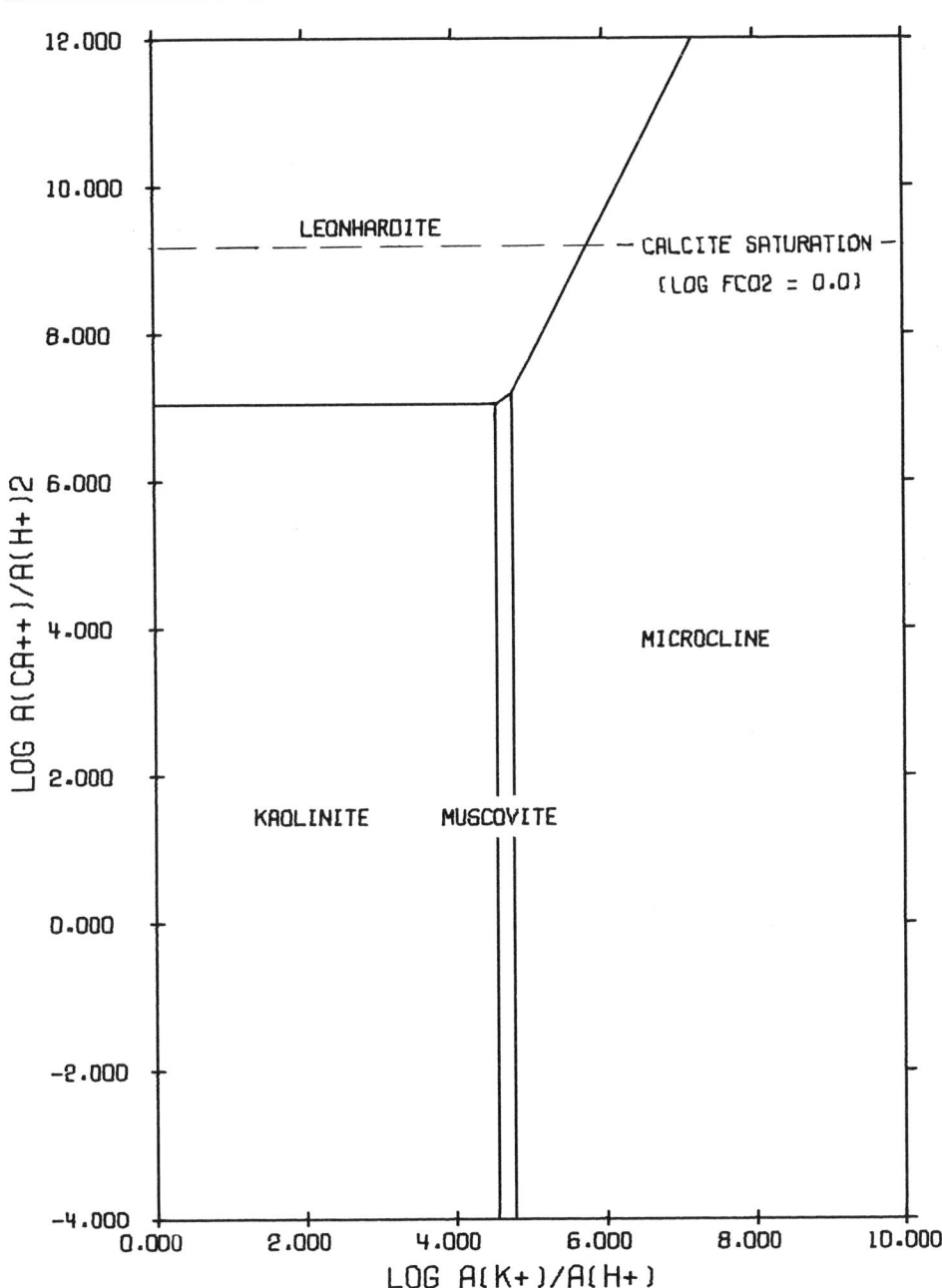

The System HCl—H$_2$O—Al$_2$O$_3$—CaO—CO$_2$—K$_2$O—SiO$_2$ at 100°C; log $a_{H_4SiO_4}$ = −3.08 = quartz saturation.

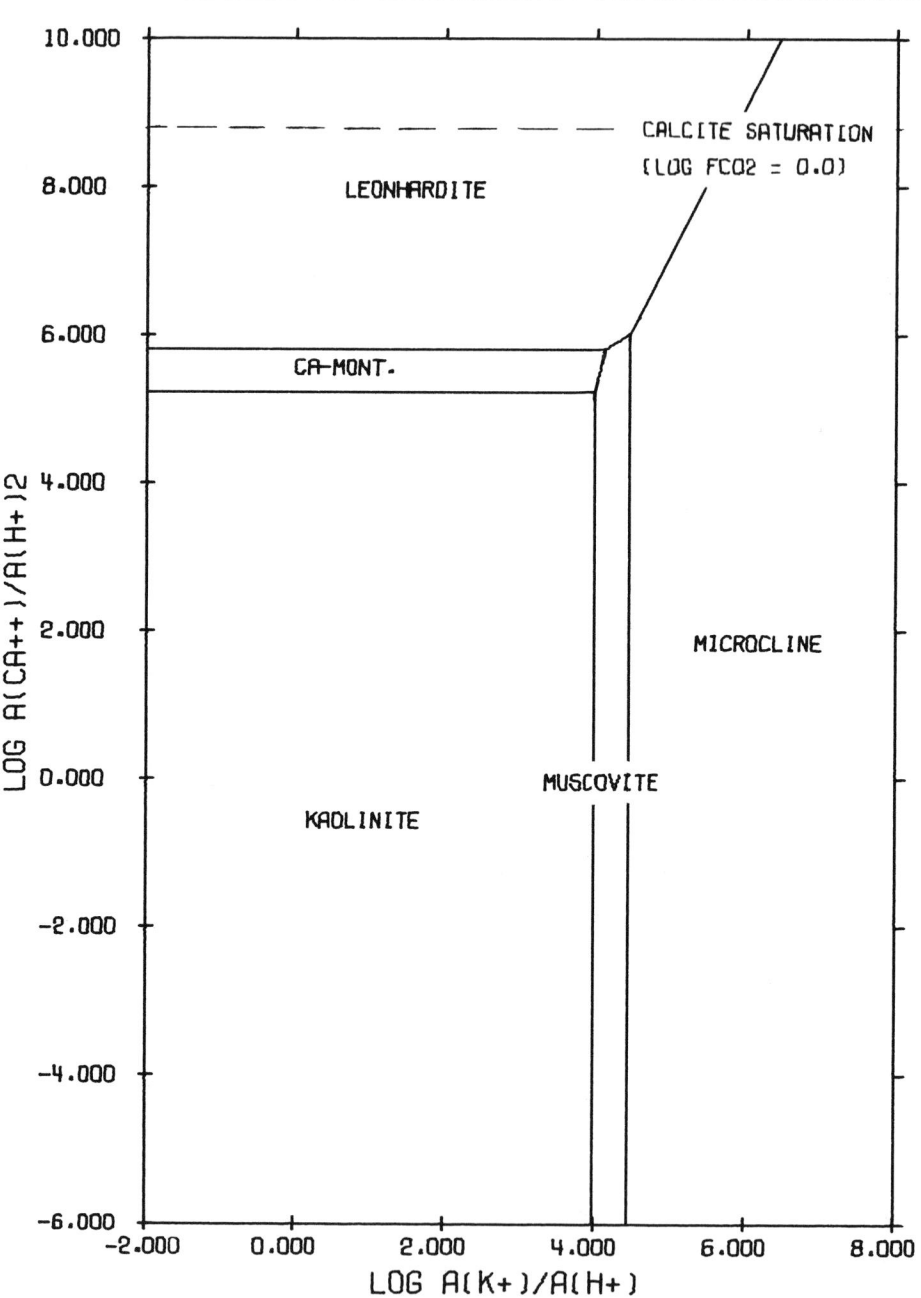

The System HCl—H_2O—Al_2O_3—CaO—CO_2—K_2O—SiO_2 at 150°C; log $a_{H_4SiO_4}$ = −2.67 = quartz saturation.

Activity Diagrams

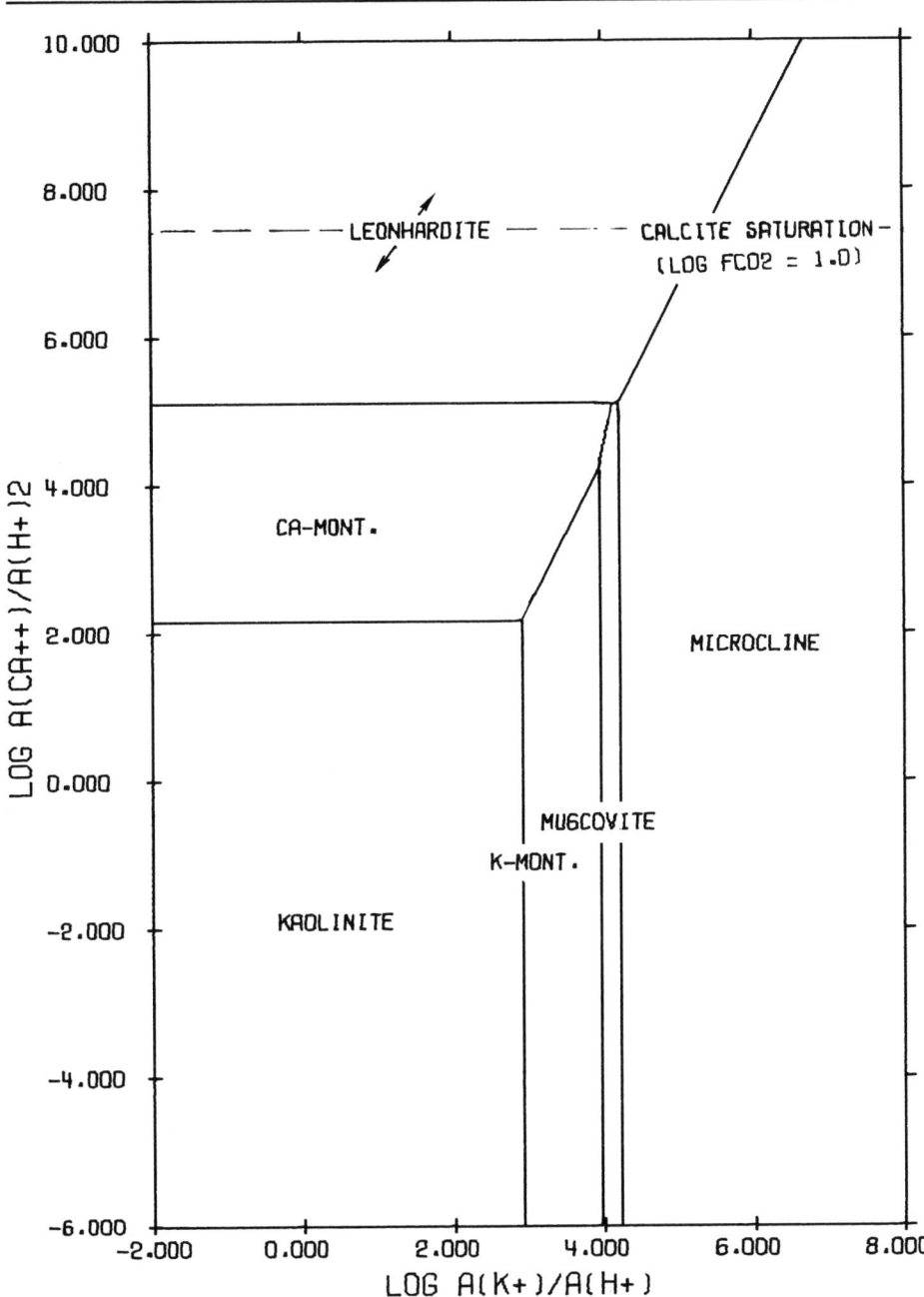

The System $HCl—H_2O—Al_2O_3—CaO—CO_2—K_2O—SiO_2$ at 200°C; $\log a_{H_4SiO_4} = -2.35 =$ quartz saturation.

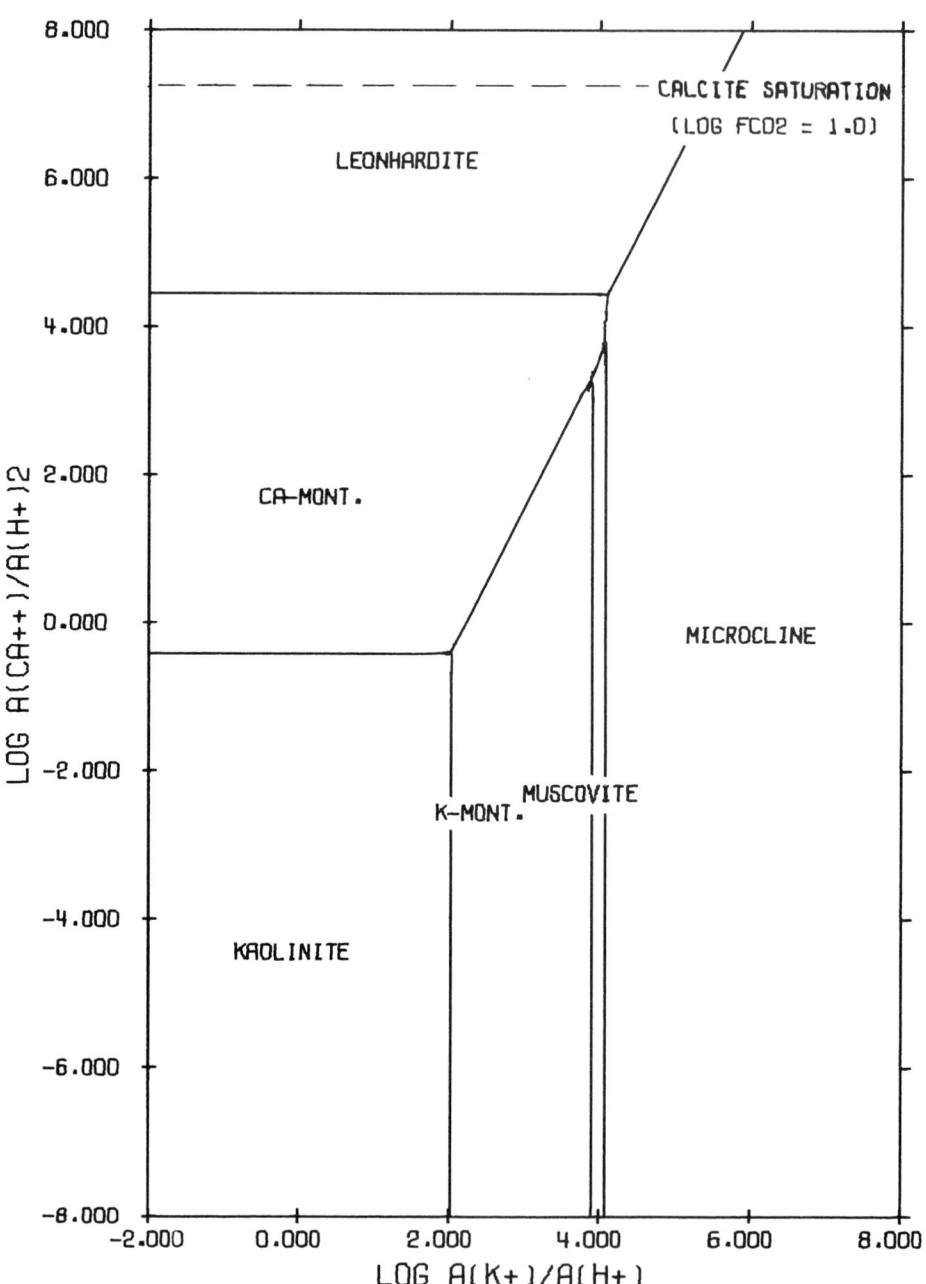

The System HCl—H_2O—Al_2O_3—CaO—CO_2—K_2O—SiO_2 at 250°C; log $a_{H_4SiO_4}$ = −2.11 = quartz saturation.

Activity Diagrams

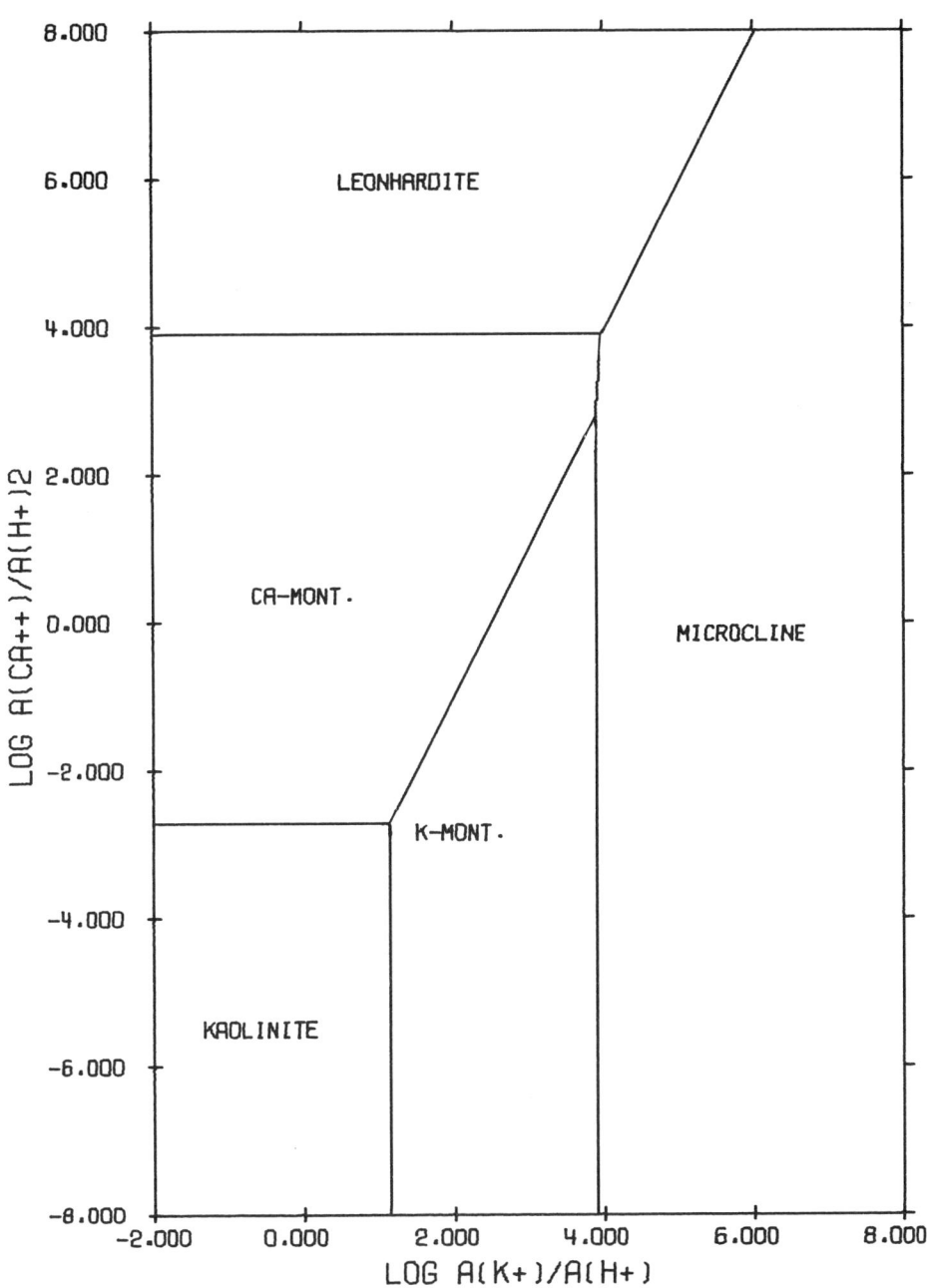

The System HCl—H_2O—Al_2O_3—CaO—CO_2—K_2O—SiO_2 at 300°C; log $a_{H_4SiO_4}$ = −1.94 = quartz saturation.

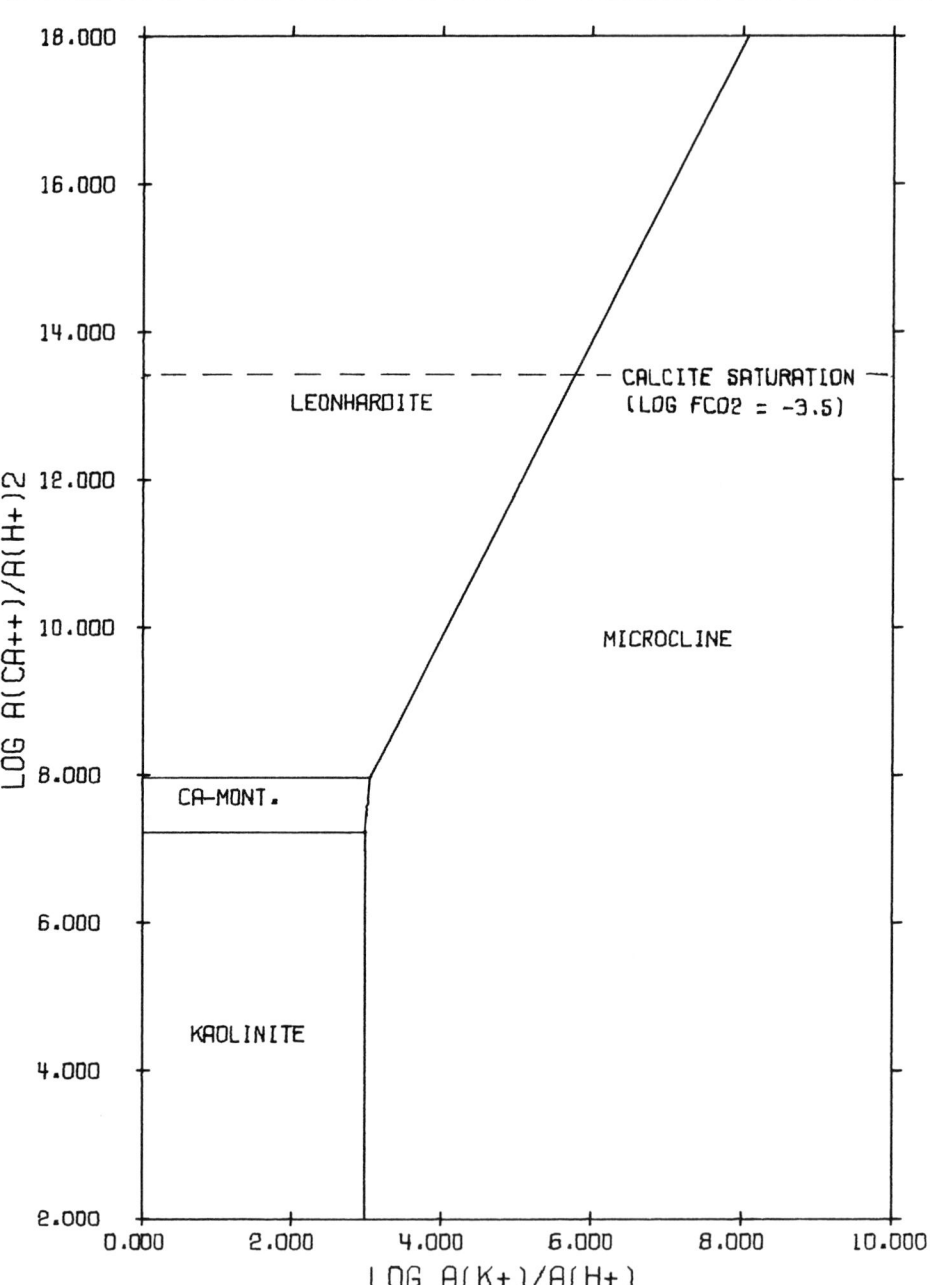

The System HCl—H$_2$O—Al$_2$O$_3$—CaO—CO$_2$—K$_2$O—SiO$_2$ at 0°C; log $a_{H_4SiO_4}$ = -3.00 = amorphous silica saturation.

Activity Diagrams

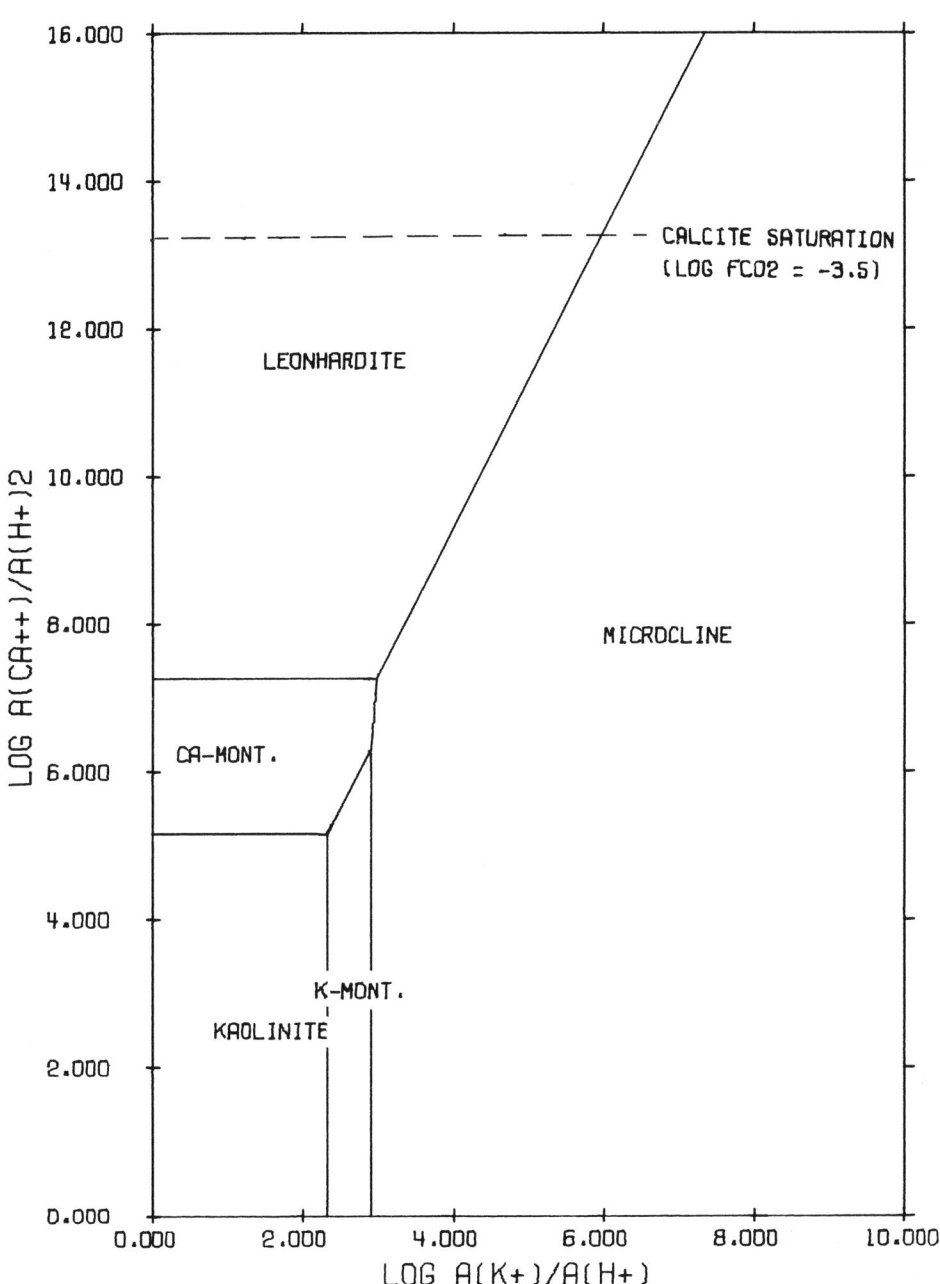

The System HCl—H$_2$O—Al$_2$O$_3$—CaO—CO$_2$—K$_2$O—SiO$_2$ at 25°C; log $a_{\text{H}_4\text{SiO}_4}$ = −2.70 = amorphous silica saturation.

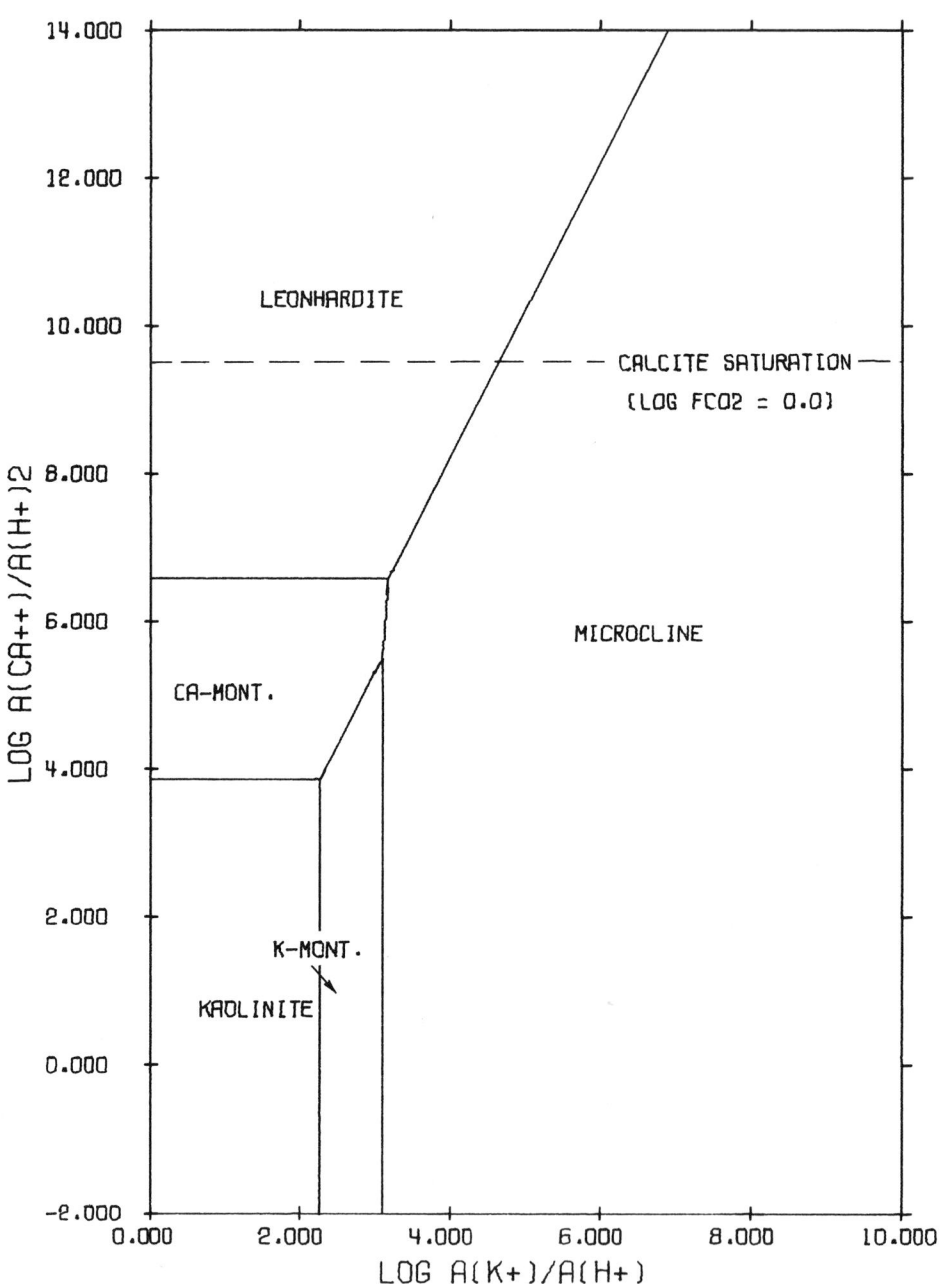

The System HCl—H$_2$O—Al$_2$O$_3$—CaO—CO$_2$—K$_2$O—SiO$_2$ at 60°C; log $a_{H_4SiO_4}$ = −2.47 = amorphous silica saturation.

Activity Diagrams

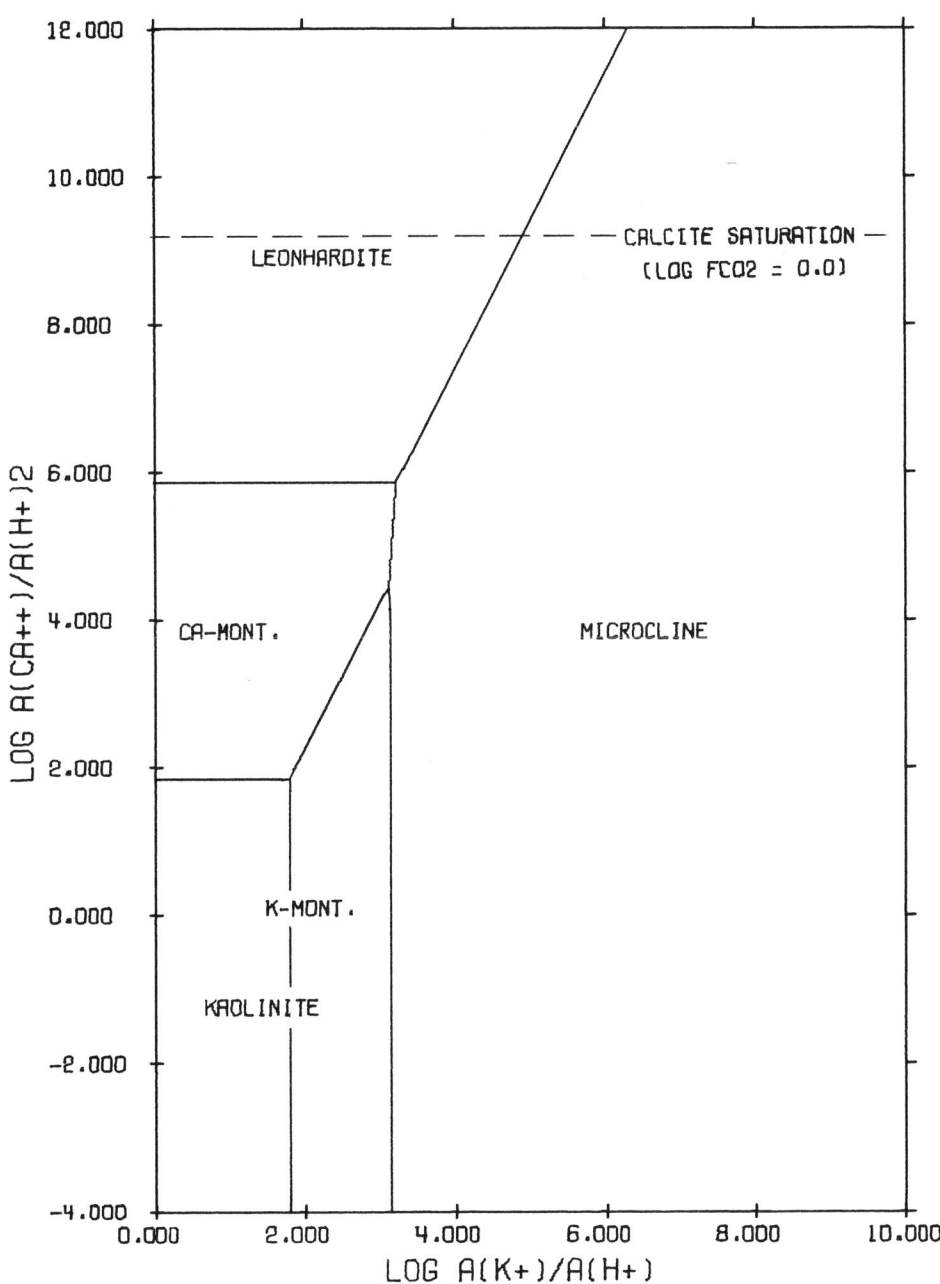

The System HCl—H_2O—Al_2O_3—CaO—CO_2—K_2O—SiO_2 at 100°C; log $a_{H_4SiO_4}$ = −2.20 = amorphous silica saturation.

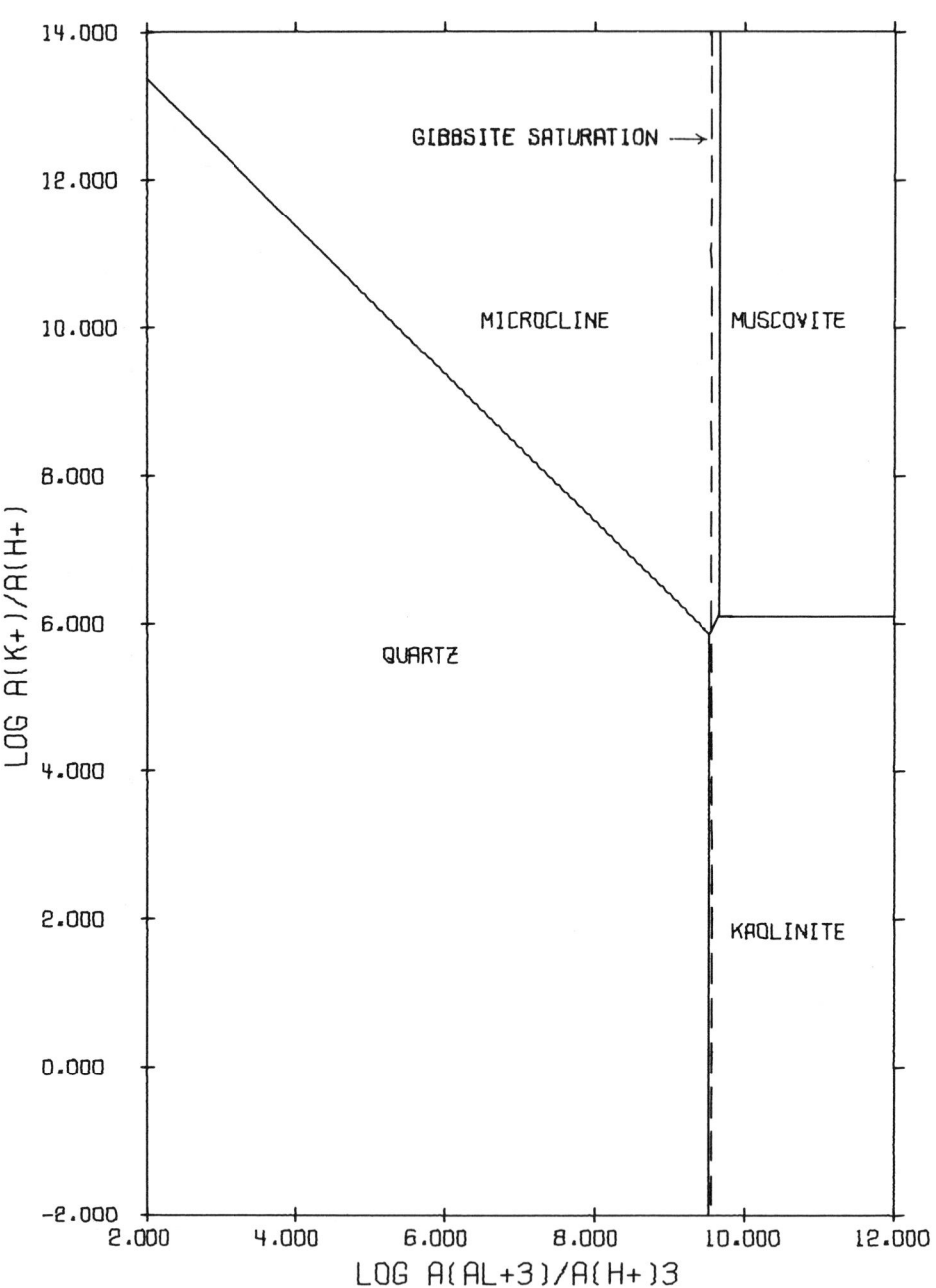

The System HCl—H$_2$O—Al$_2$O$_3$—K$_2$O—SiO$_2$ at 0°C.

Activity Diagrams

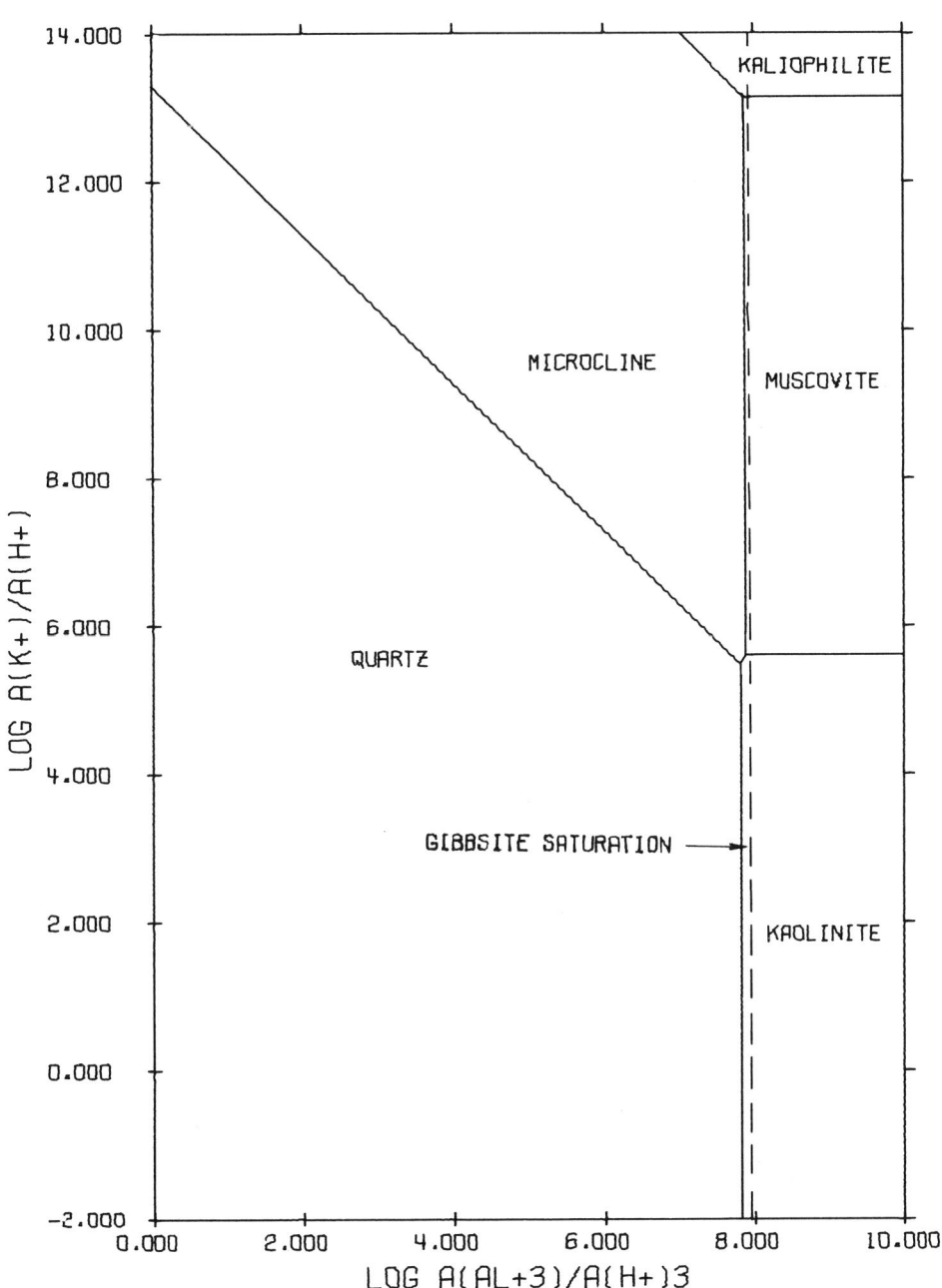

The System HCl—H_2O—Al_2O_3—K_2O—SiO_2 at 25°C.

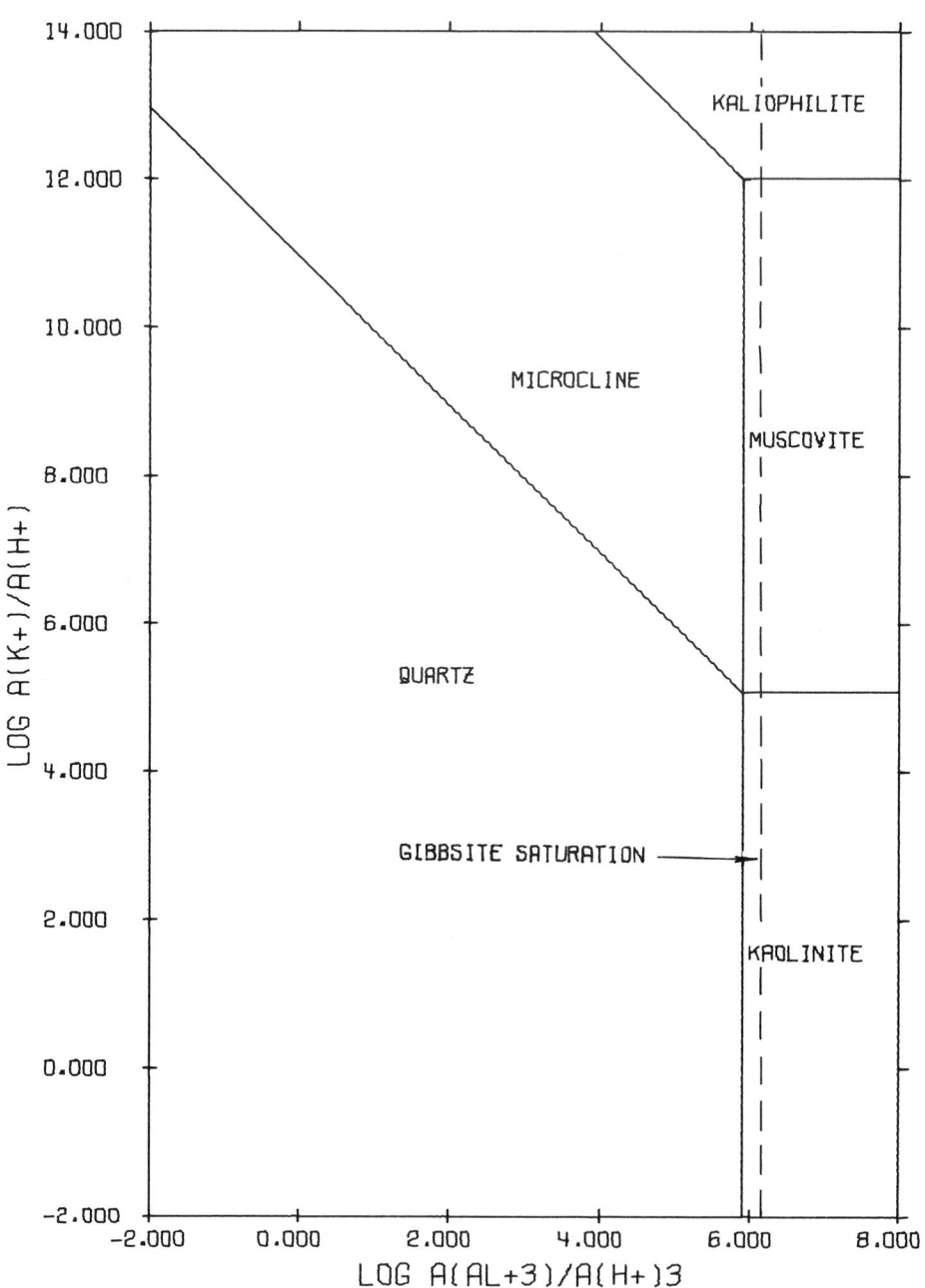

The System HCl—H₂O—Al₂O₃—K₂O—SiO₂ at 60°C.

Activity Diagrams

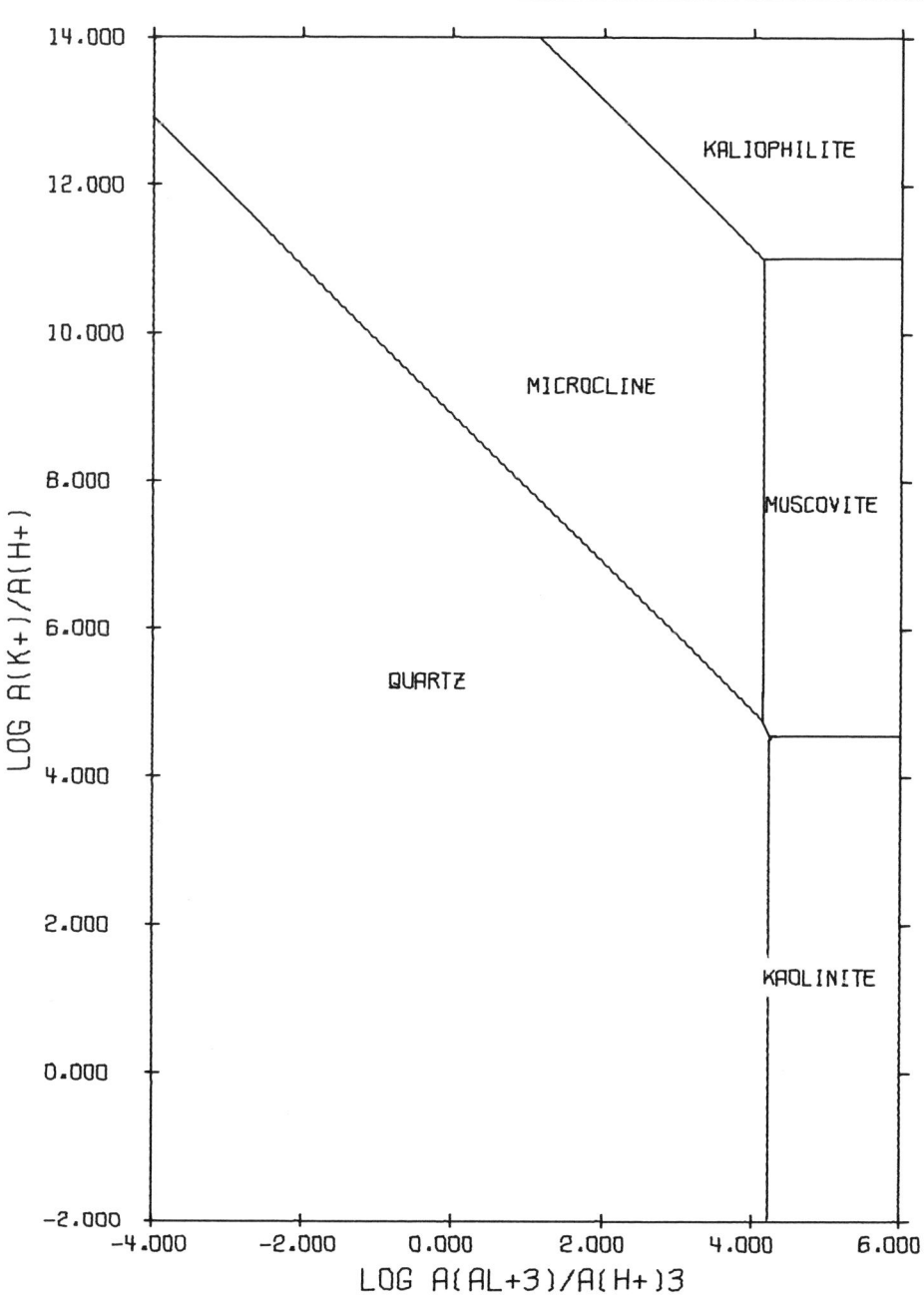

The System HCl—H₂O—Al₂O₃—K₂O—SiO₂ at 100°C.

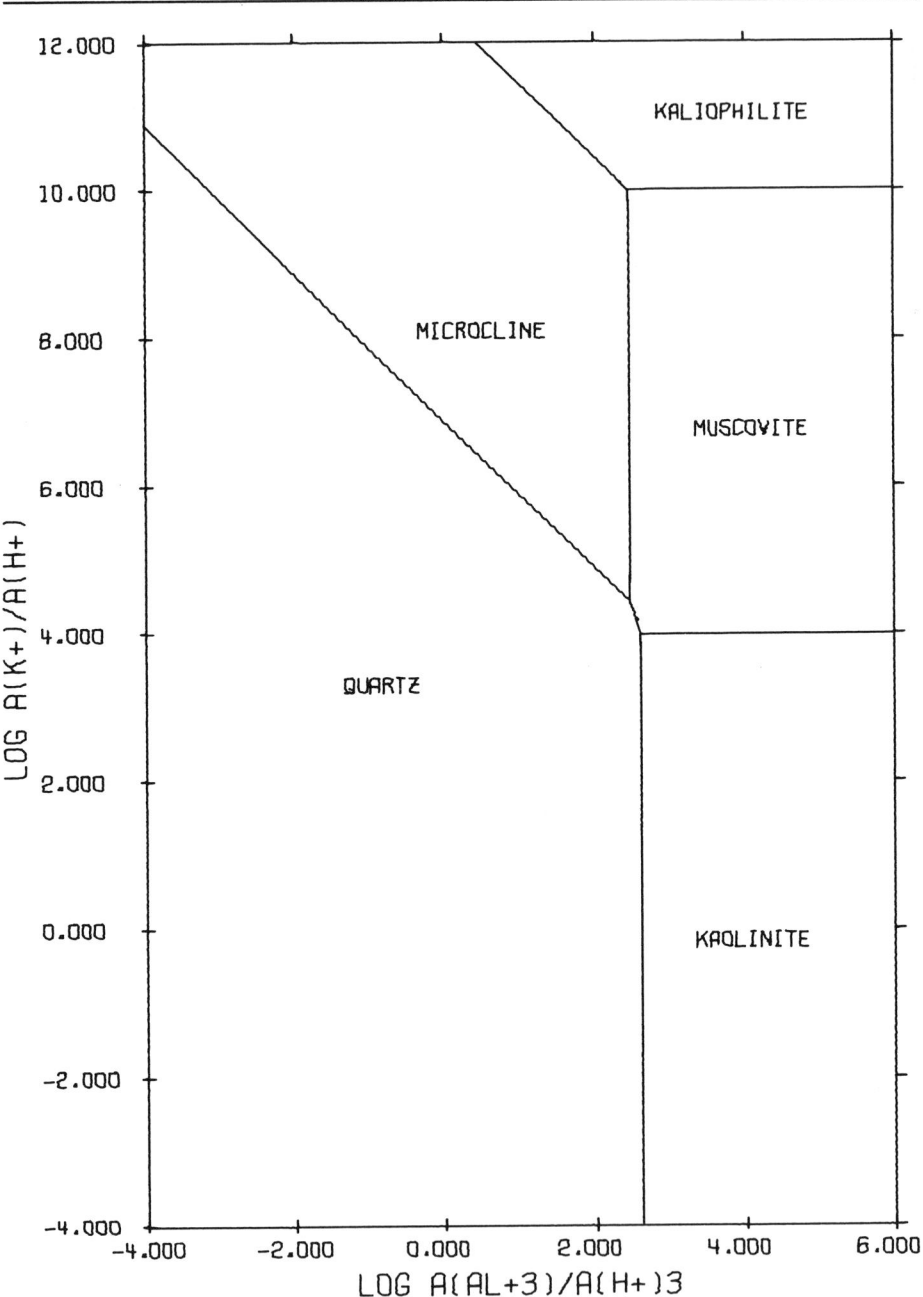

The System HCl—H$_2$O—Al$_2$O$_3$—K$_2$O—SiO$_2$ at 150°C.

Activity Diagrams

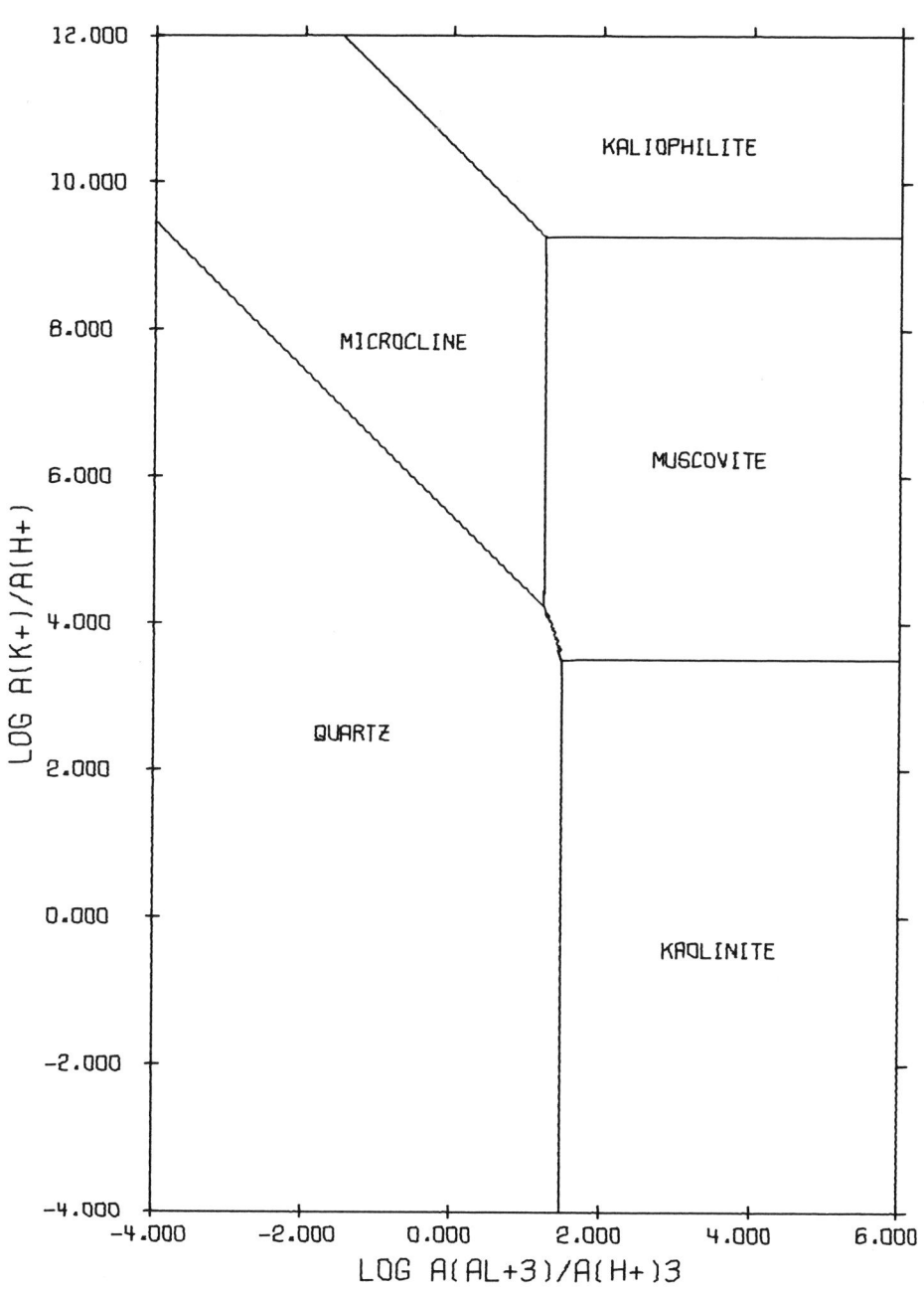

The System HCl—H$_2$O—Al$_2$O$_3$—K$_2$O—SiO$_2$ at 200°C.

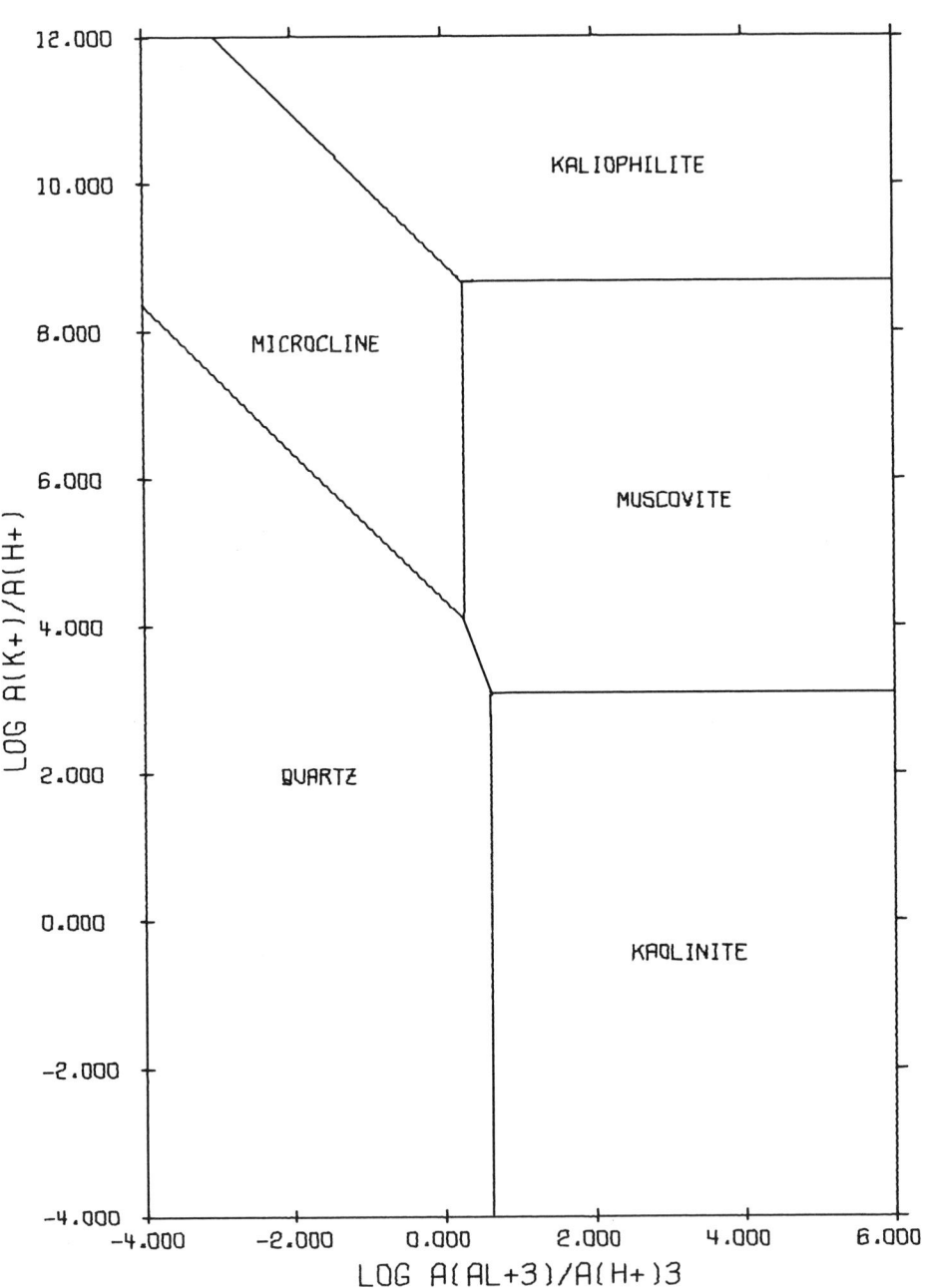

The System HCl—H$_2$O—Al$_2$O$_3$—K$_2$O—SiO$_2$ at 250°C.

Activity Diagrams

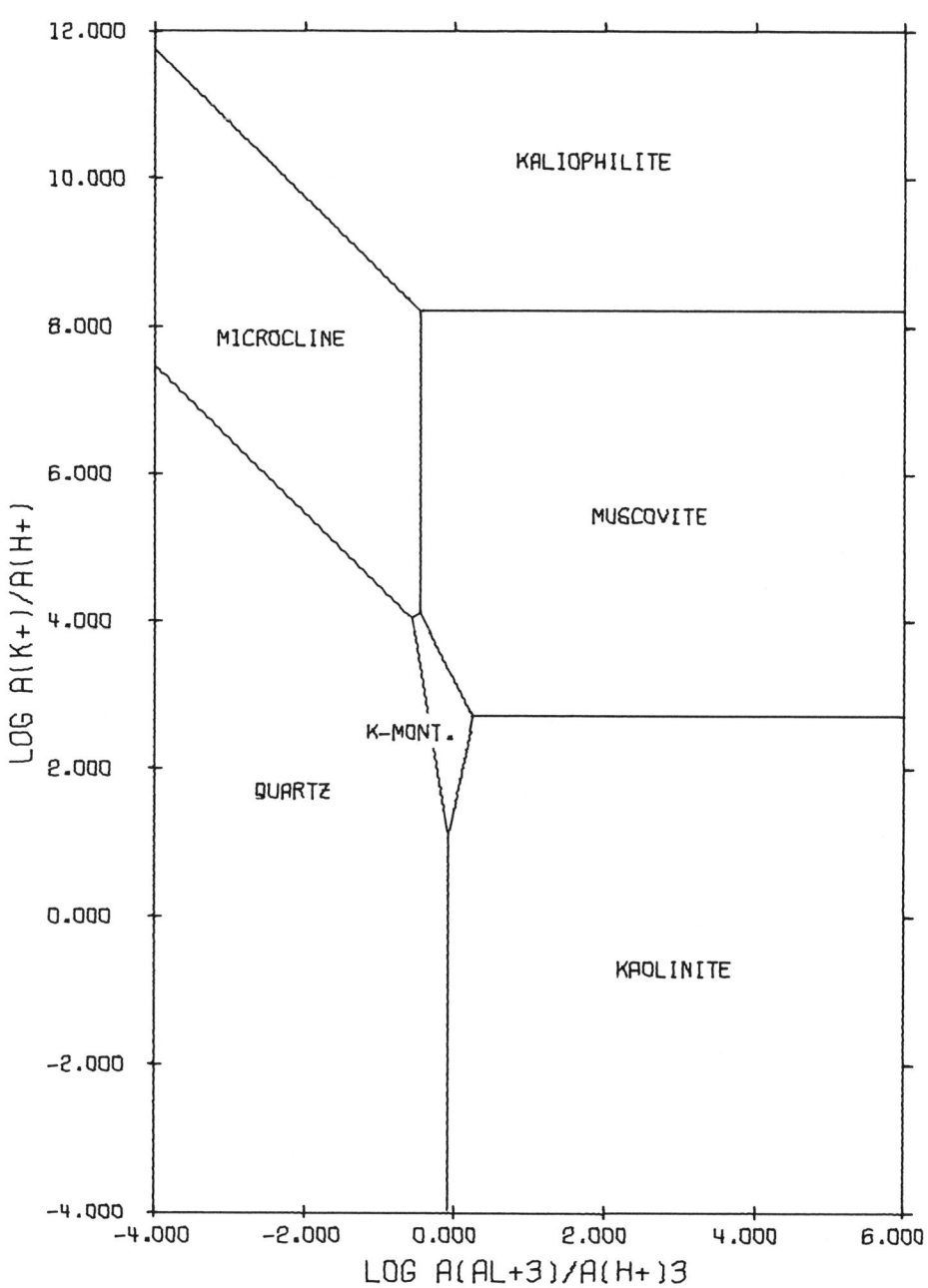

The System HCl—H_2O—Al_2O_3—K_2O—SiO_2 at 300°C.

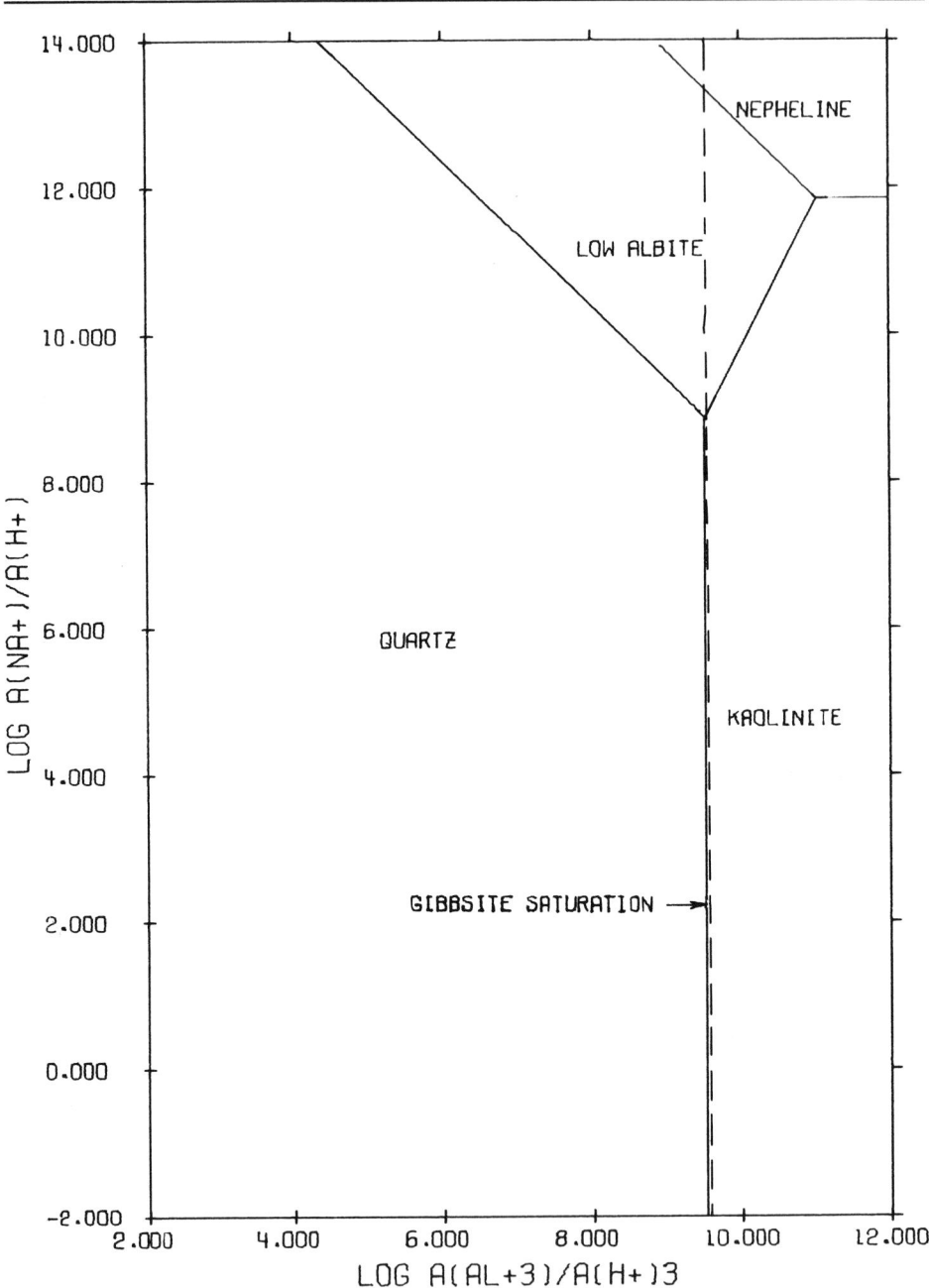

The System HCl—H₂O—Al₂O₃—Na₂O—SiO₂ at 0°C.

Activity Diagrams

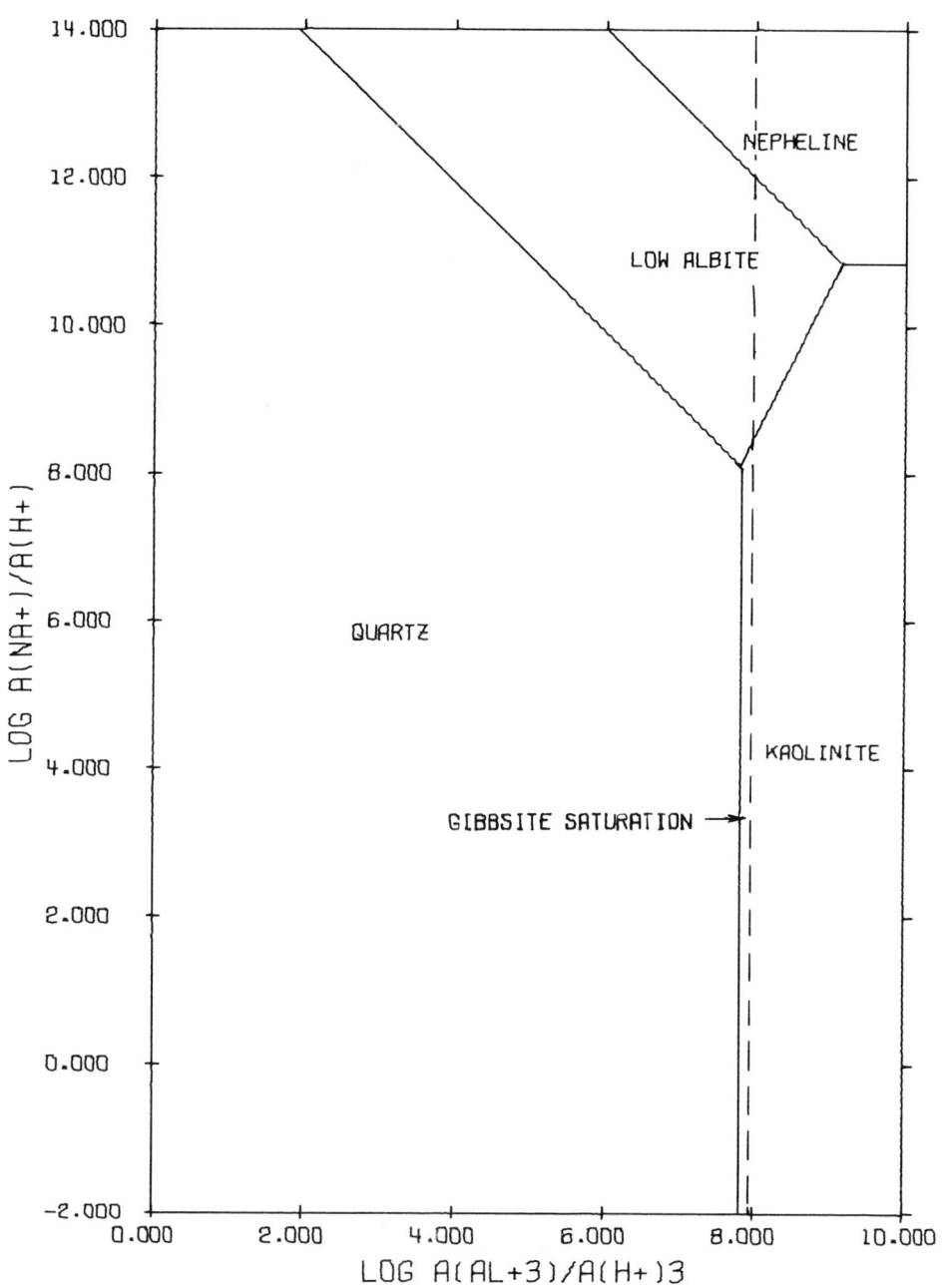

The System HCl—H$_2$O—Al$_2$O$_3$—Na$_2$O—SiO$_2$ at 25°C.

Activity Diagrams

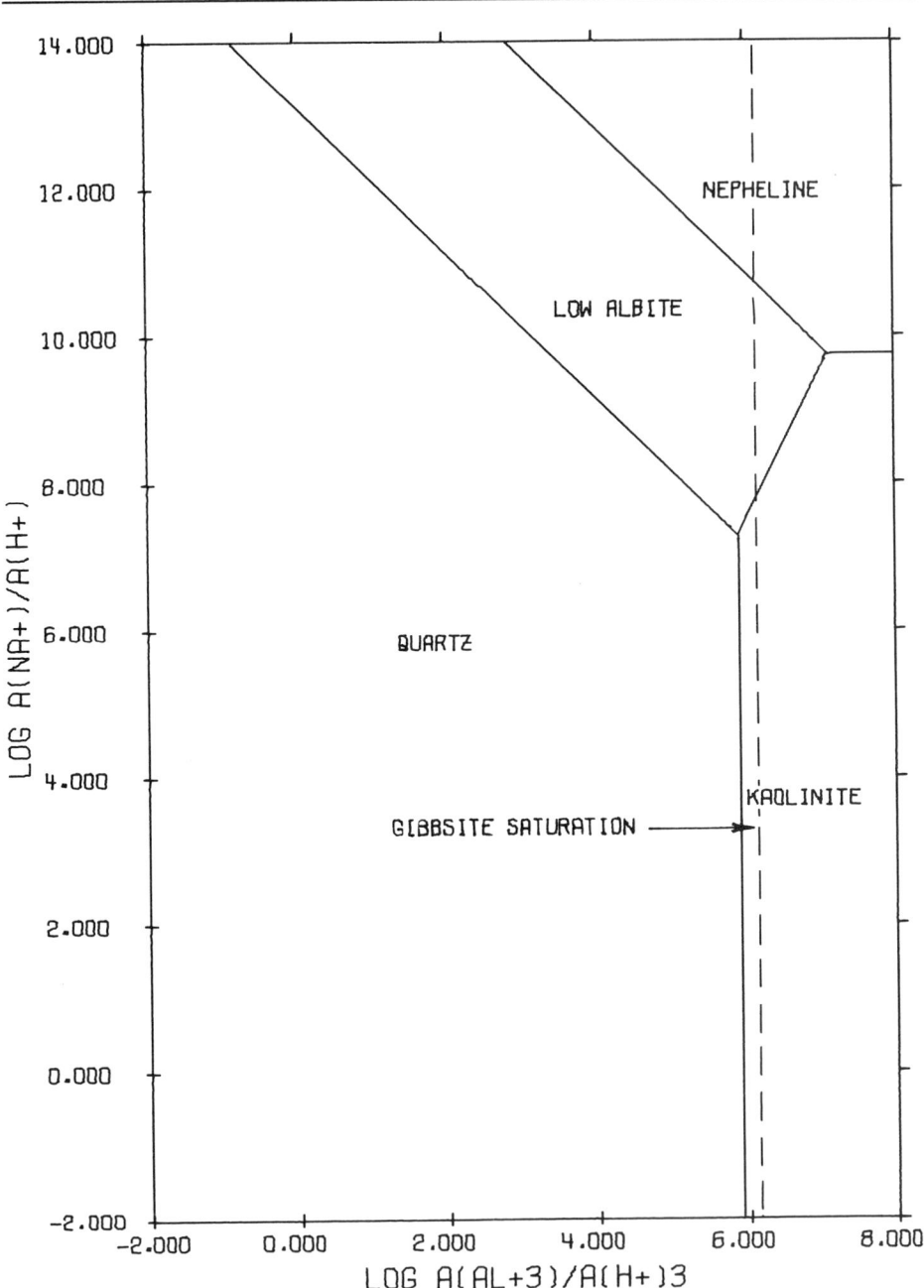

The System HCl—H$_2$O—Al$_2$O$_3$—Na$_2$O—SiO$_2$ at 60°C.

Activity Diagrams

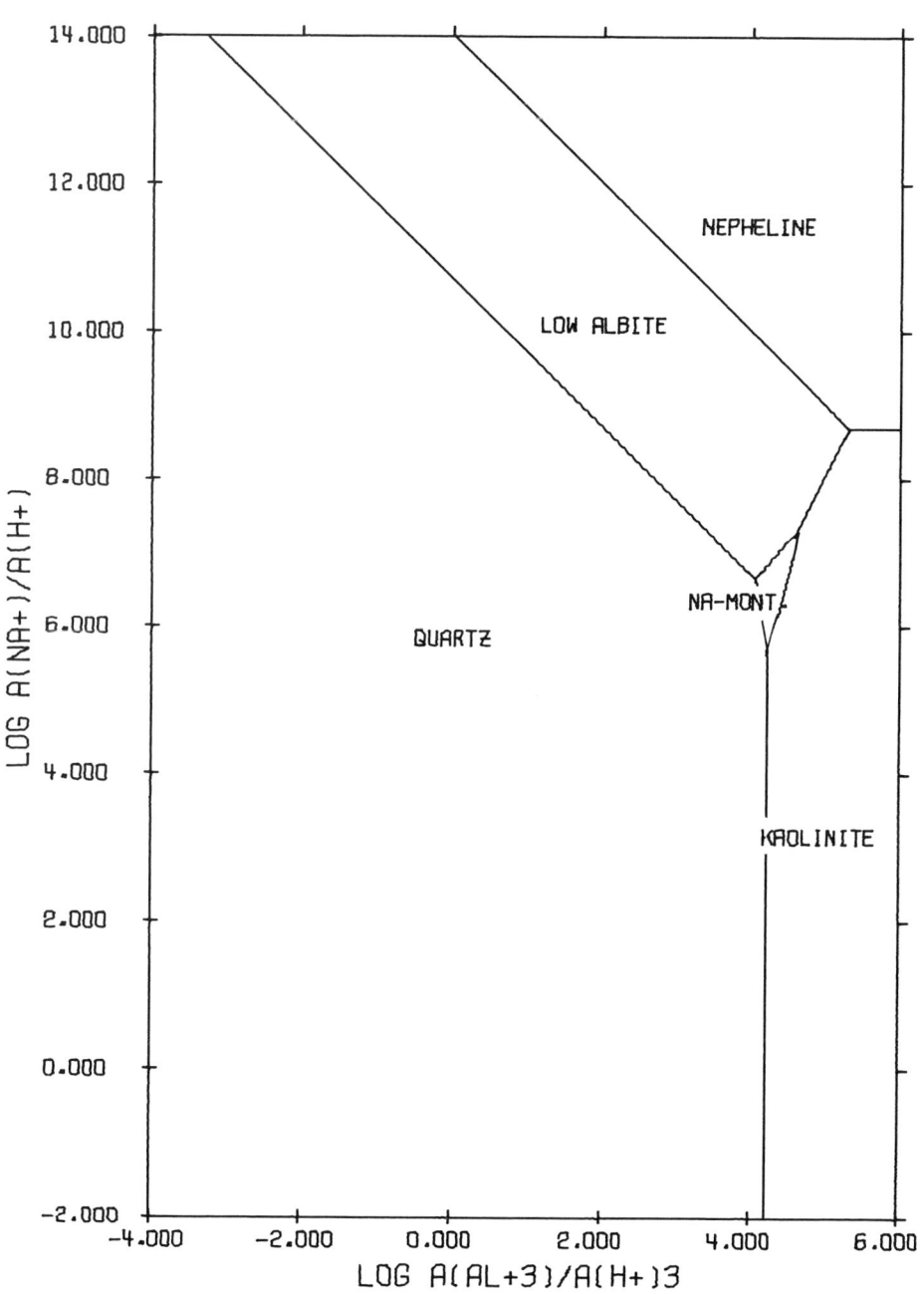

The System HCl—H$_2$O—Al$_2$O$_3$—Na$_2$O—SiO$_2$ at 100°C.

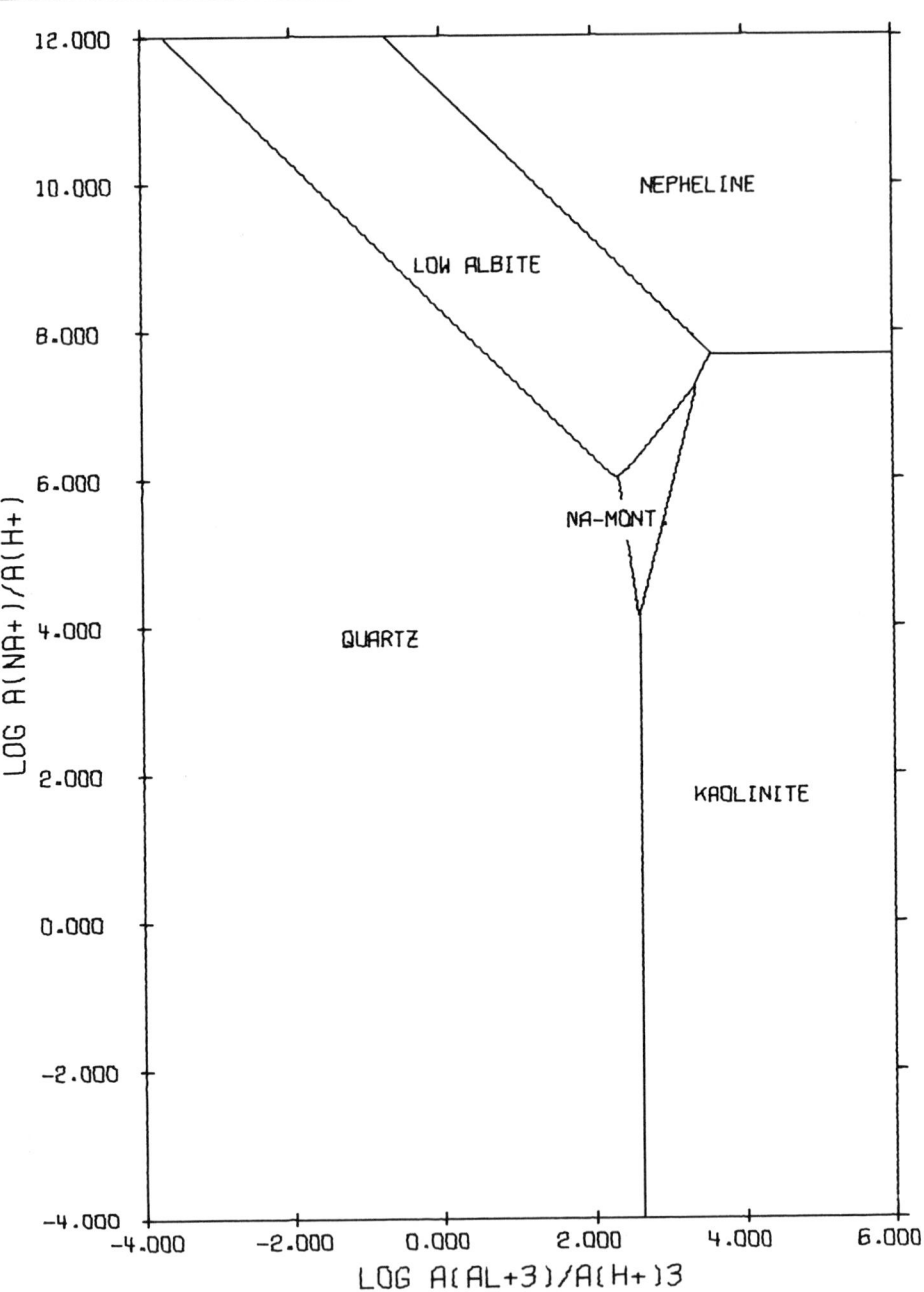

The System HCl—H_2O—Al_2O_3—Na_2O—SiO_2 at 150°C.

Activity Diagrams

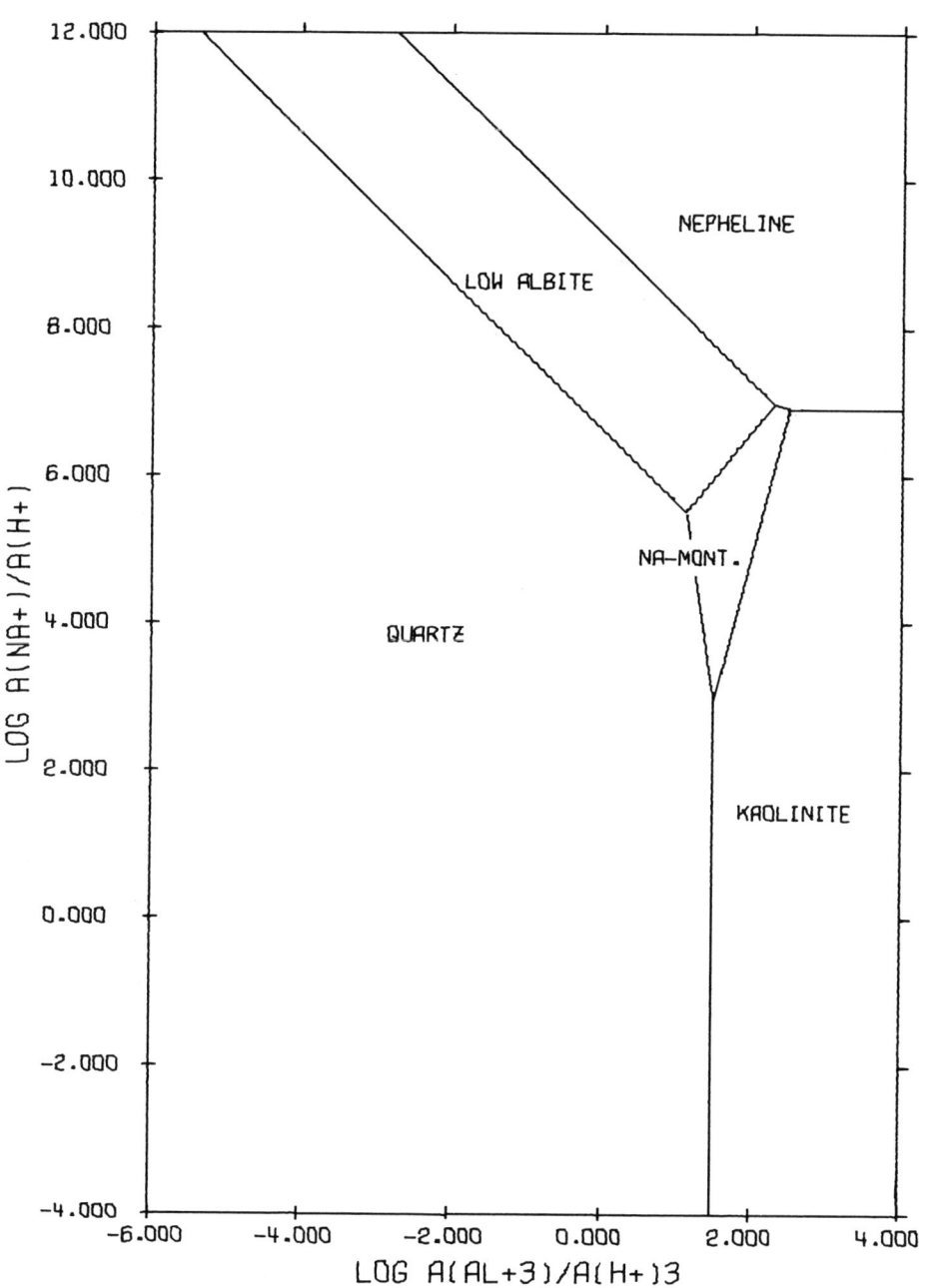

The System HCl—H$_2$O—Al$_2$O$_3$—Na$_2$O—SiO$_2$ at 200°C.

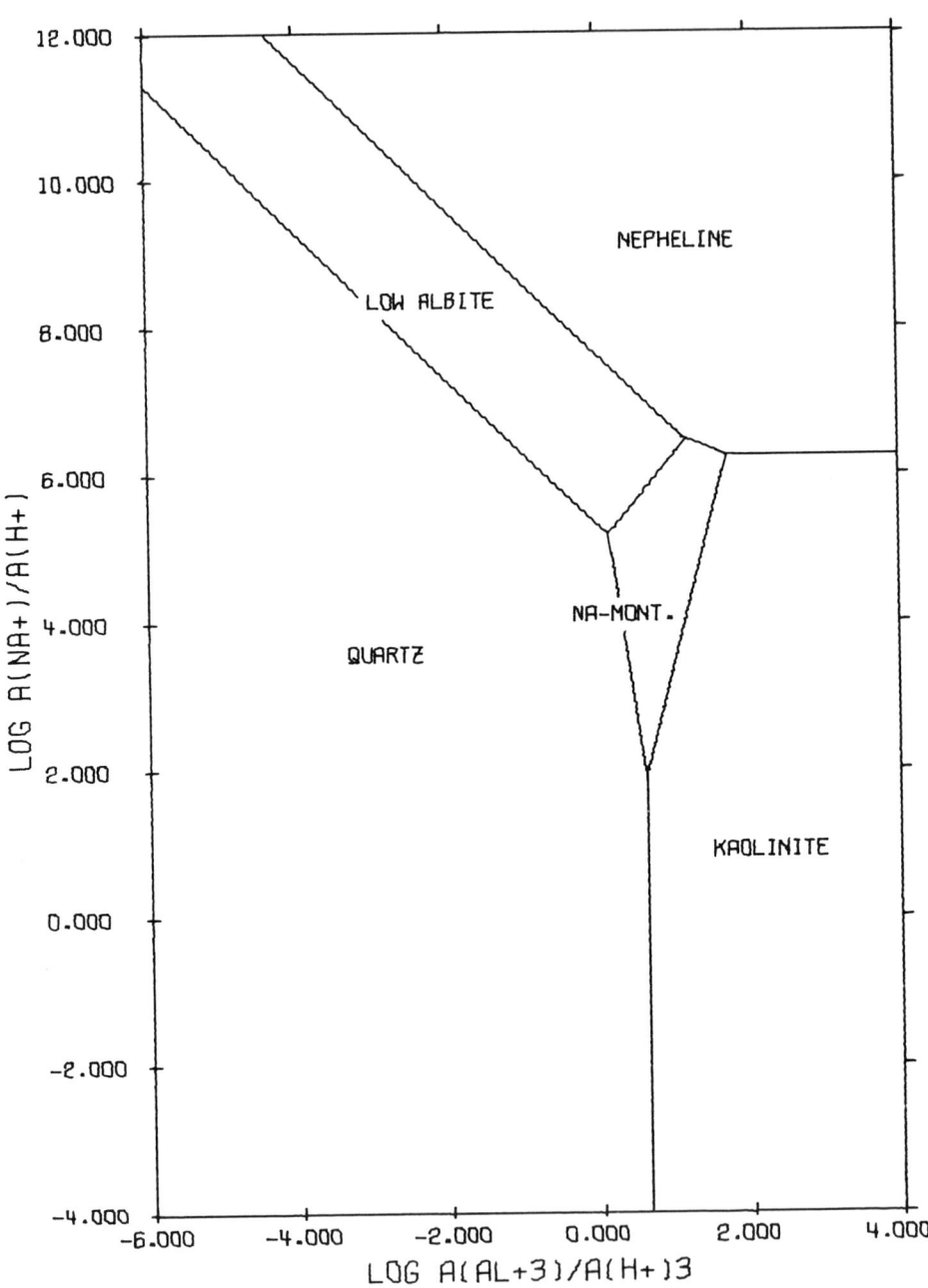

The System HCl—H₂O—Al₂O₃—Na₂O—SiO₂ at 250°C.

Activity Diagrams

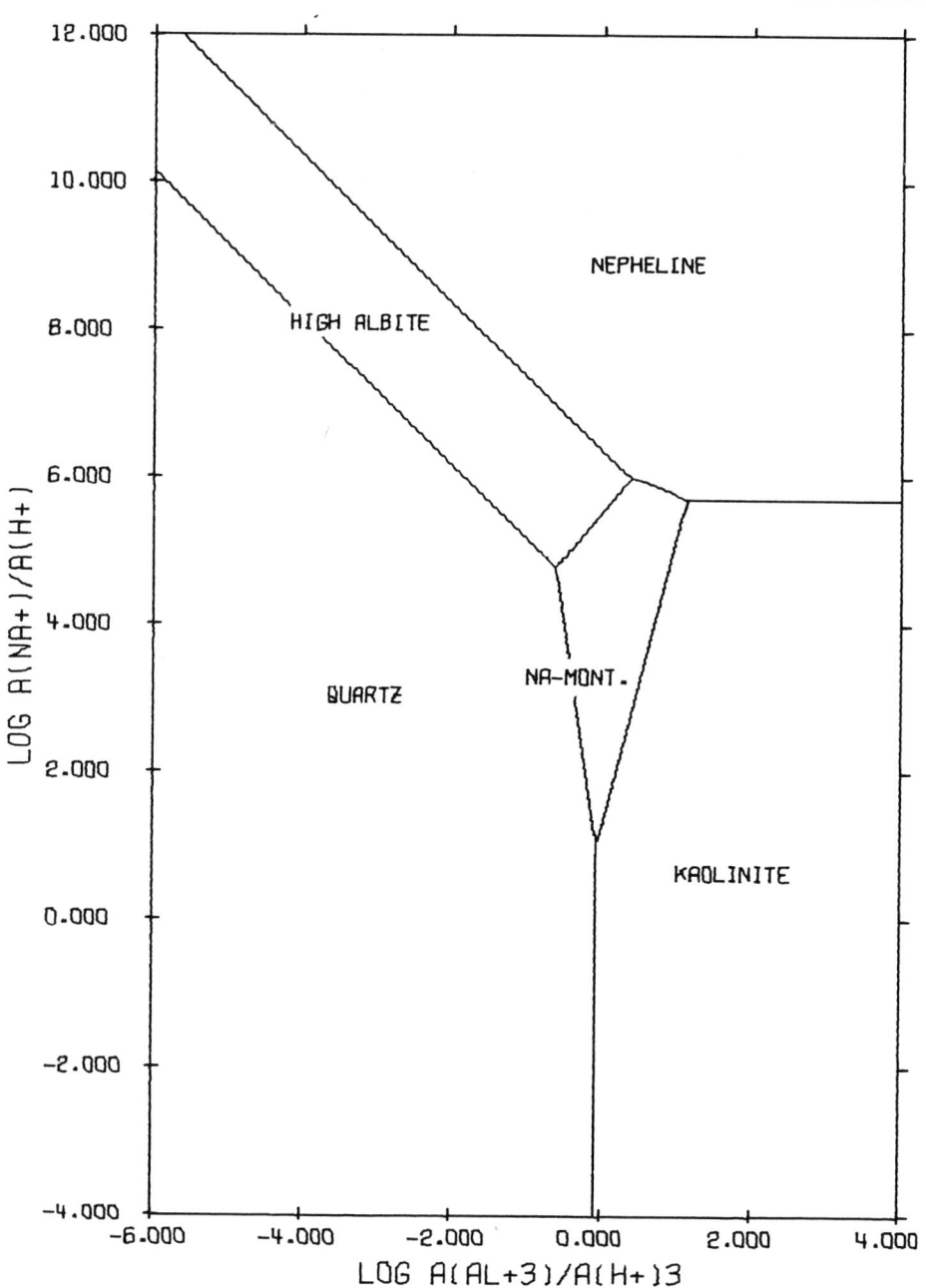

The System HCl—H$_2$O—Al$_2$O$_3$—Na$_2$O—SiO$_2$ at 300°C.

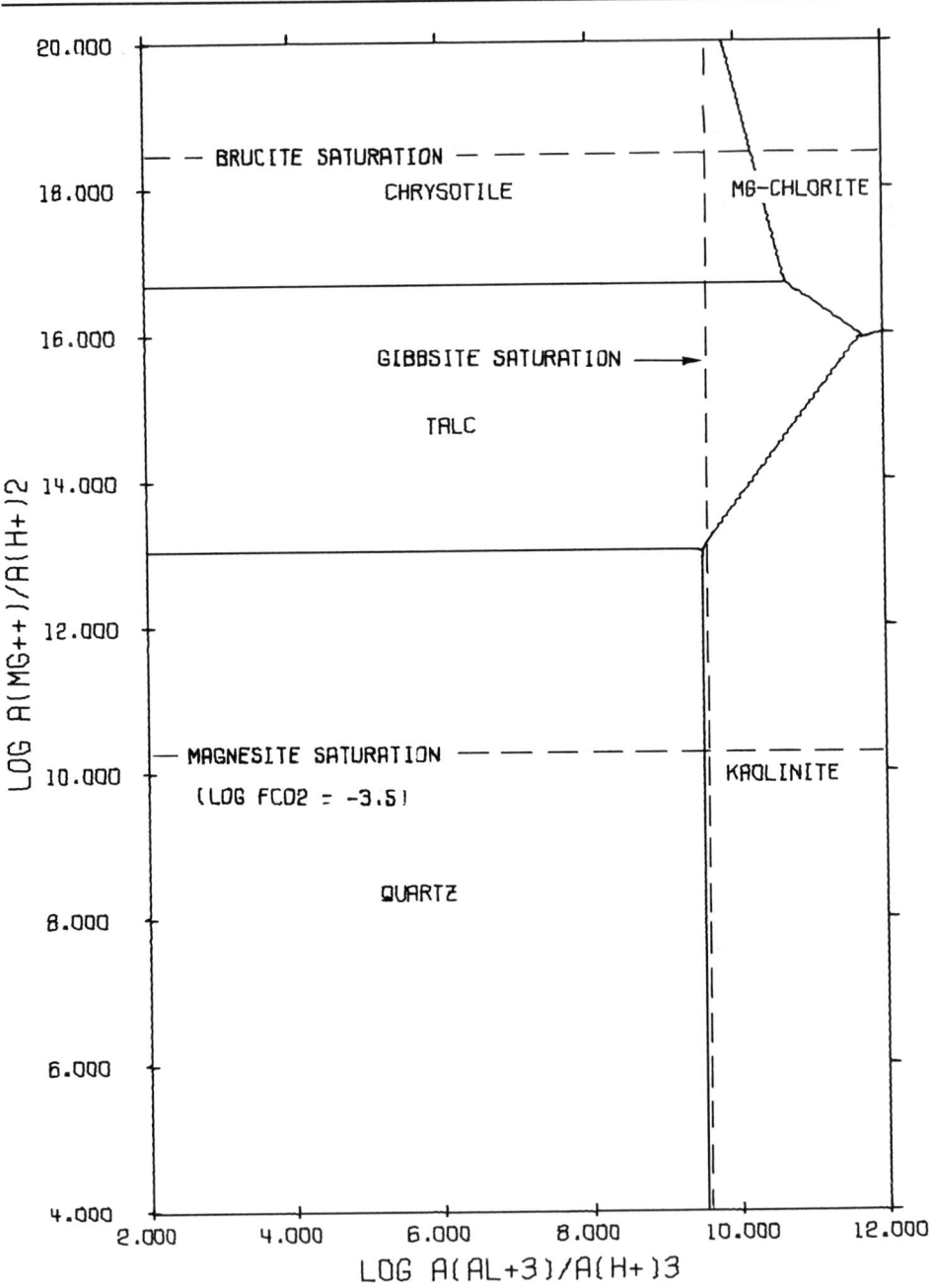

The System HCl—H$_2$O—Al$_2$O$_3$—CO$_2$—MgO—SiO$_2$ at 0°C.

Activity Diagrams

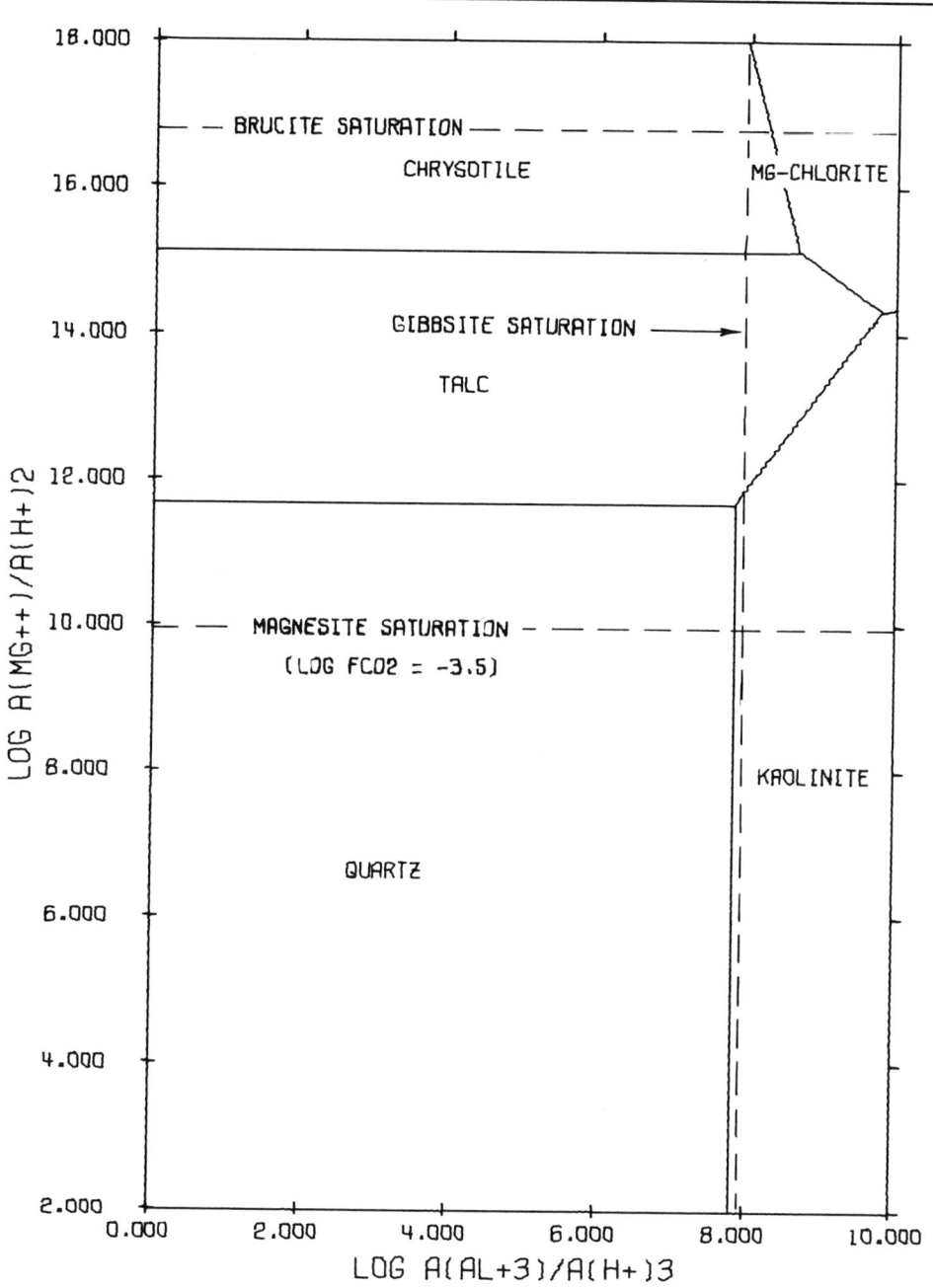

The System HCl—H$_2$O—Al$_2$O$_3$—CO$_2$—MgO—SiO$_2$ at 25°C.

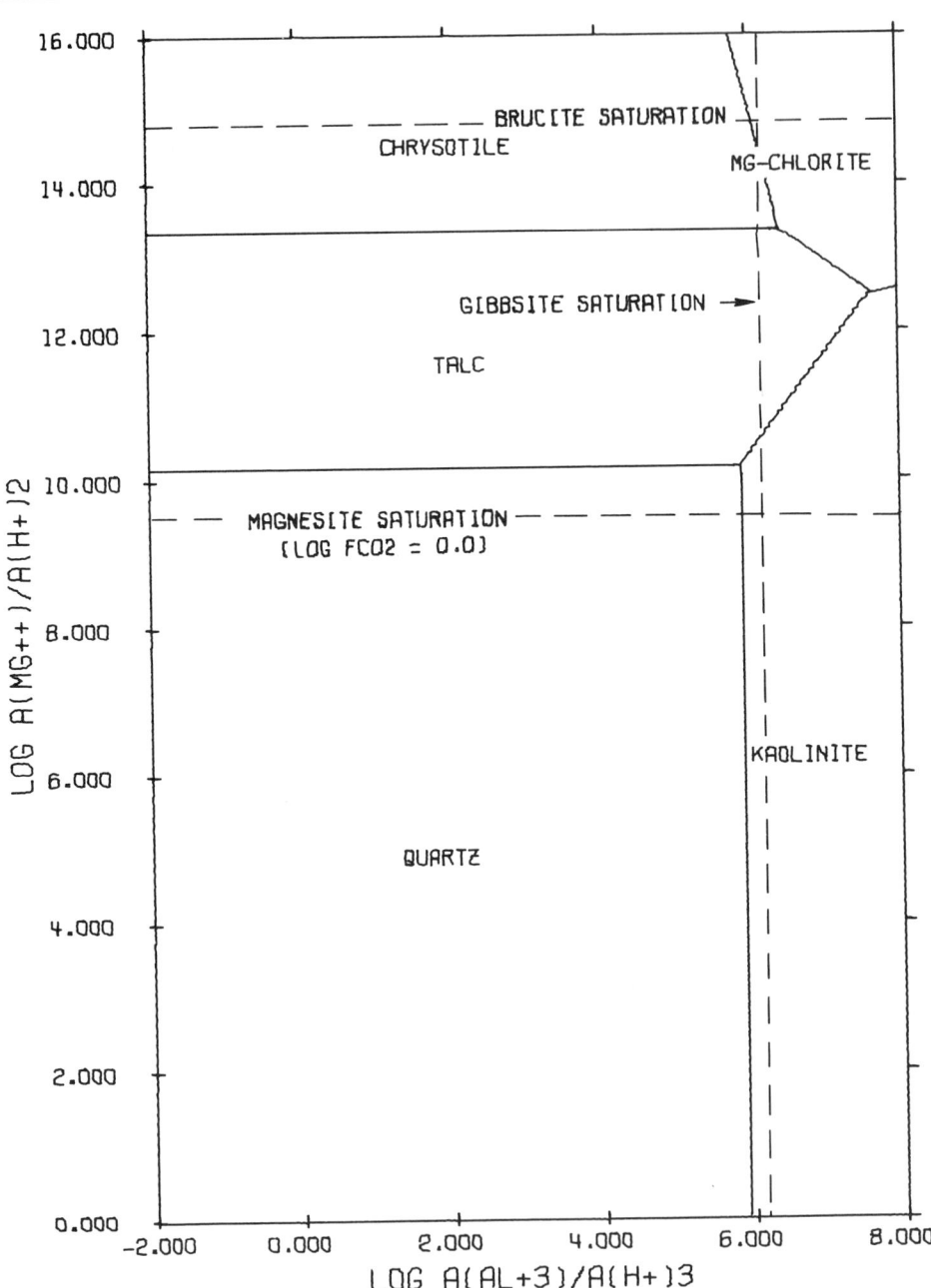

The System HCl—H$_2$O—Al$_2$O$_3$—CO$_2$—MgO—SiO$_2$ at 60°C.

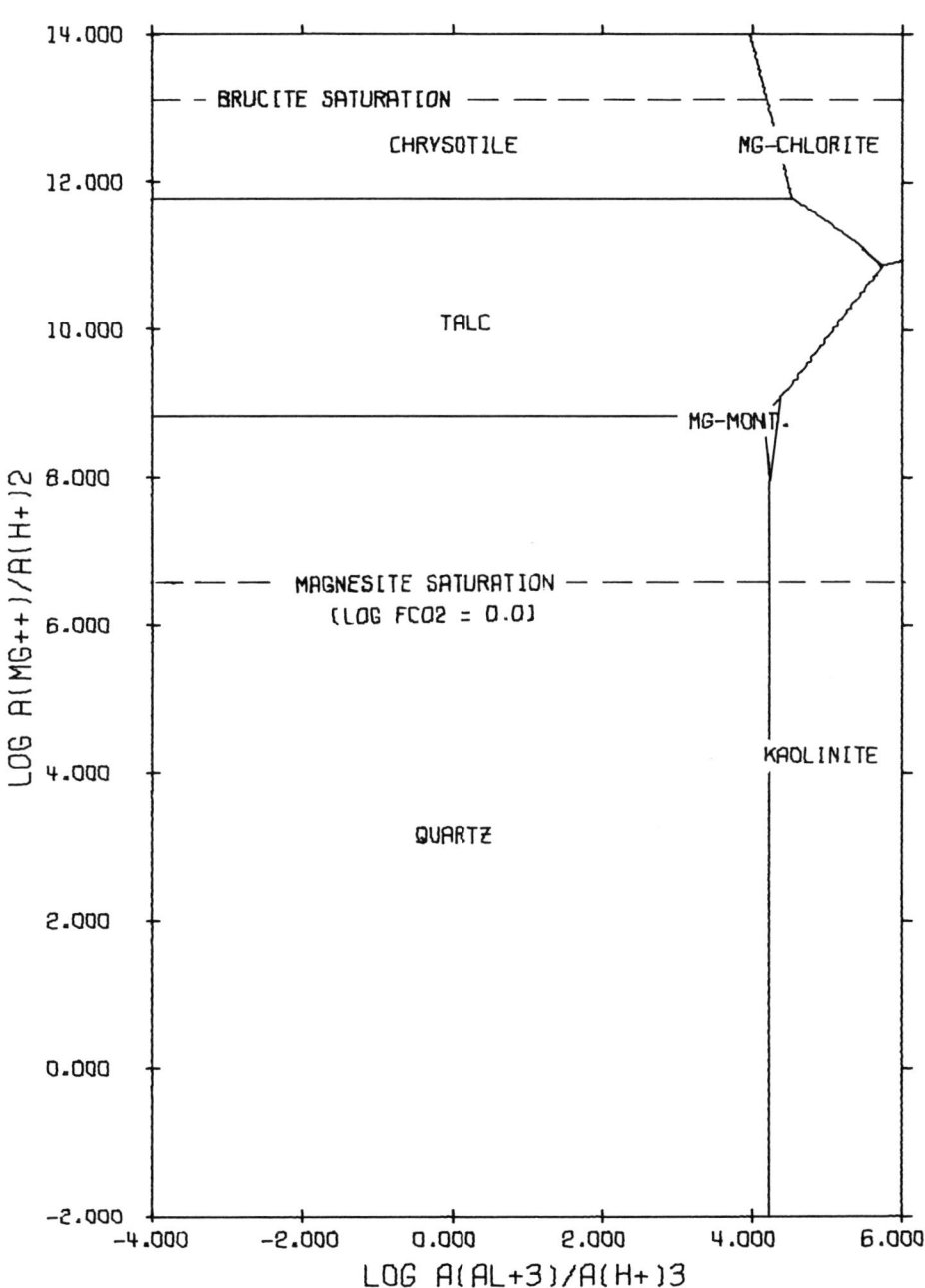

The System HCl—H_2O—Al_2O_3—CO_2—MgO—SiO_2 at 100°C.

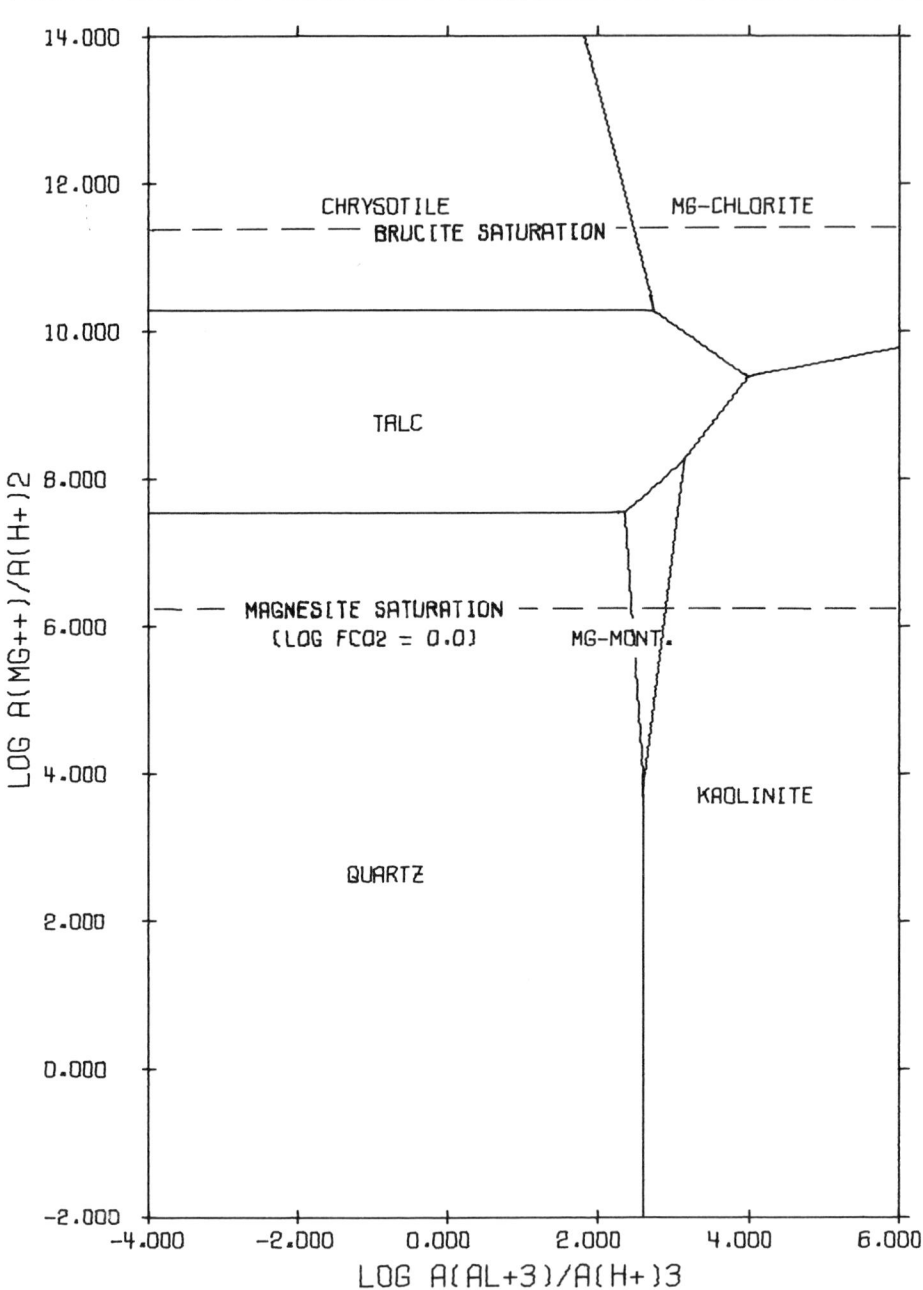

The System HCl—H$_2$O—Al$_2$O$_3$—CO$_2$—MgO—SiO$_2$ at 150°C.

Activity Diagrams

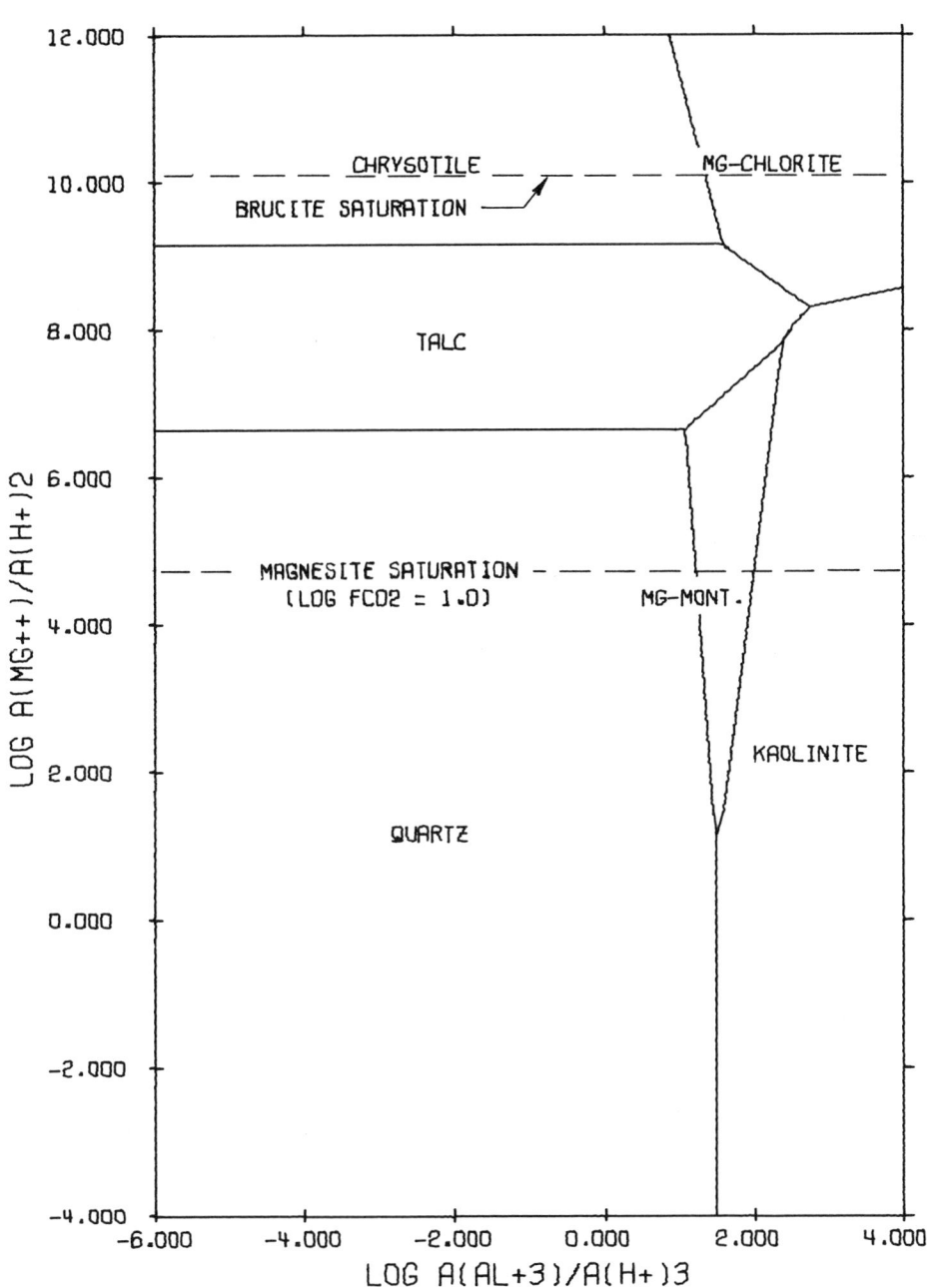

The System HCl—H$_2$O—Al$_2$O$_3$—CO$_2$—MgO—SiO$_2$ at 200°C.

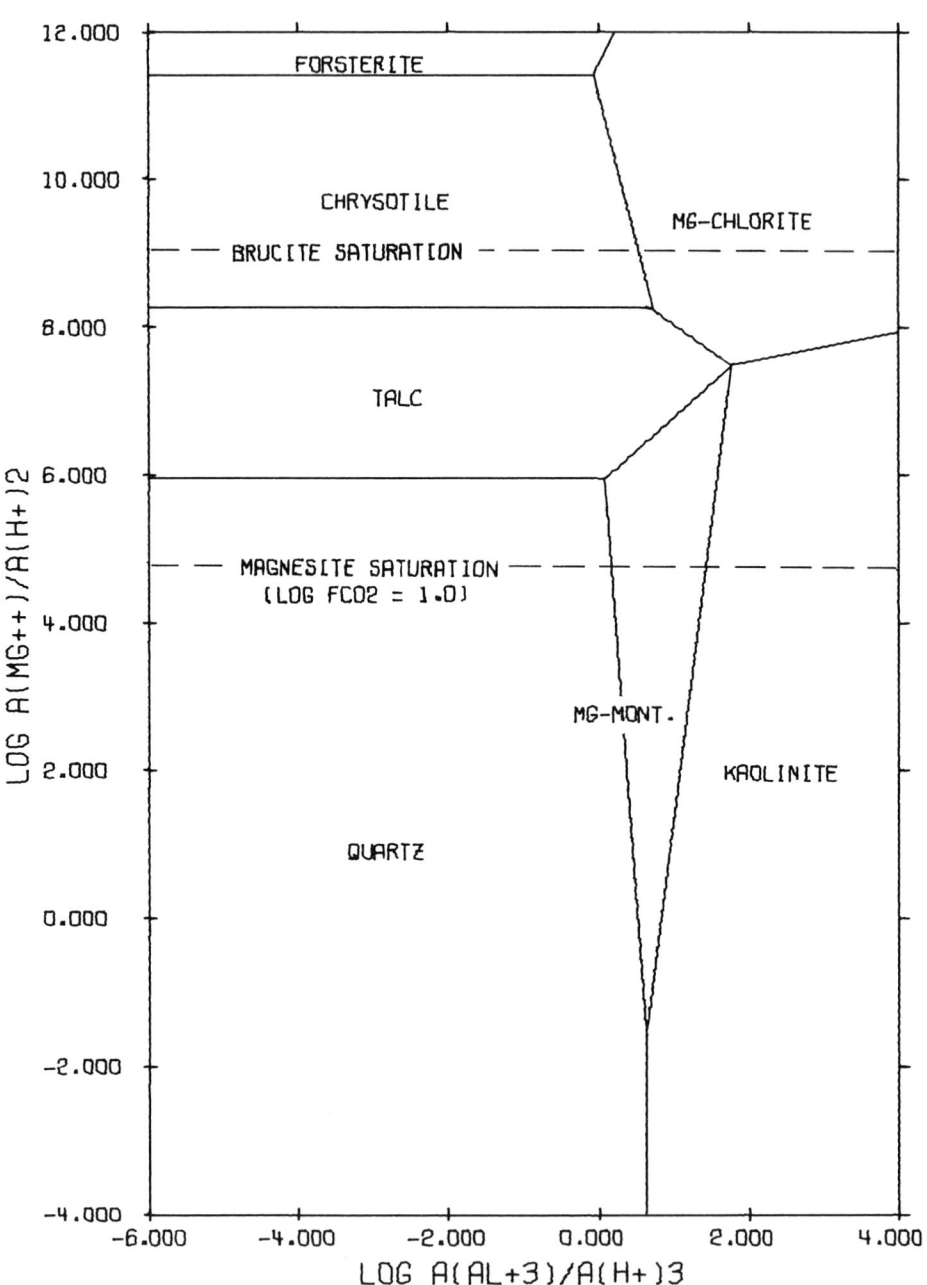

The System HCl—H_2O—Al_2O_3—CO_2—MgO—SiO_2 at 250°C.

Activity Diagrams

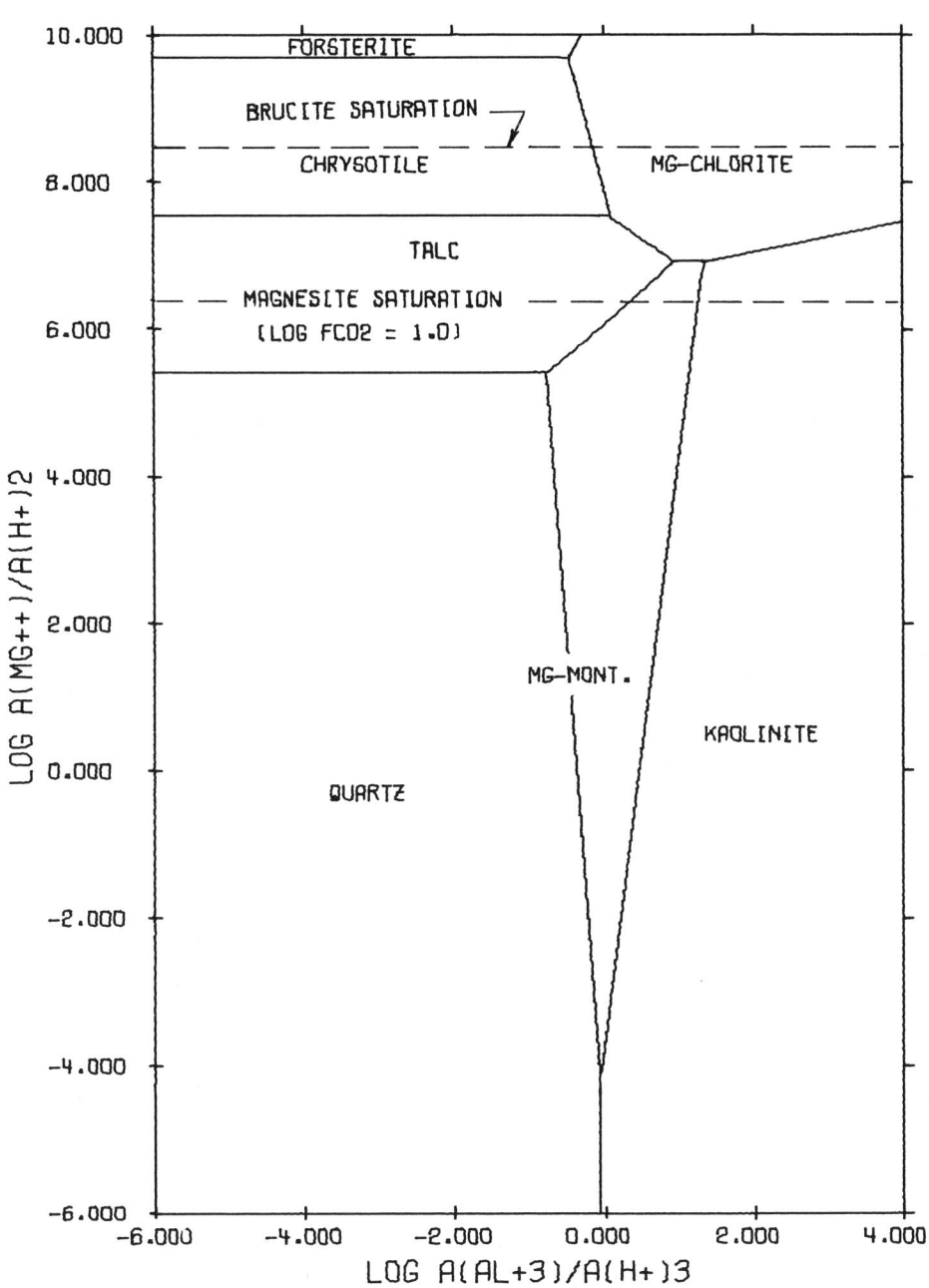

The System HCl—H$_2$O—Al$_2$O$_3$—CO$_2$—MgO—SiO$_2$ at 300°C.

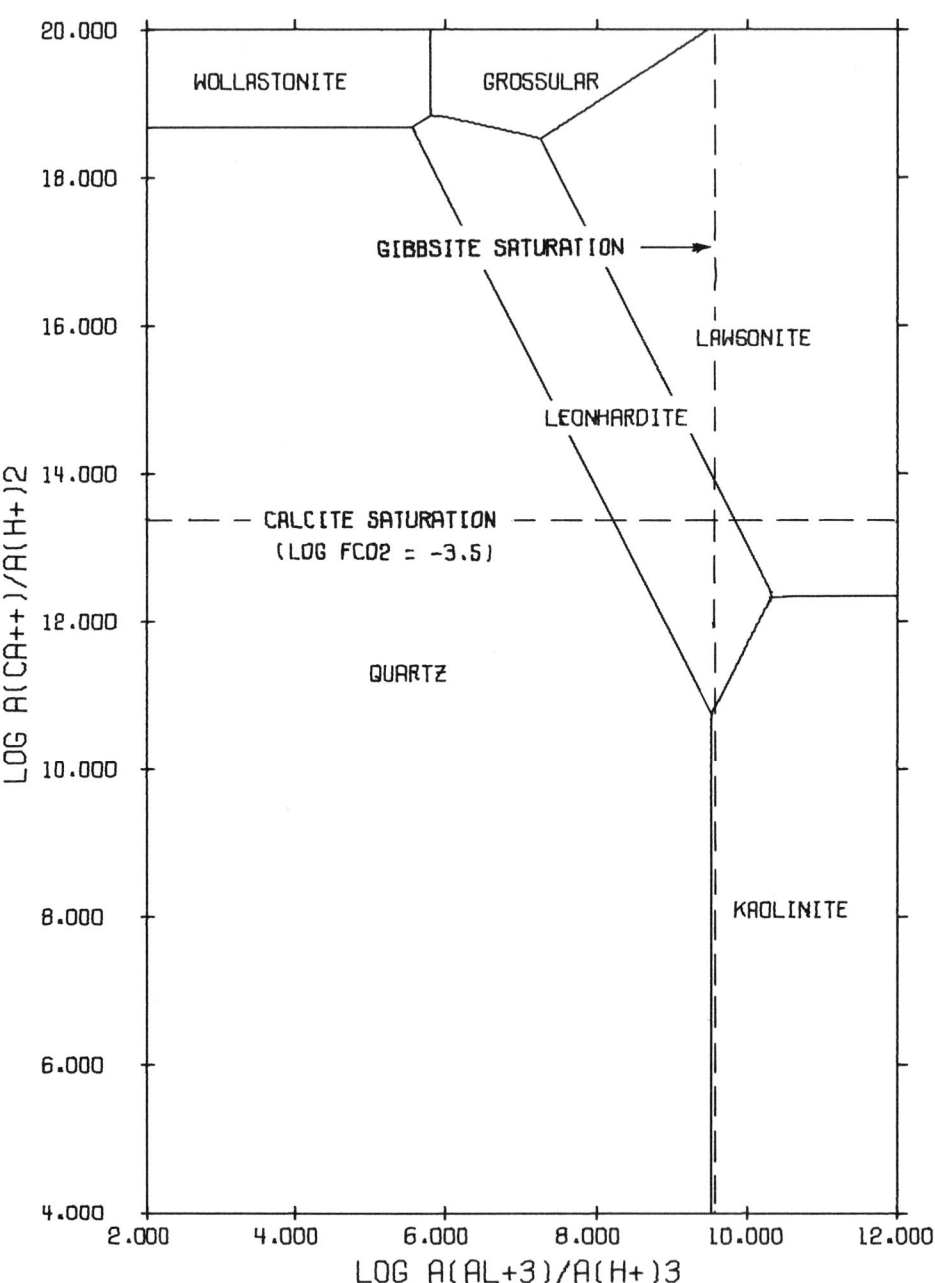

The System HCl—H$_2$O—Al$_2$O$_3$—CaO—CO$_2$—SiO$_2$ at 0°C.

Activity Diagrams

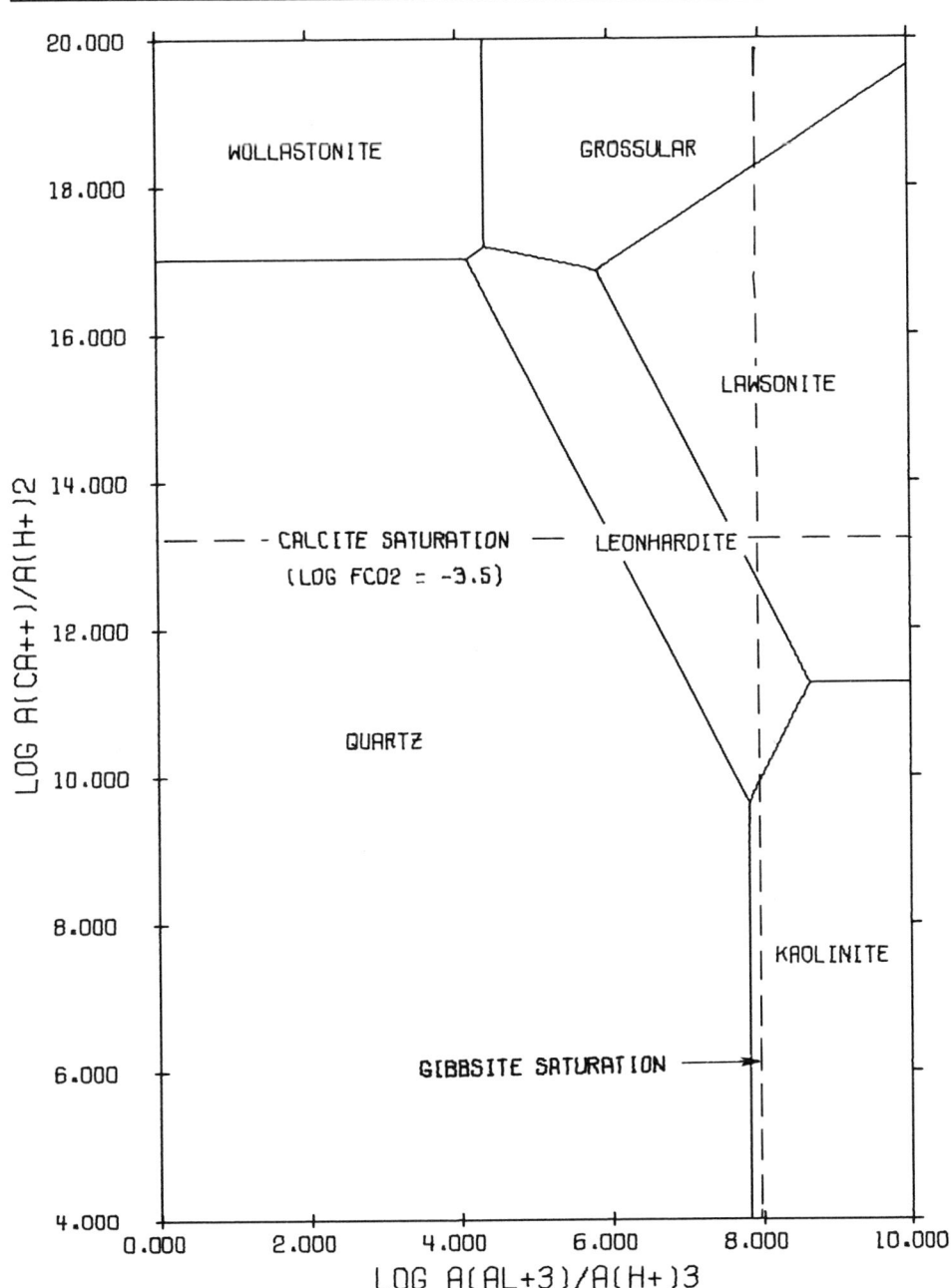

The System HCl—H$_2$O—Al$_2$O$_3$—CaO—CO$_2$—SiO$_2$ at 25°C.

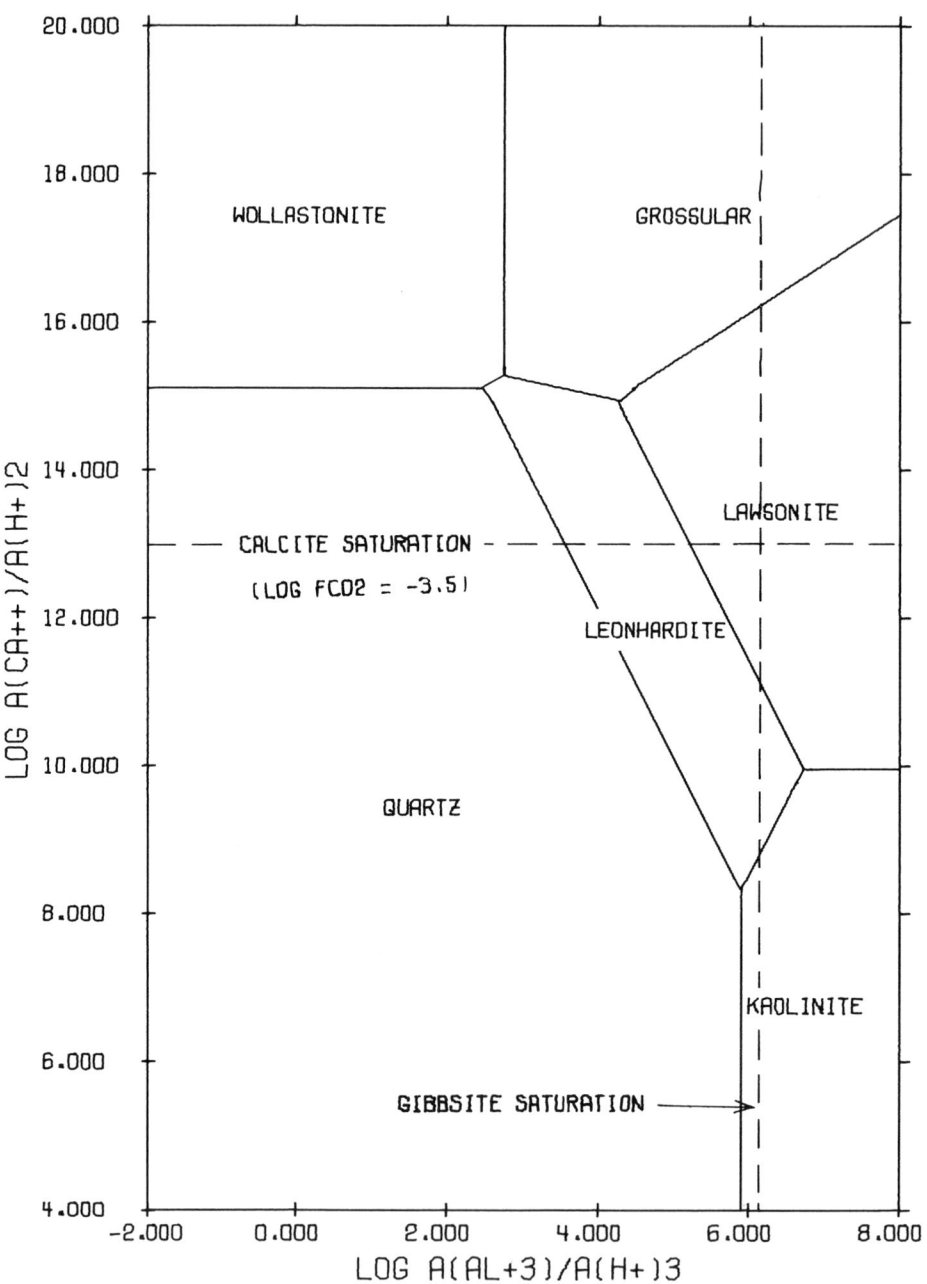

The System HCl—H$_2$O—Al$_2$O$_3$—CaO—CO$_2$—SiO$_2$ at 60°C.

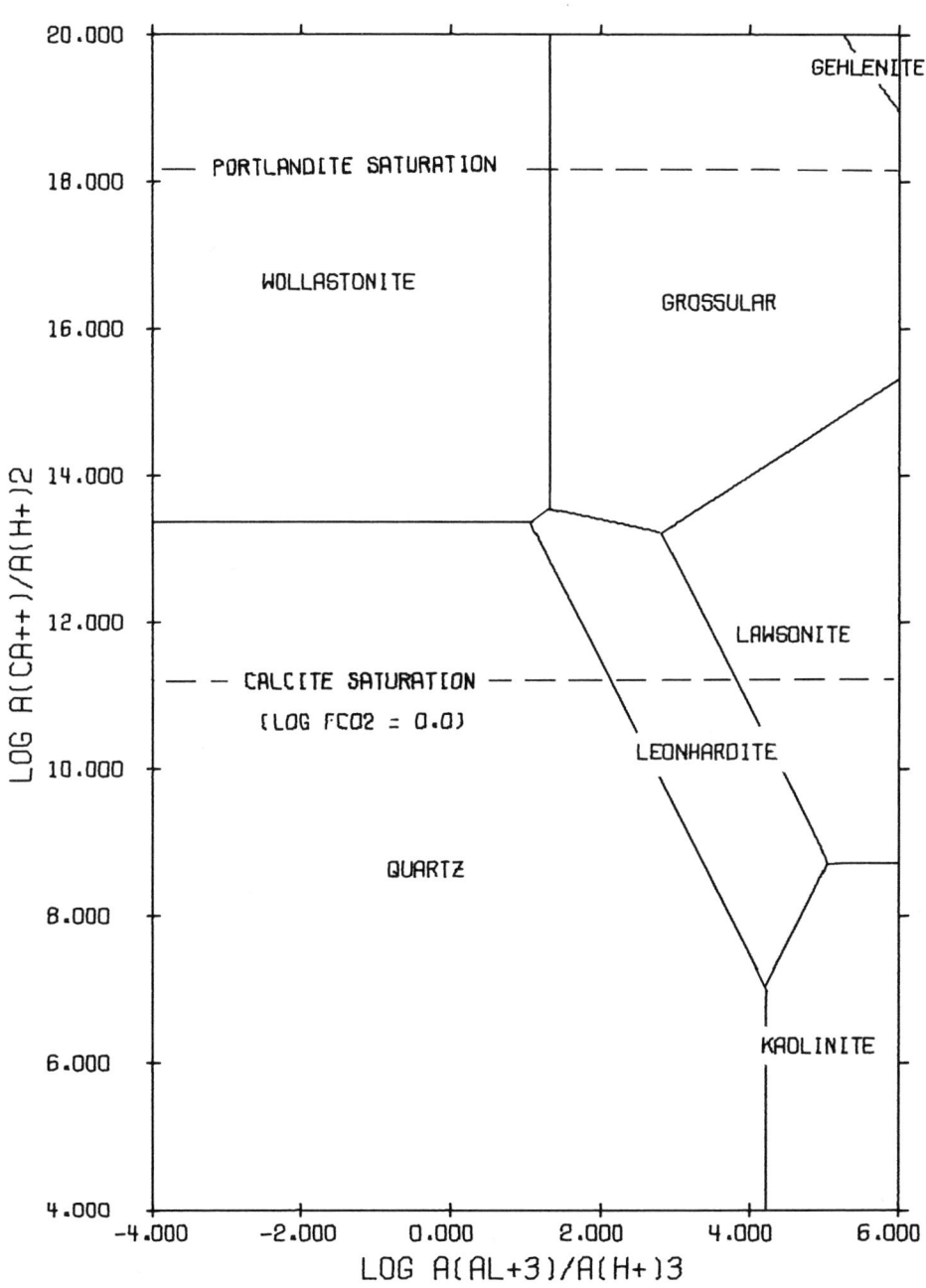

The System HCl—H_2O—Al_2O_3—CaO—CO_2—SiO_2 at 100°C.

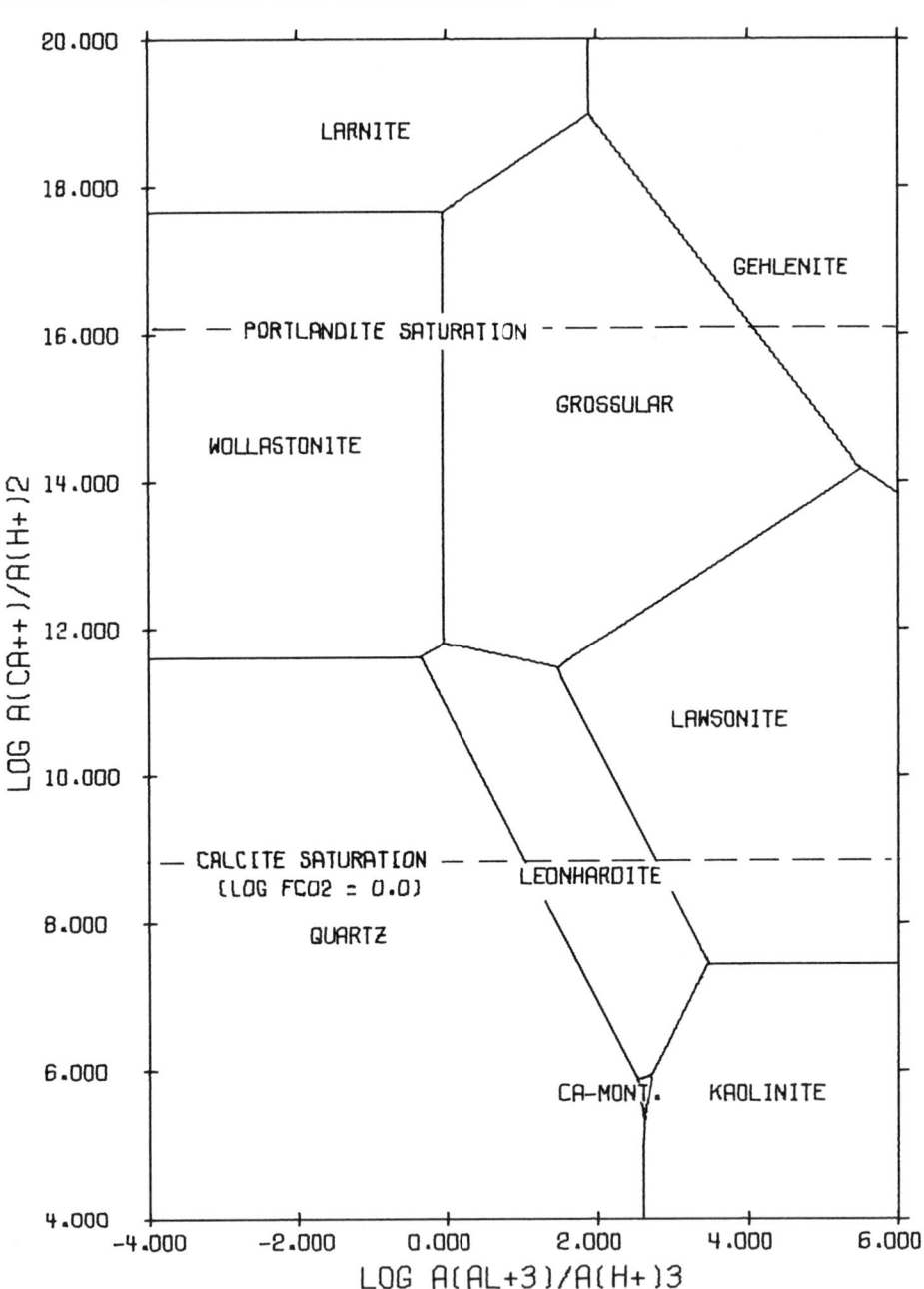

The System HCl—H$_2$O—Al$_2$O$_3$—CaO—CO$_2$—SiO$_2$ at 150°C.

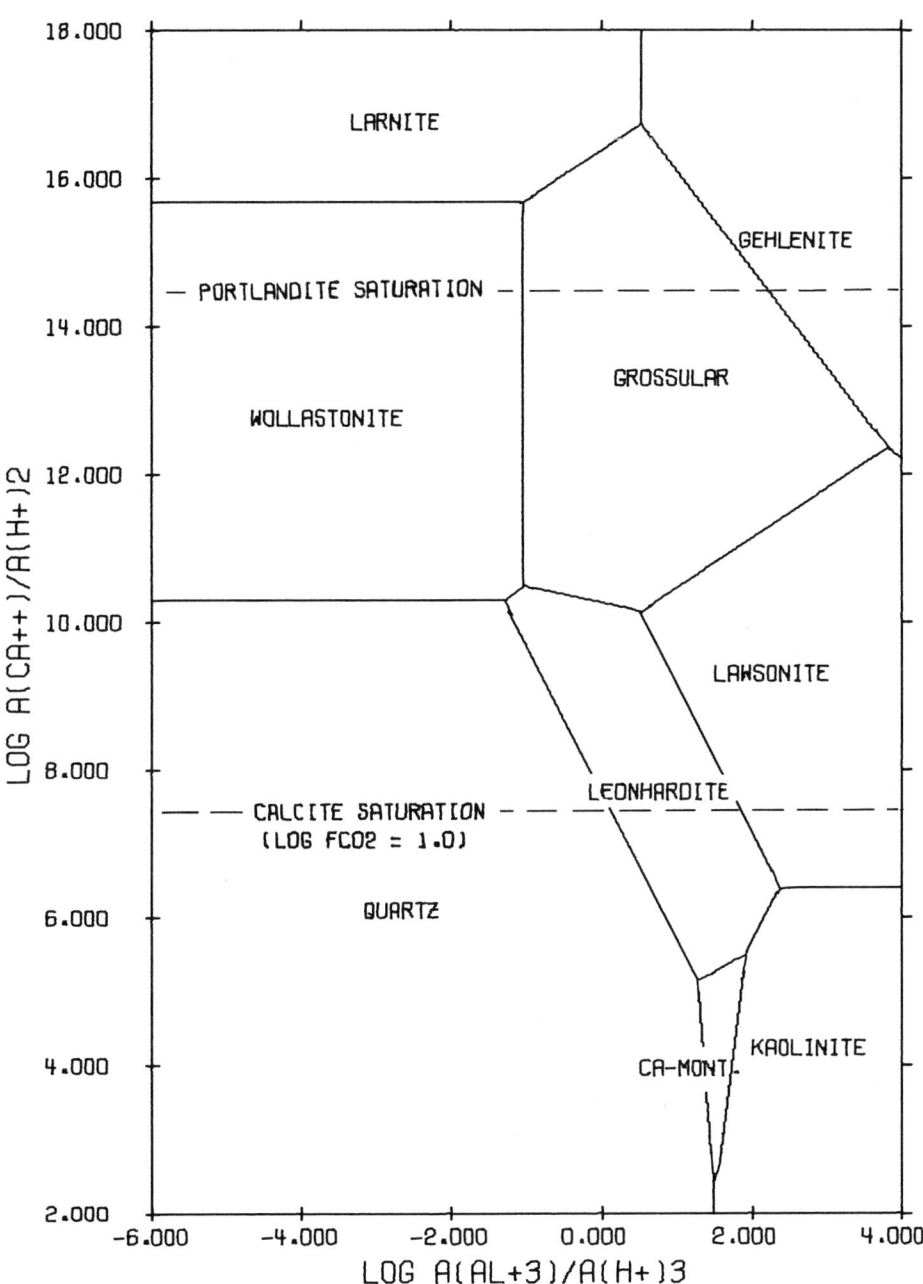

The System HCl—H$_2$O—Al$_2$O$_3$—CaO—CO$_2$—SiO$_2$ at 200°C.

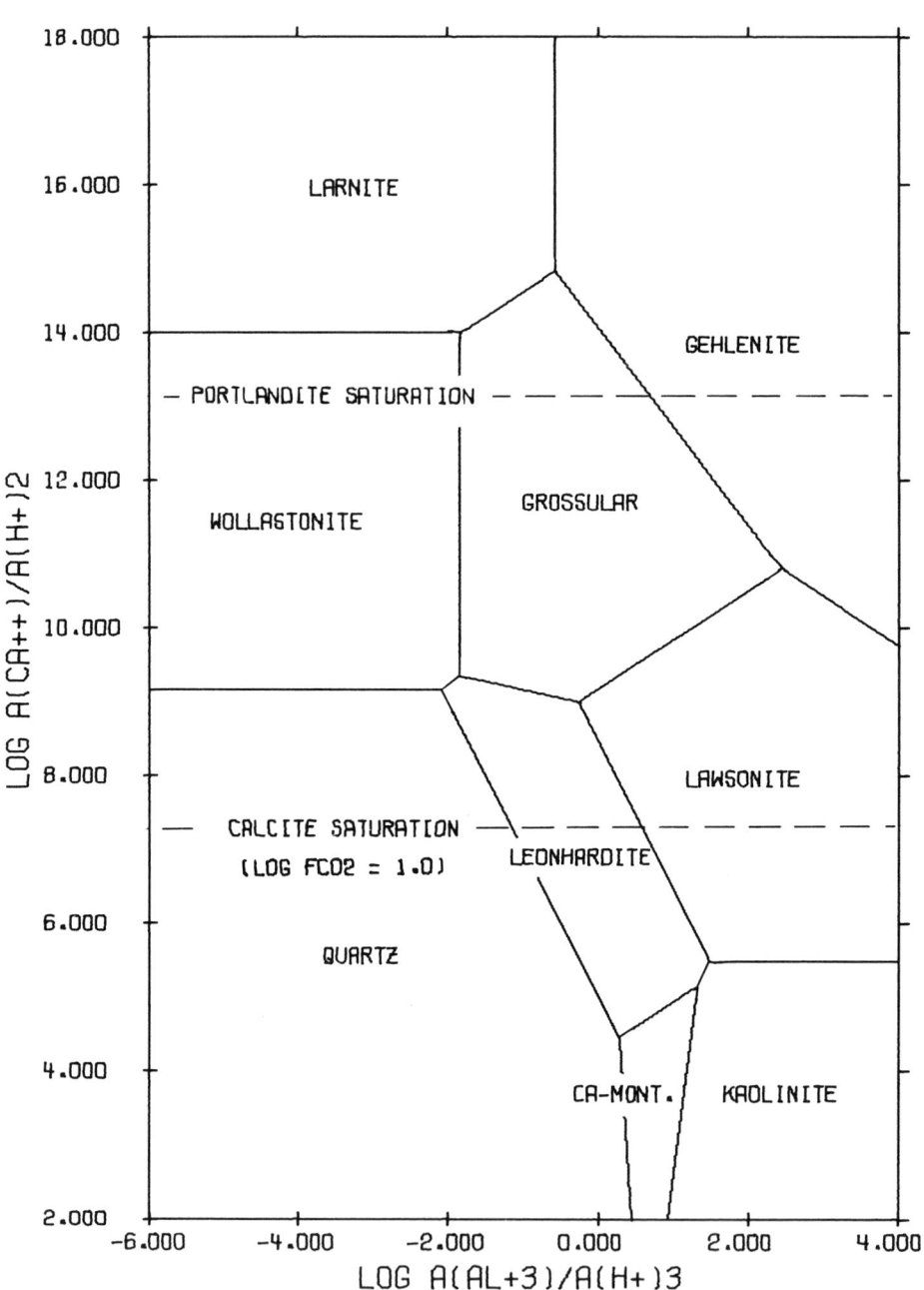

The System HCl—H$_2$O—Al$_2$O$_3$—CaO—CO$_2$—SiO$_2$ at 250°C.

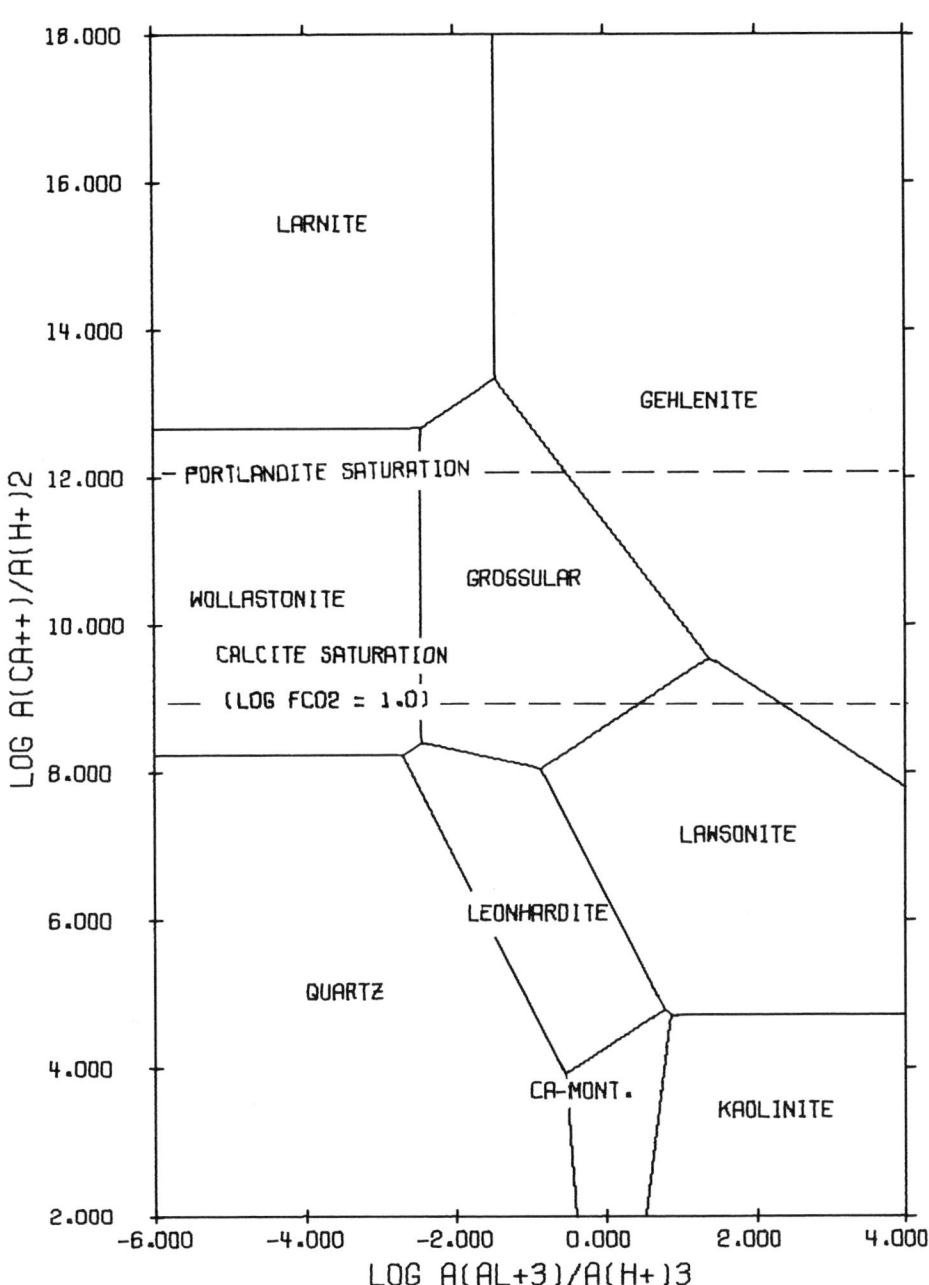

The System HCl—H$_2$O—Al$_2$O$_3$—CaO—CO$_2$—SiO$_2$ at 300°C.

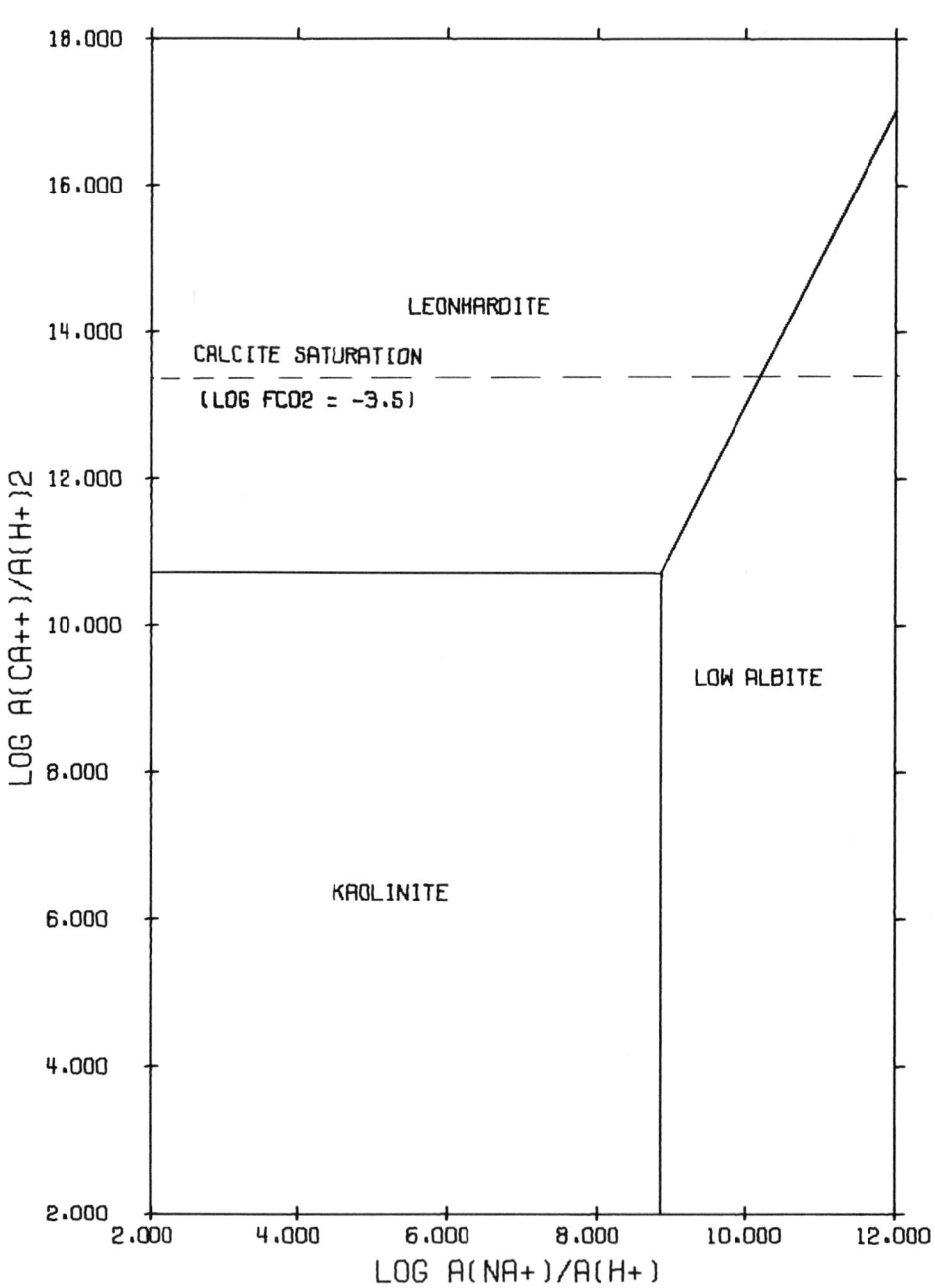

The System HCl—H$_2$O—Al$_2$O$_3$—CaO—CO$_2$—Na$_2$O—SiO$_2$ at 0°C; log $a_{H_4SiO_4}$ = −4.44 = quartz saturation.

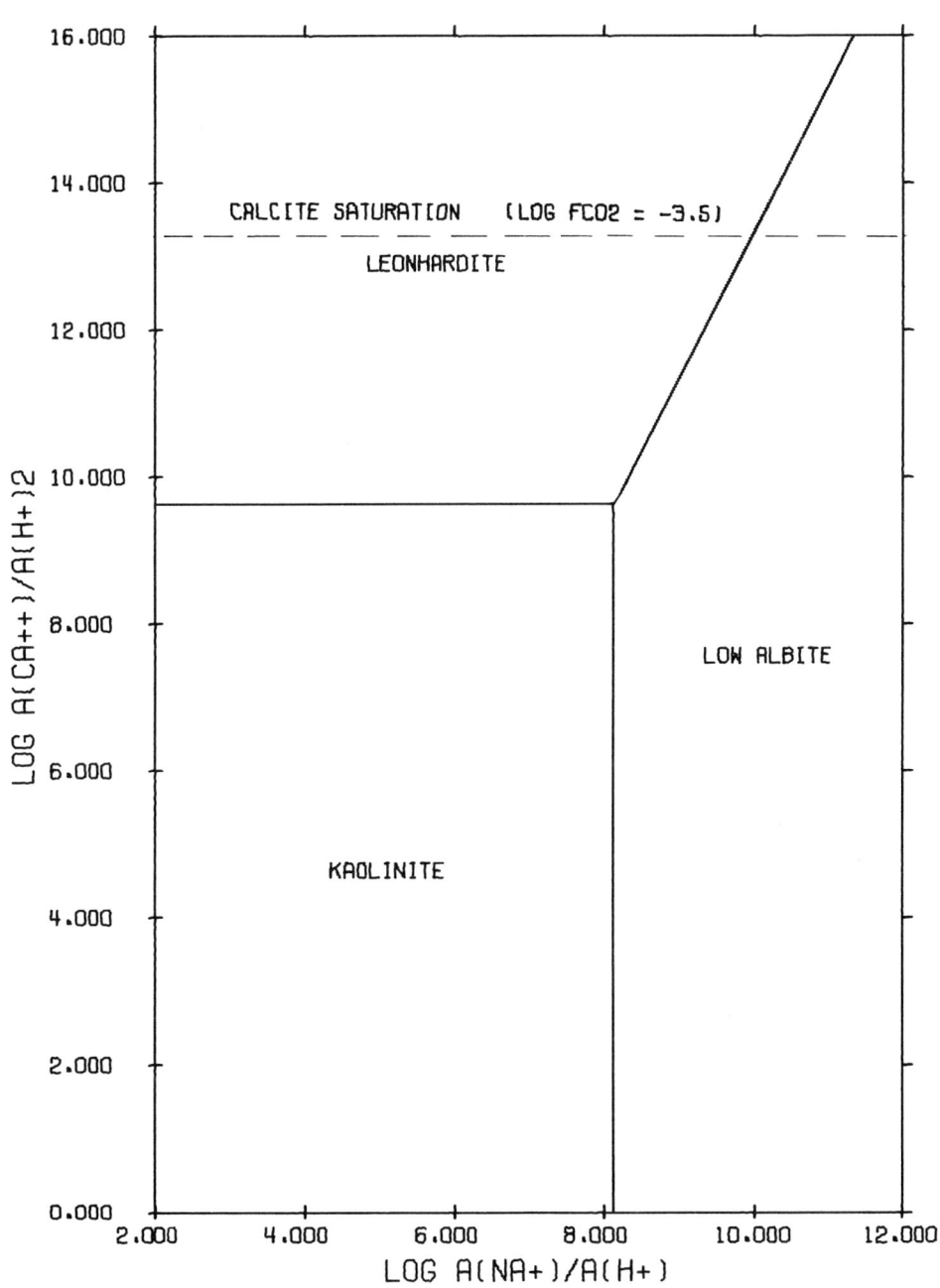

The System HCl—H$_2$O—Al$_2$O$_3$—CaO—CO$_2$—Na$_2$O—SiO$_2$ at 25°C; log $a_{H_4SiO_4}$ = −4.00 = quartz saturation.

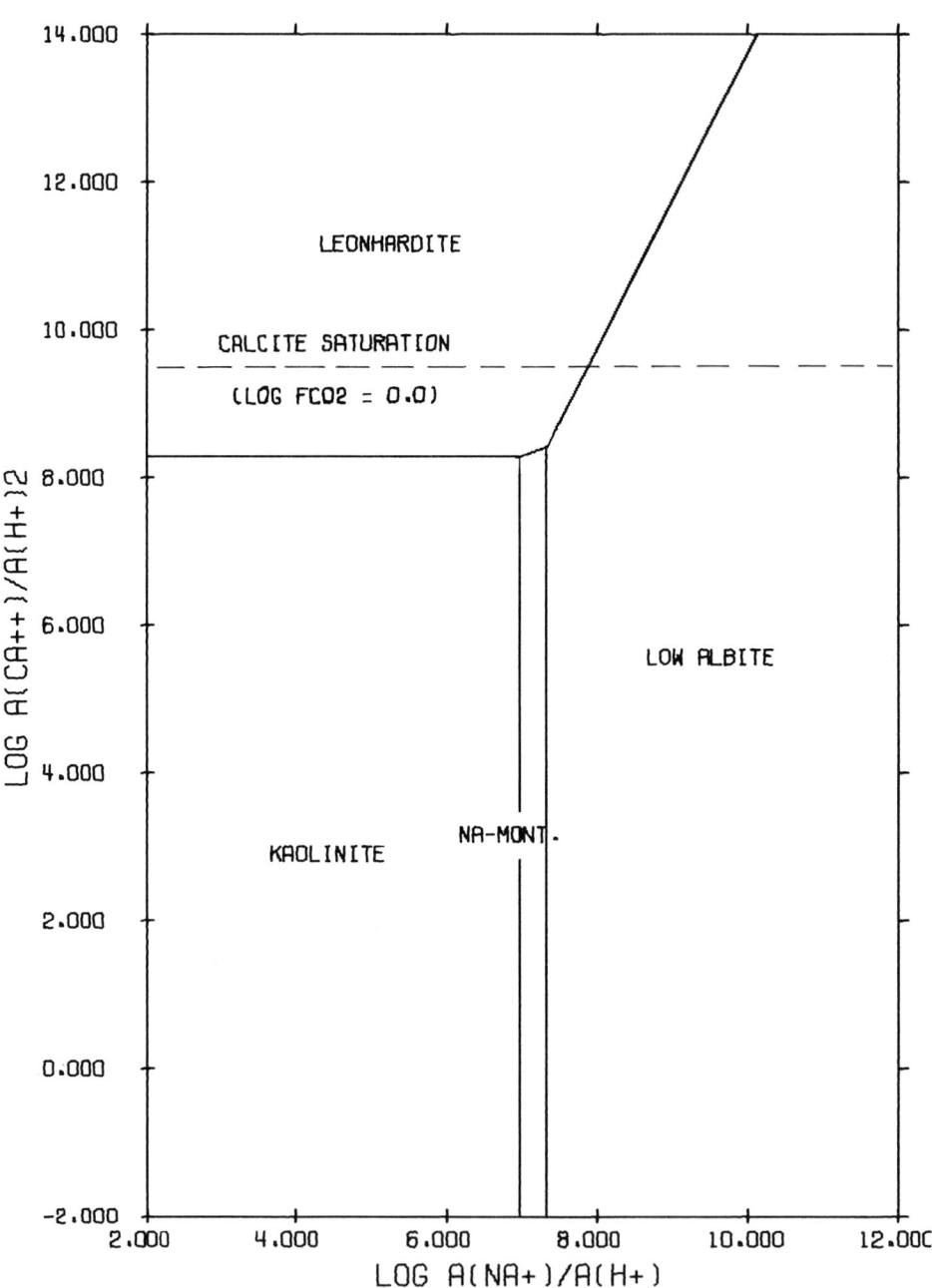

The System HCl—H$_2$O—Al$_2$O$_3$—CaO—CO$_2$—Na$_2$O—SiO$_2$ at 60°C; log $a_{\text{H}_4\text{SiO}_4}$ = −3.52 = quartz saturation.

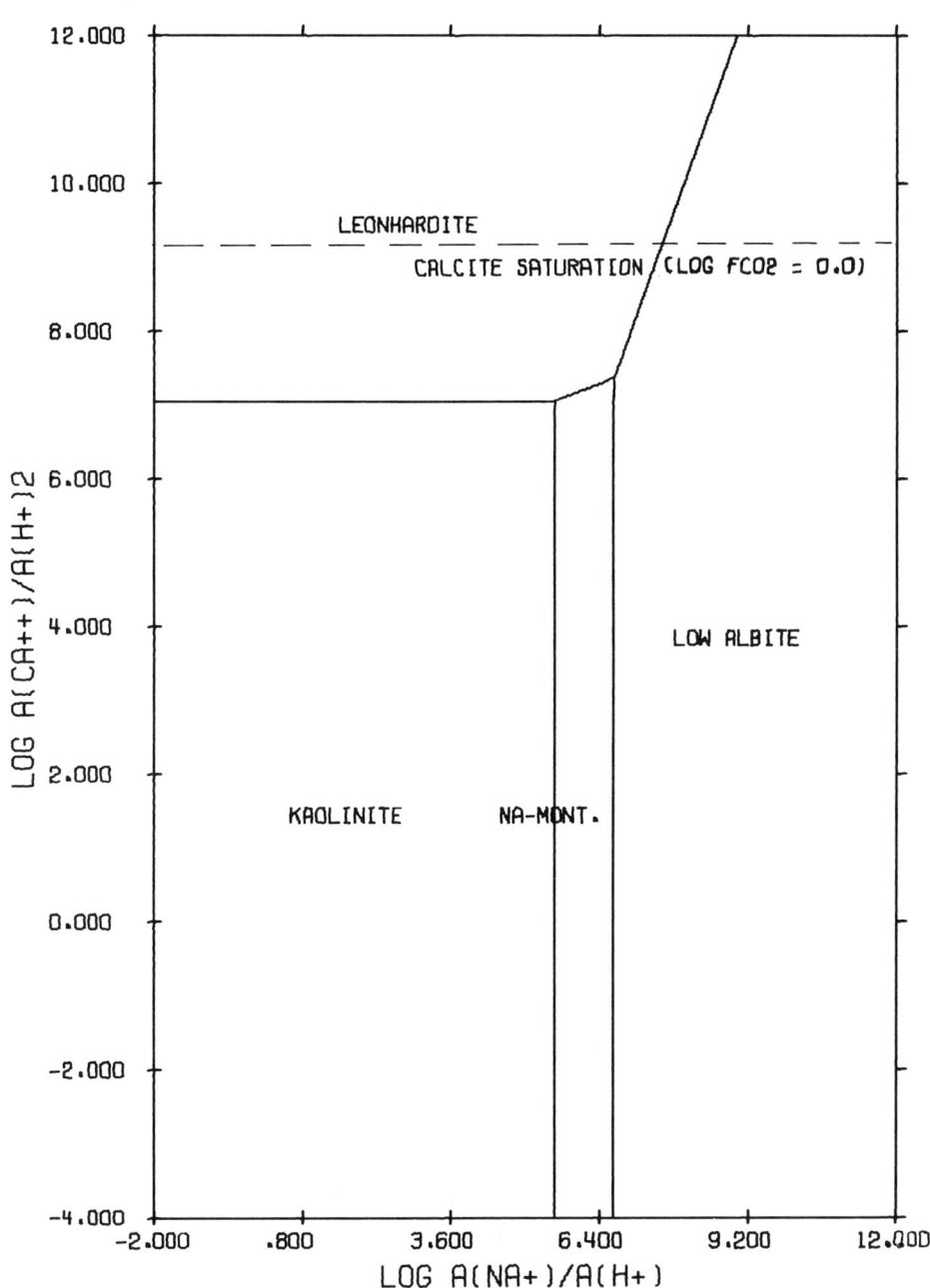

The System HCl—H_2O—Al_2O_3—CaO—CO_2—Na_2O—SiO_2 at 100°C; log $a_{H_4SiO_4}$ = −3.08 = quartz saturation.

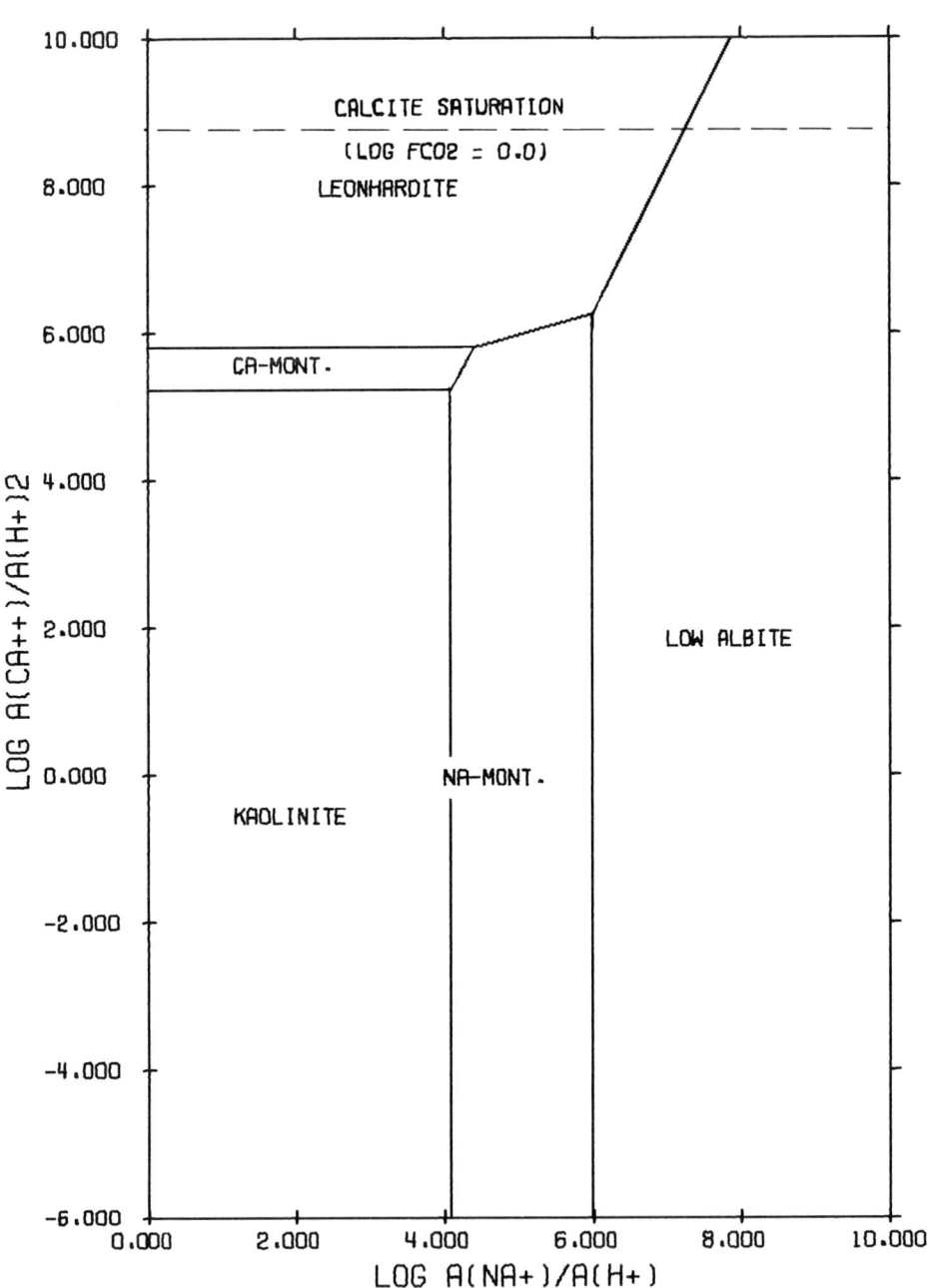

The System HCl—H$_2$O—Al$_2$O$_3$—CaO—CO$_2$—Na$_2$O—SiO$_2$ at 150°C; log $a_{H_4SiO_4}$ = −2.67 = quartz saturation.

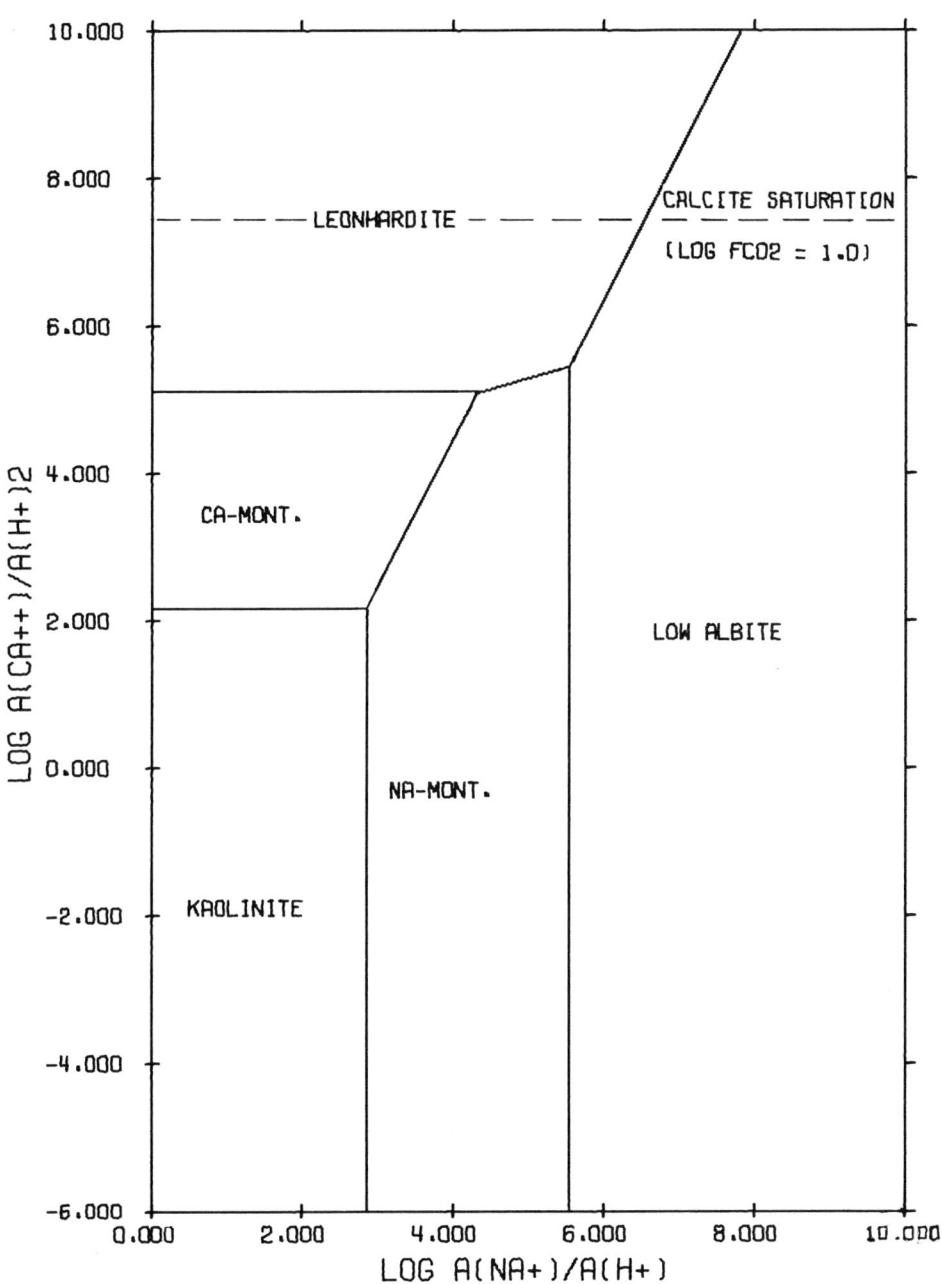

The System $HCl—H_2O—Al_2O_3—CaO—CO_2—Na_2O—SiO_2$ at 200°C; $\log a_{H_4SiO_4} = -2.35 =$ quartz saturation.

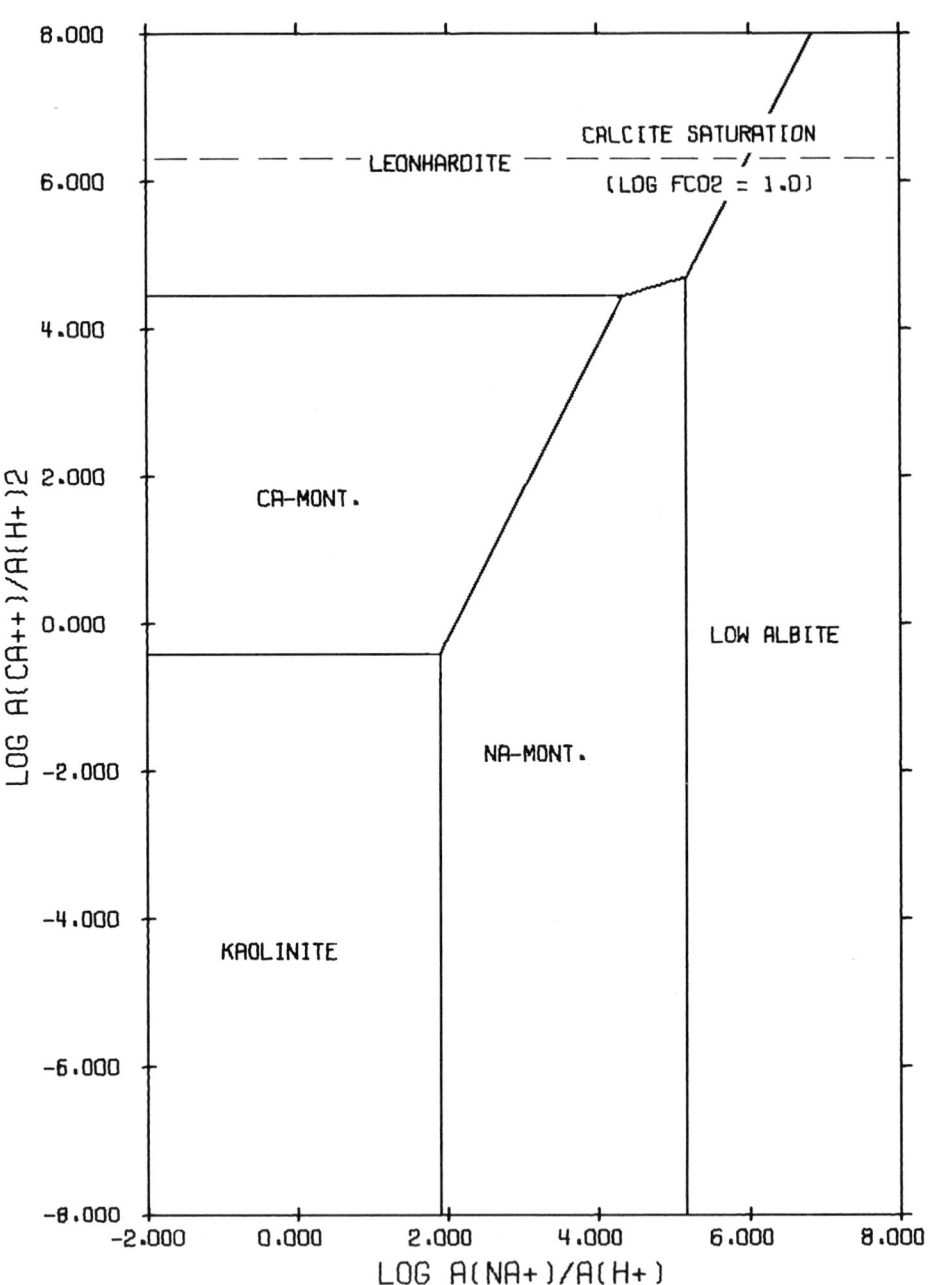

The System HCl—H_2O—Al_2O_3—CaO—CO_2—Na_2O—SiO_2 at 250°C; log $a_{H_4SiO_4}$ = −2.11 = quartz saturation.

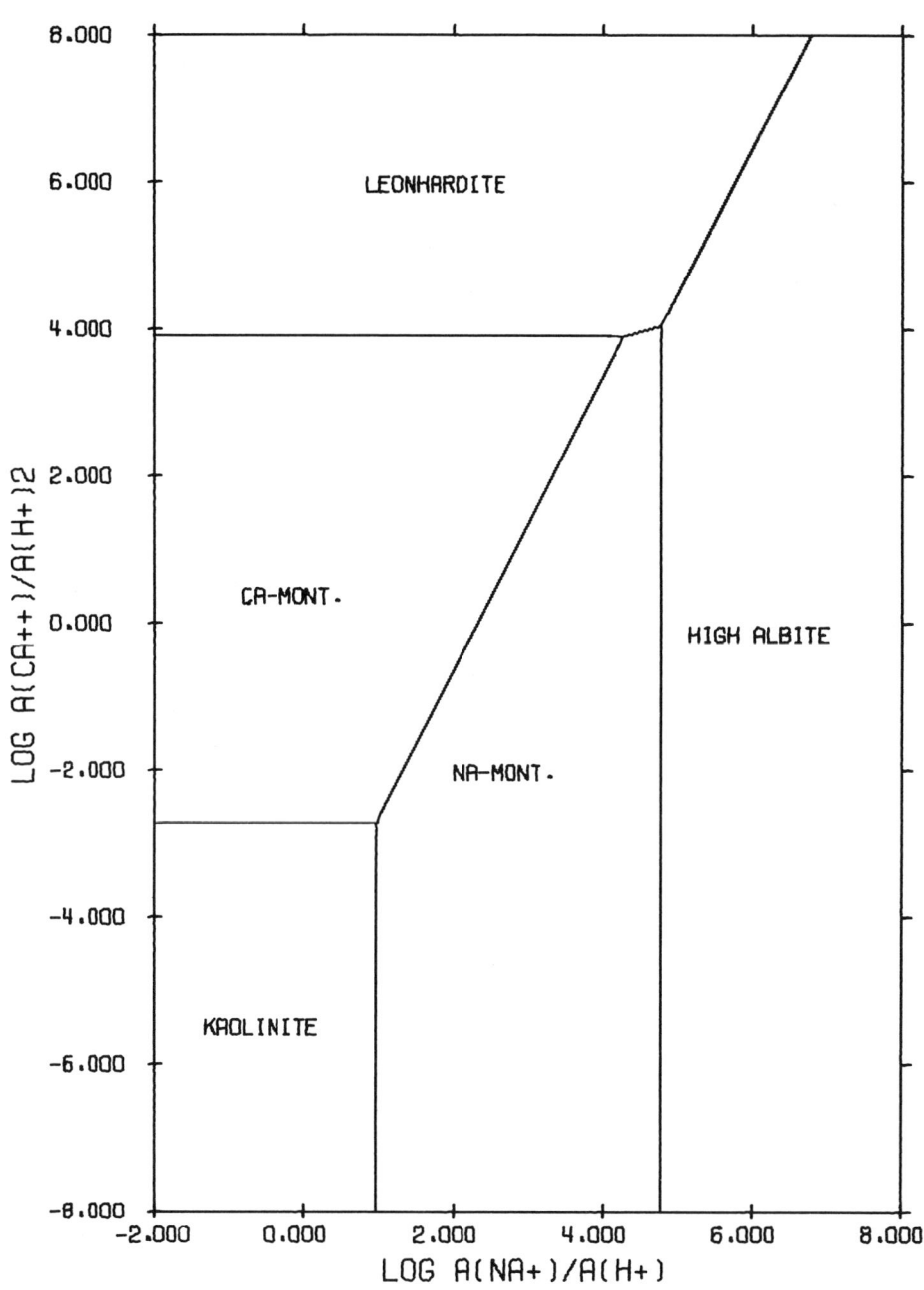

The System HCl—H$_2$O—Al$_2$O$_3$—CaO—CO$_2$—Na$_2$O—SiO$_2$ at 300°C; log $a_{\text{H}_4\text{SiO}_4}$ = −1.94 = quartz saturation.

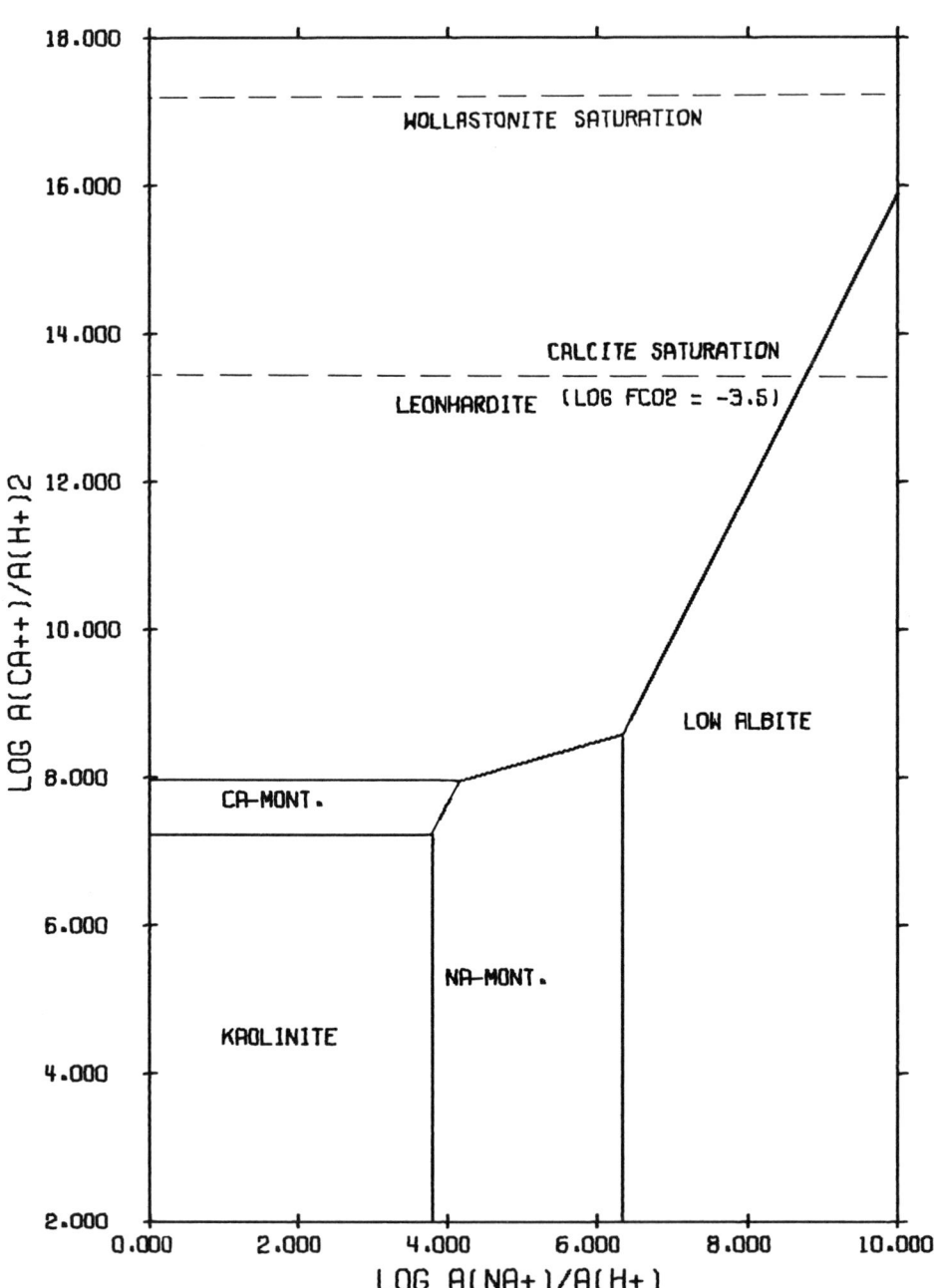

The System HCl—H$_2$O—Al$_2$O$_3$—CaO—CO$_2$—Na$_2$O—SiO$_2$ at 0°C; log $a_{H_4SiO_4}$ = −3.00 = amorphous silica saturation.

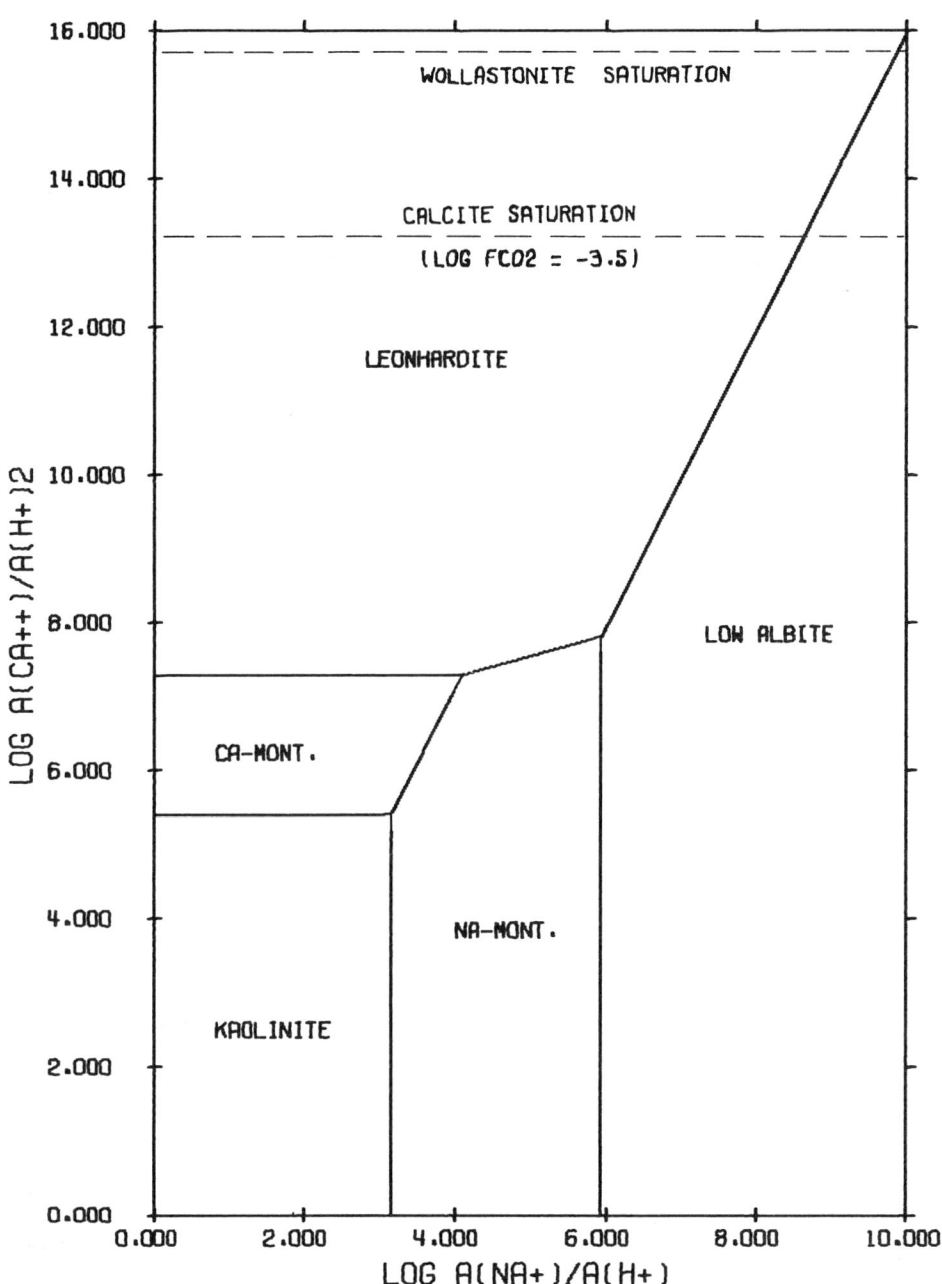

The System HCl—H$_2$O—Al$_2$O$_3$—CaO—CO$_2$—Na$_2$O—SiO$_2$ at 25°C; log $a_{\text{H}_4\text{SiO}_4}$ = −2.70 = amorphous silica saturation.

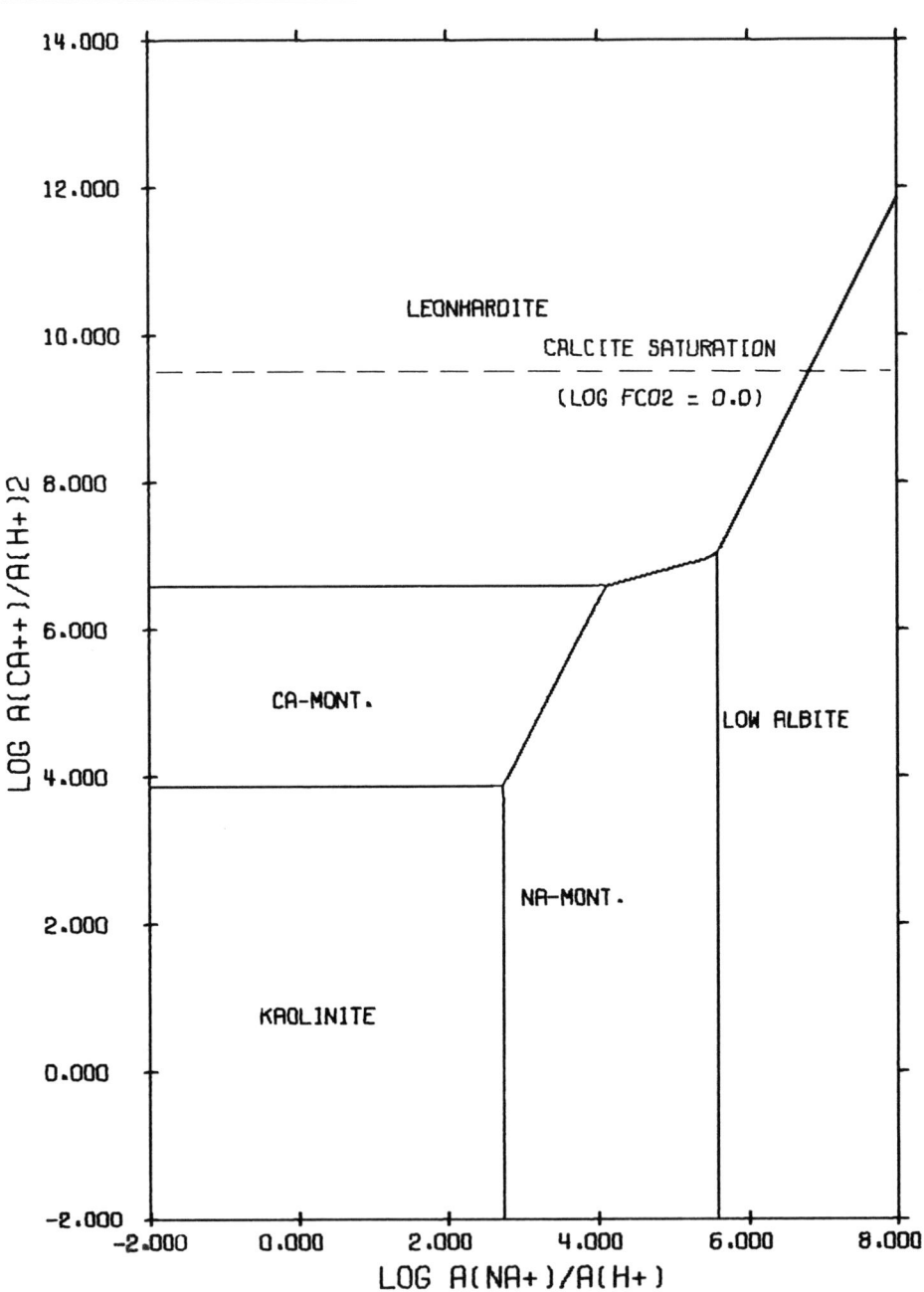

The System HCl—H$_2$O—Al$_2$O$_3$—CaO—CO$_2$—Na$_2$O—SiO$_2$ at 60°C; log $a_{H_4SiO_4}$ = -2.47 = amorphous silica saturation.

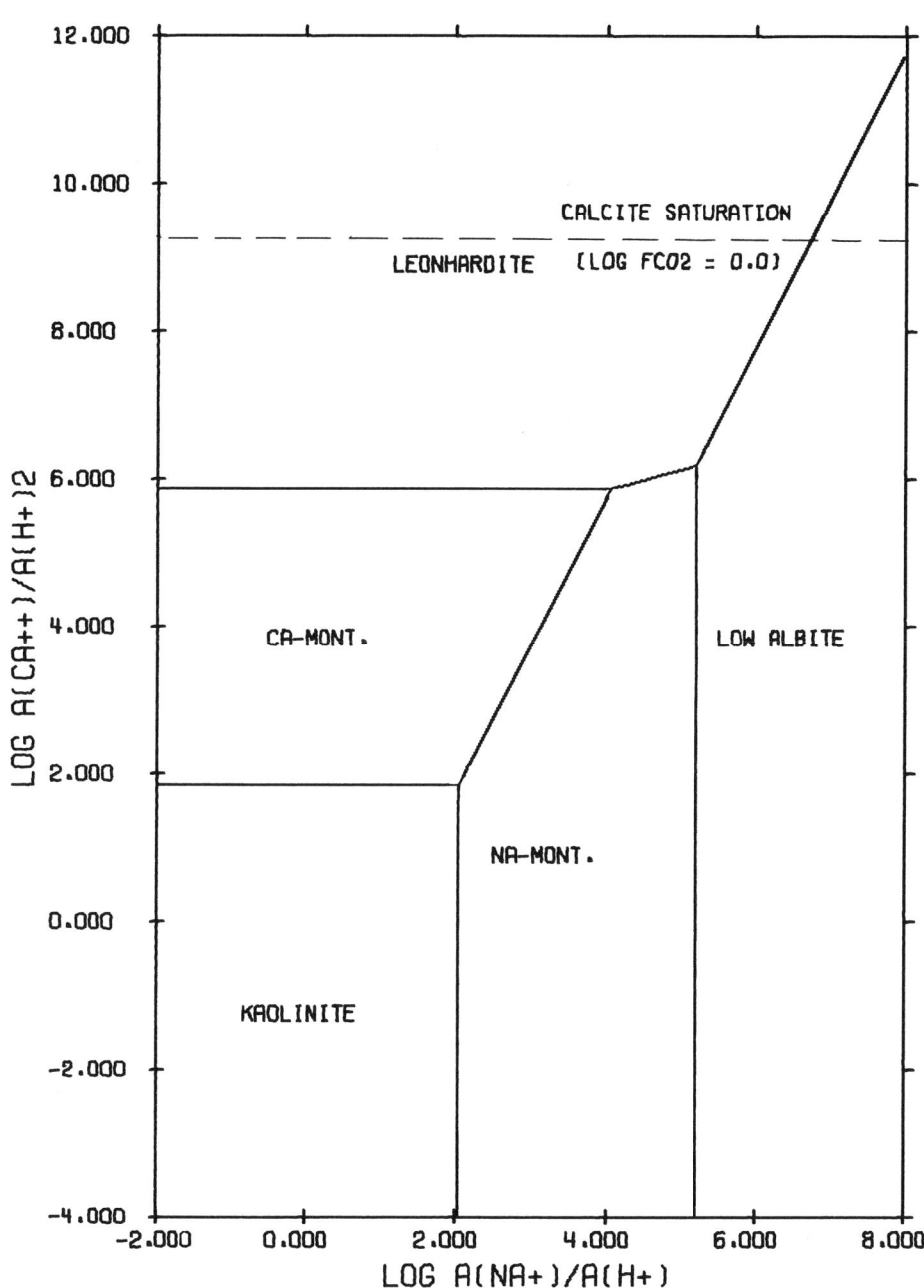

The System HCl—H$_2$O—Al$_2$O$_3$—CaO—CO$_2$—Na$_2$O—SiO$_2$ at 100°C; log $a_{H_4SiO_4}$ = −2.20 = amorphous silica saturation.

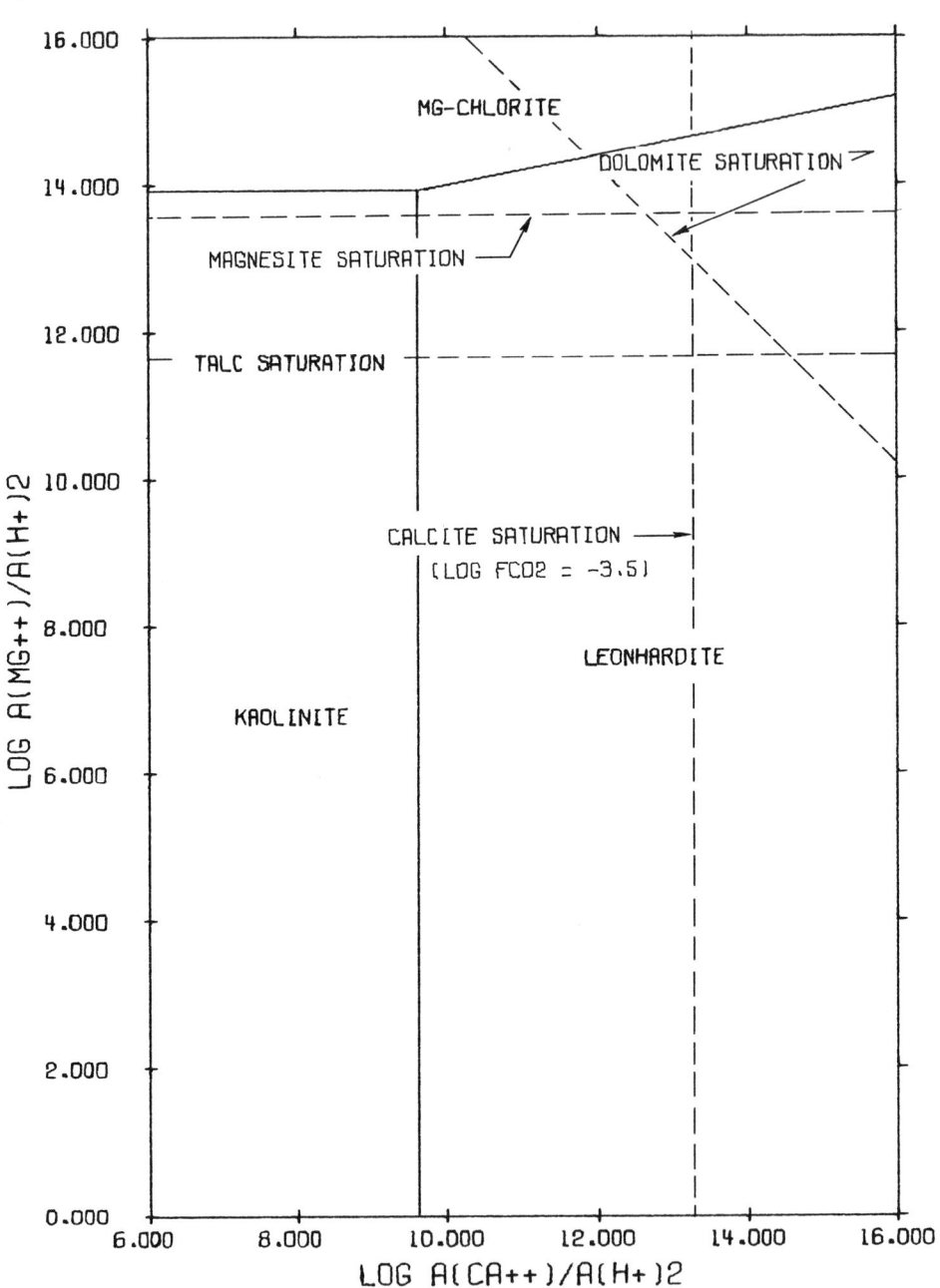

The System HCl—H$_2$O—Al$_2$O$_3$—CaO—CO$_2$—MgO—SiO$_2$ at 25°C; log $a_{H_4SiO_4}$ = −4.00 = quartz saturation.

Activity Diagrams

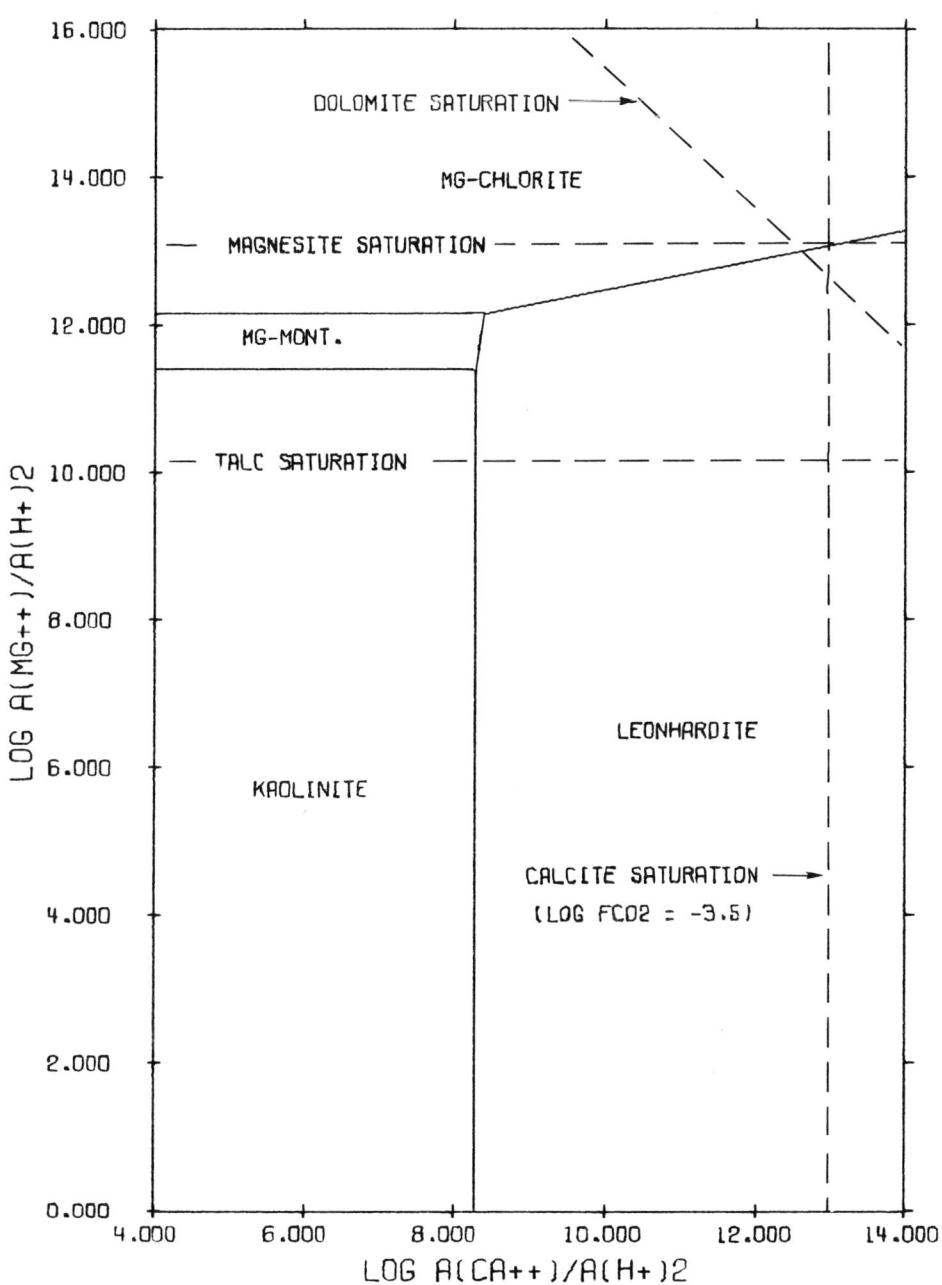

The System HCl—H$_2$O—Al$_2$O$_3$—CaO—CO$_2$—MgO—SiO$_2$ at 60°C; log $a_{\text{H}_4\text{SiO}_4}$ = −3.52 = quartz saturation.

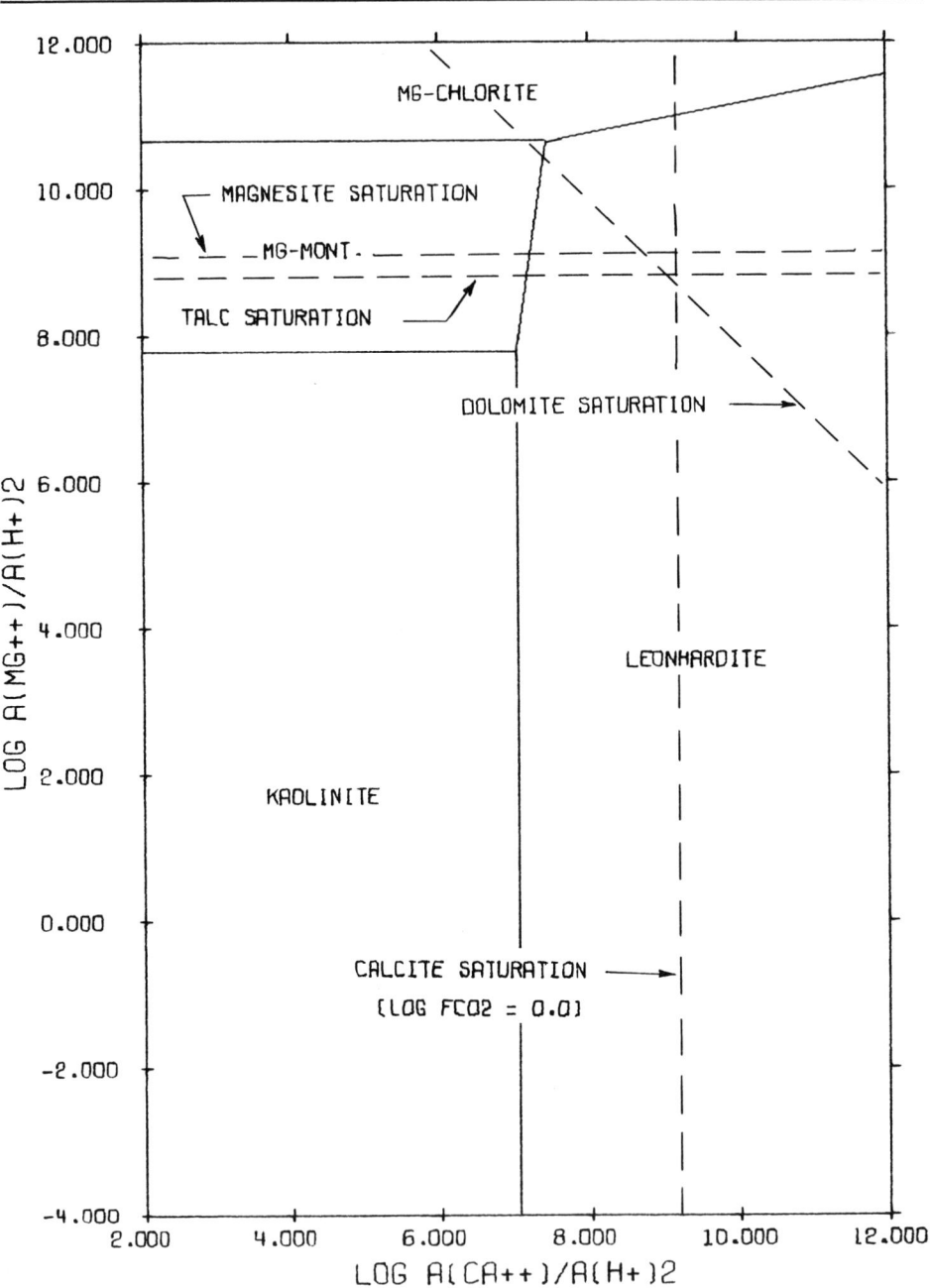

The System HCl—H$_2$O—Al$_2$O$_3$—CaO—CO$_2$—MgO—SiO$_2$ at 100°C; log $a_{H_4SiO_4}$ = −3.08 = quartz saturation.

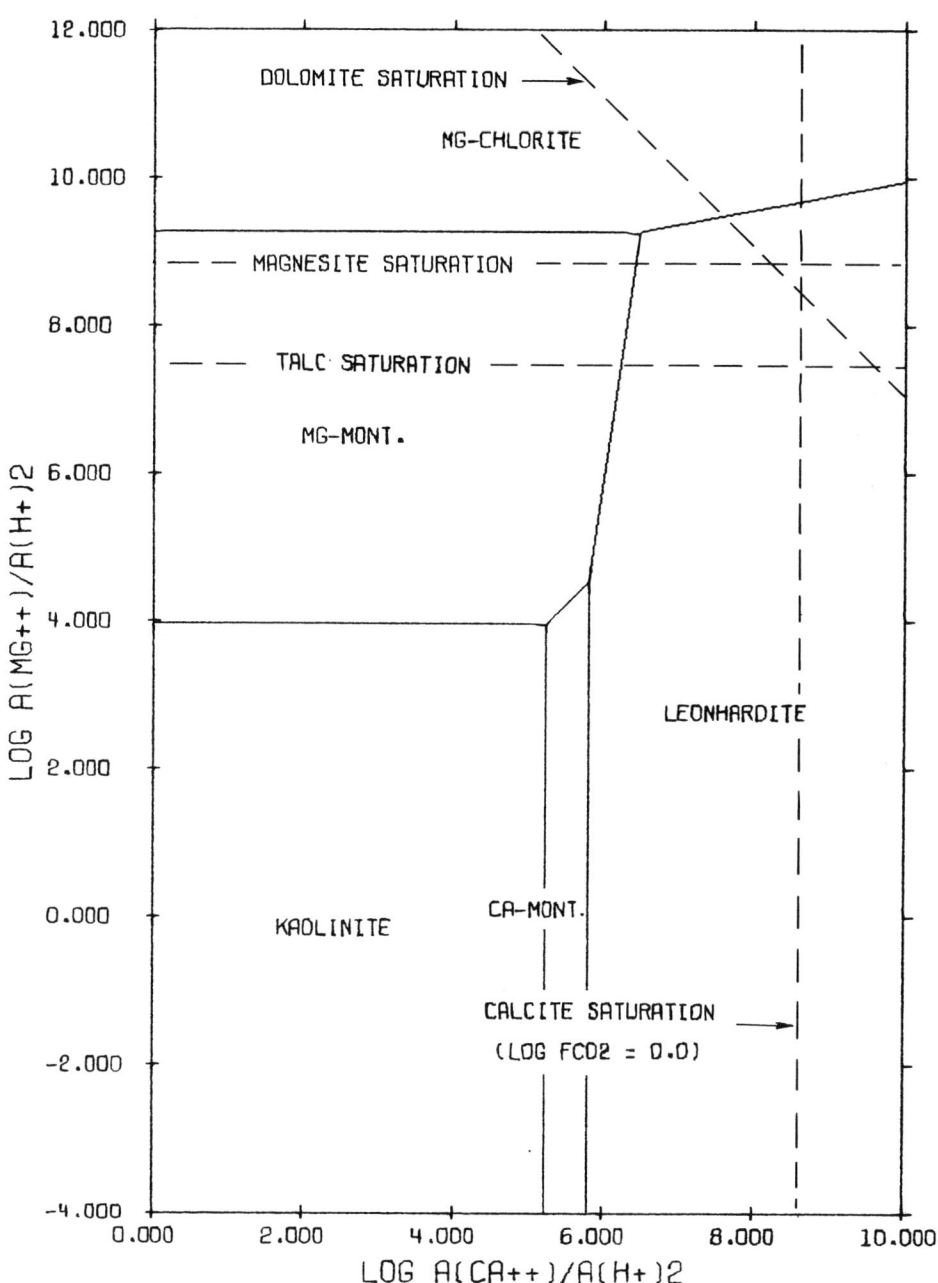

The System HCl—H$_2$O—Al$_2$O$_3$—CaO—CO$_2$—MgO—SiO$_2$ at 150°C; log $a_{H_4SiO_4}$ = −2.67 = quartz saturation.

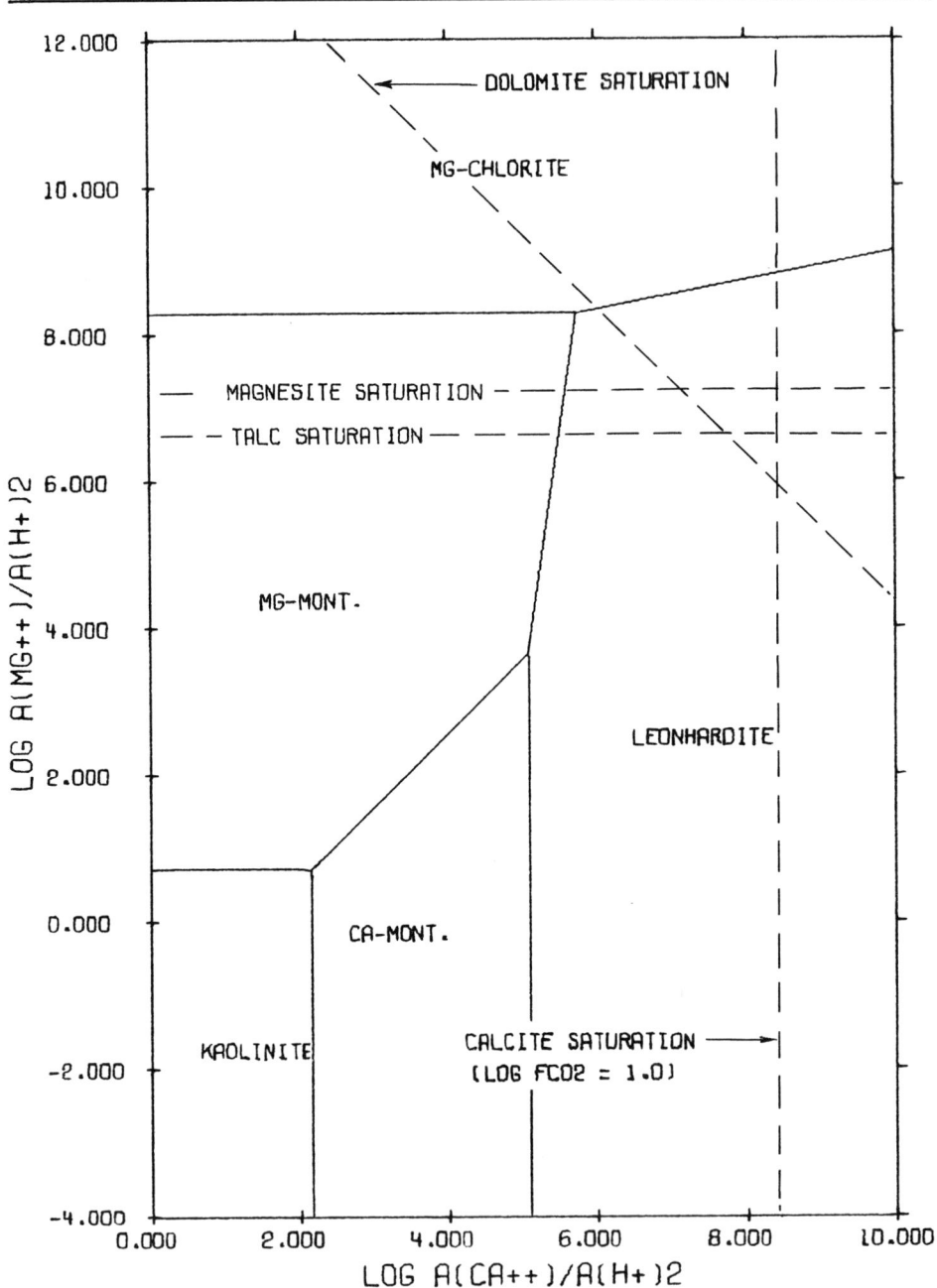

The System HCl—H_2O—Al_2O_3—CaO—CO_2—MgO—SiO_2 at 200°C; log $a_{H_4SiO_4}$ = −2.35 = quartz saturation.

Activity Diagrams

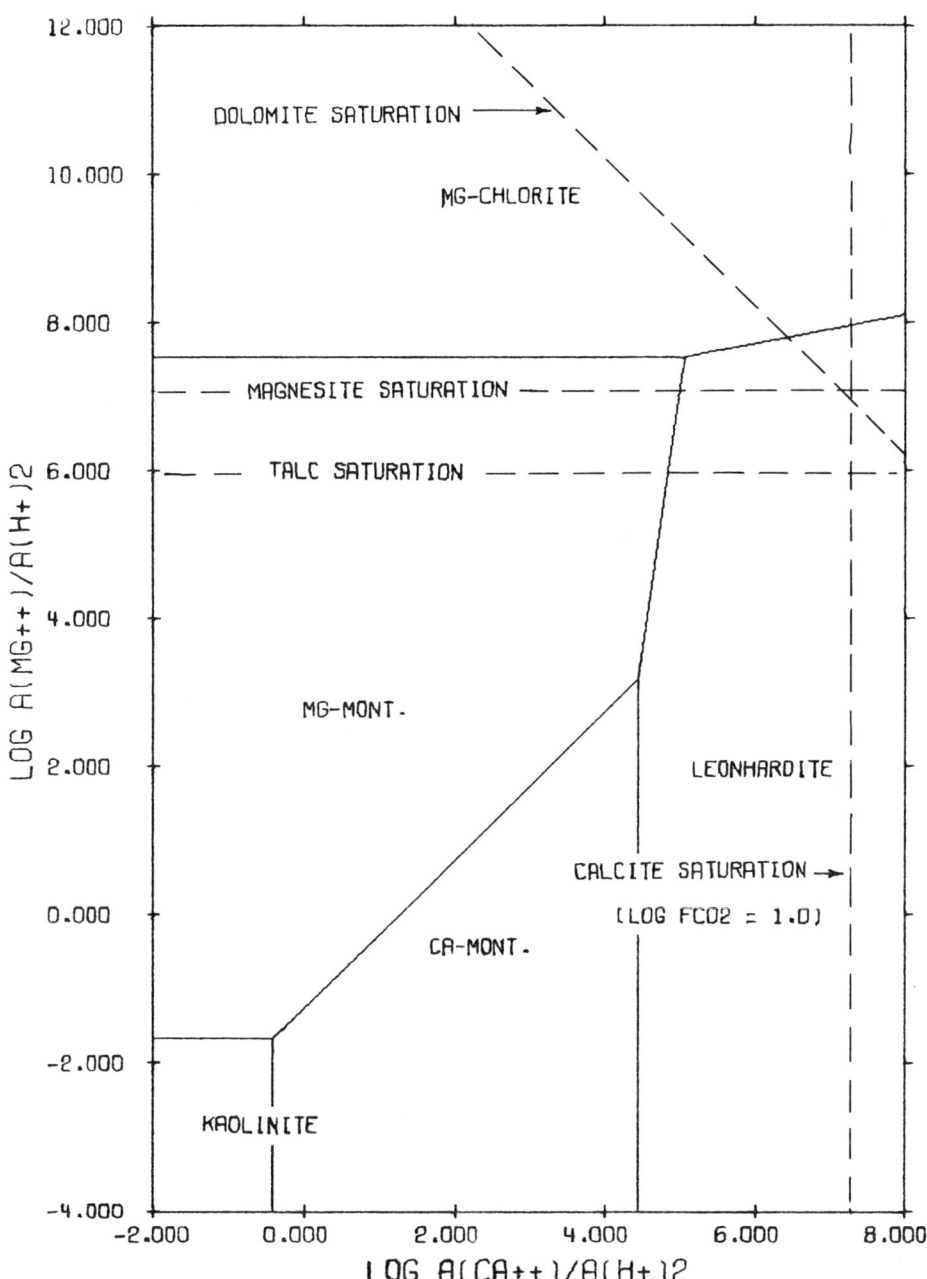

The System $HCl—H_2O—Al_2O_3—CaO—CO_2—MgO—SiO_2$ at 250°C; $\log a_{H_4SiO_4} = -2.11 =$ quartz saturation.

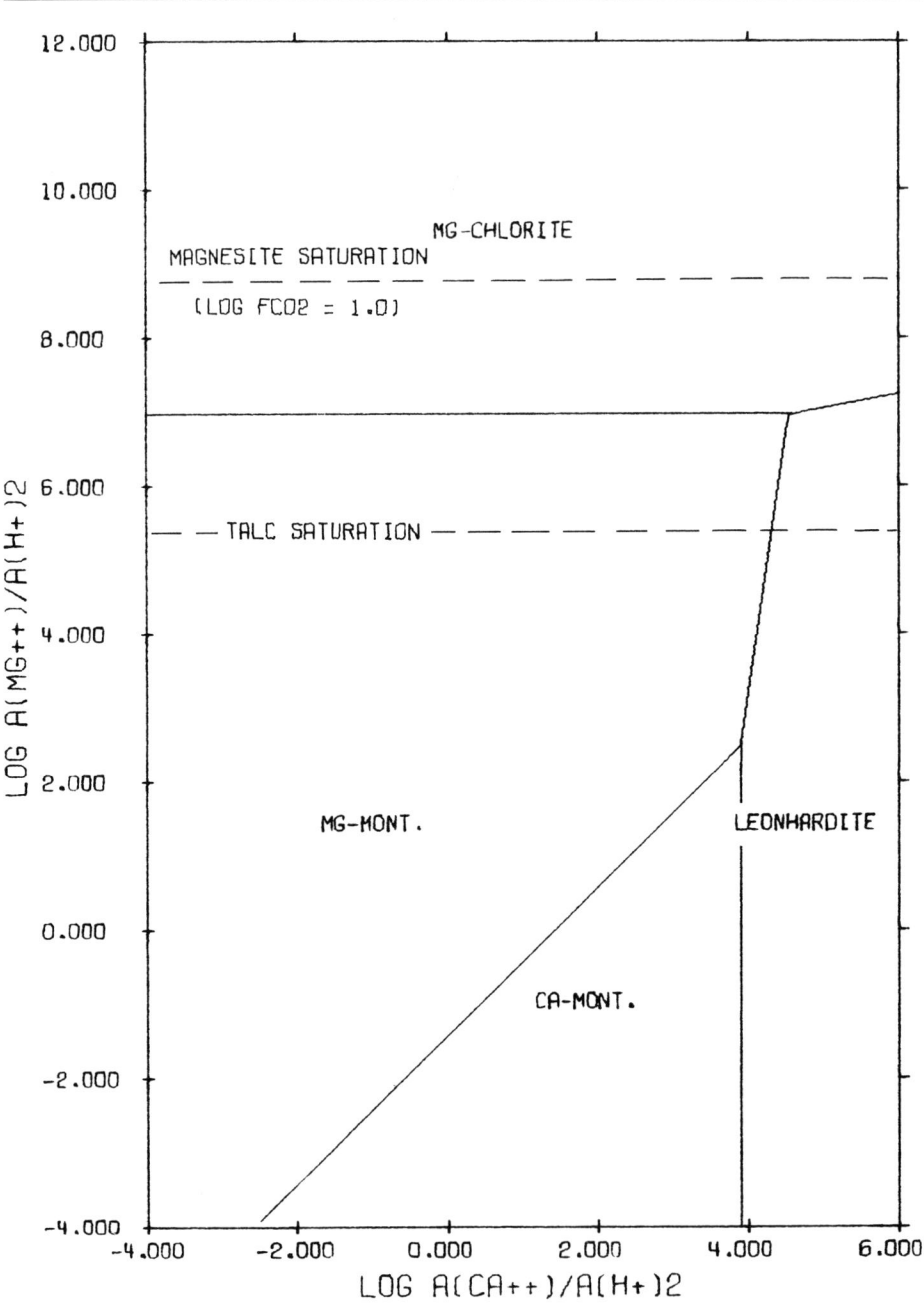

The System HCl—H_2O—Al_2O_3—CaO—CO_2—MgO—SiO_2 at 300°C; log $a_{H_4SiO_4}$ = -1.94 = quartz saturation.

Activity Diagrams

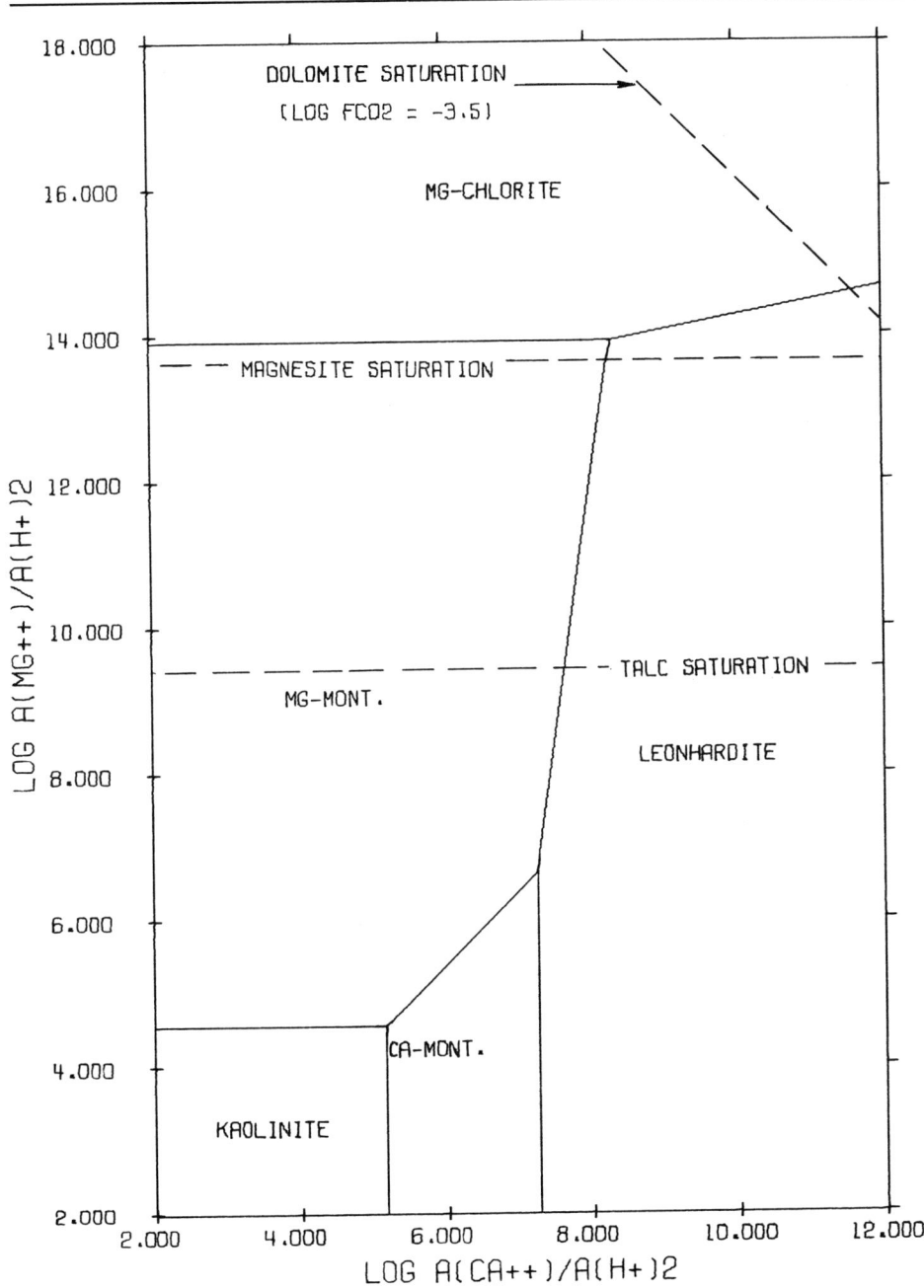

The System HCl—H$_2$O—Al$_2$O$_3$—CaO—CO$_2$—MgO—SiO$_2$ at 25°C; log $a_{H_4SiO_4}$ = -2.70 = amorphous silica saturation.

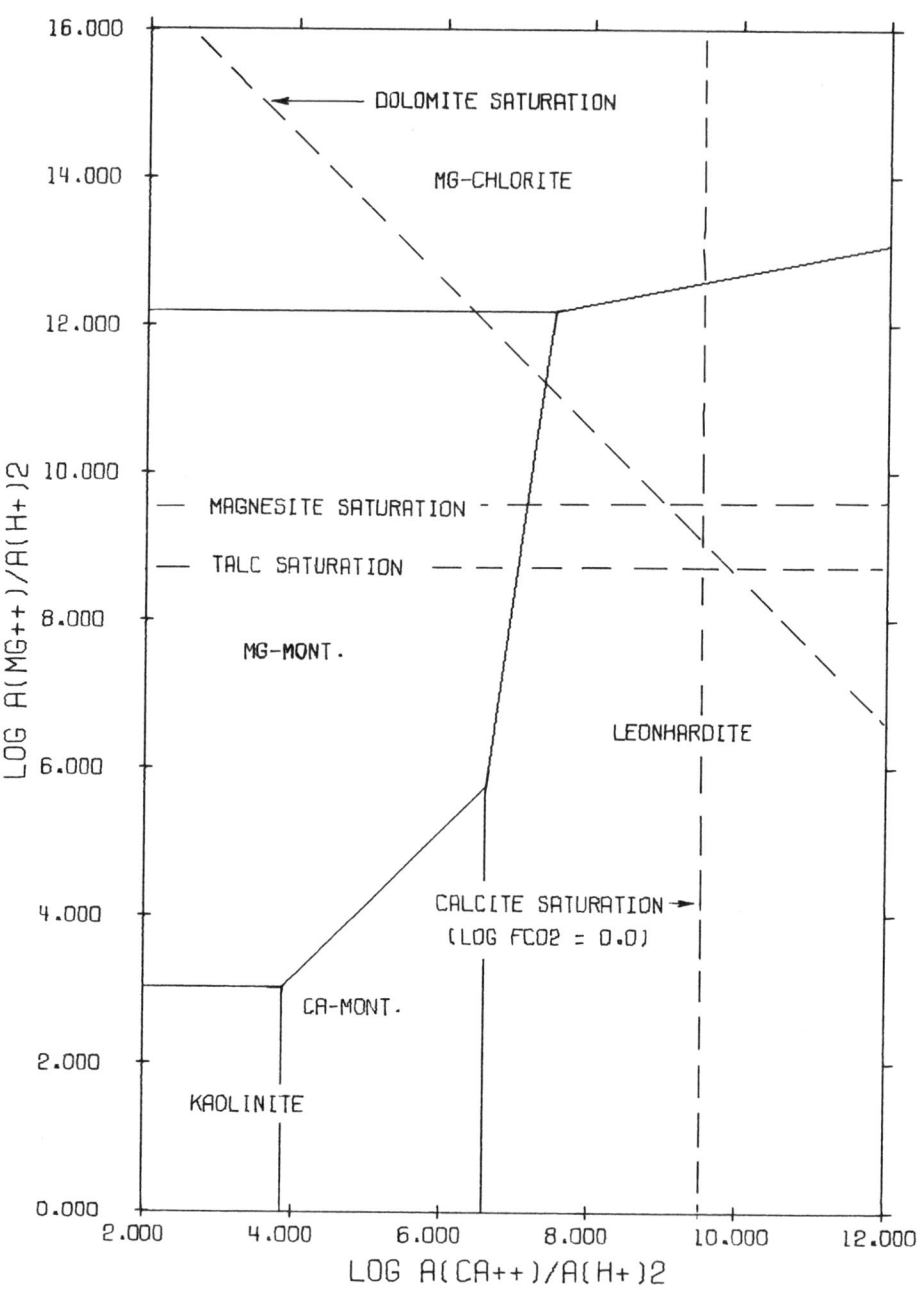

The System HCl—H$_2$O—Al$_2$O$_3$—CaO—CO$_2$—MgO—SiO$_2$ at 60°C; log $a_{H_4SiO_4}$ = −2.47 = amorphous silica saturation.

Activity Diagrams

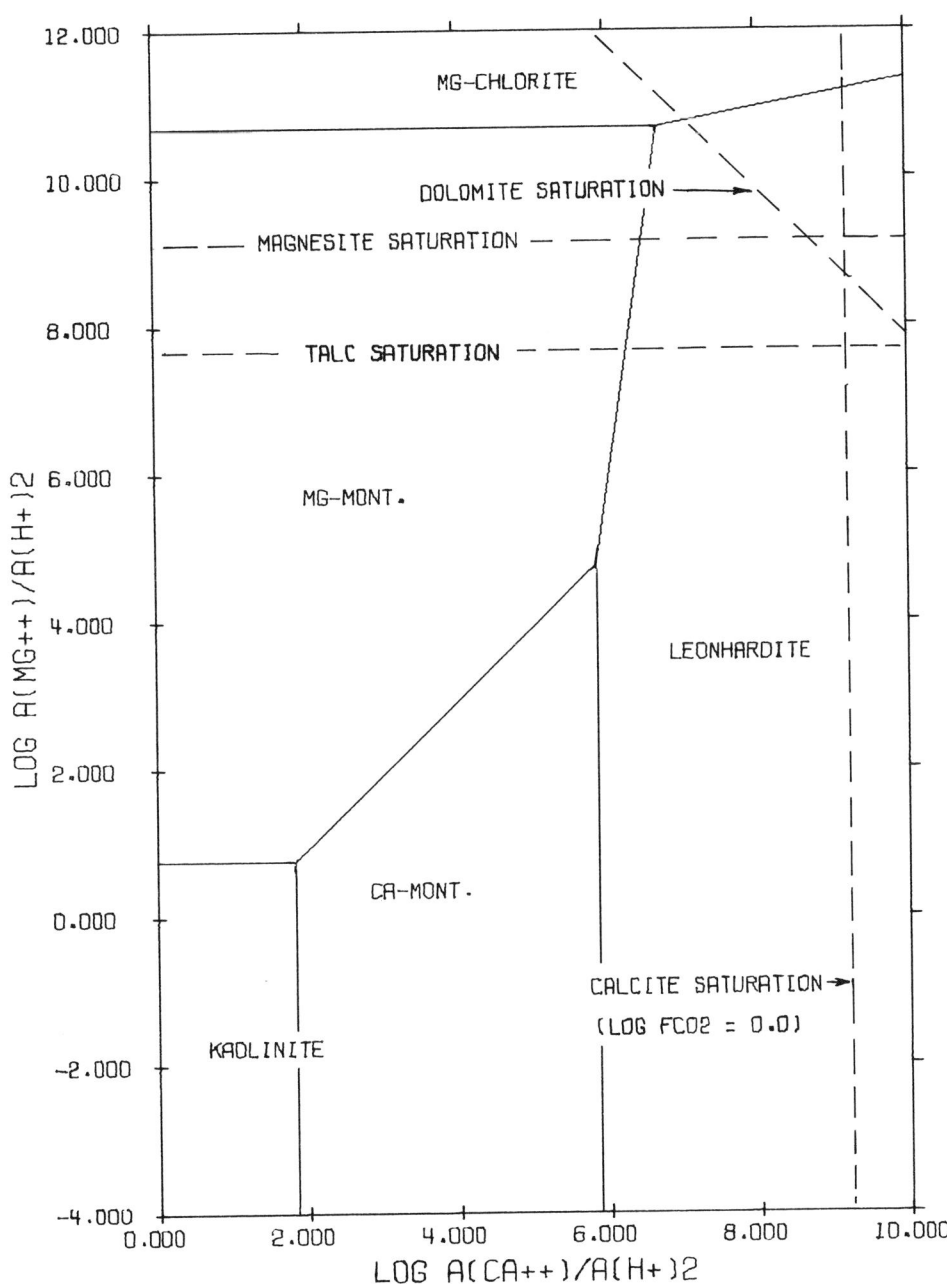

The System HCl—H$_2$O—Al$_2$O$_3$—CaO—CO$_2$—MgO—SiO$_2$ at 100°C; log $a_{H_4SiO_4}$ = −2.20 = amorphous silica saturation.

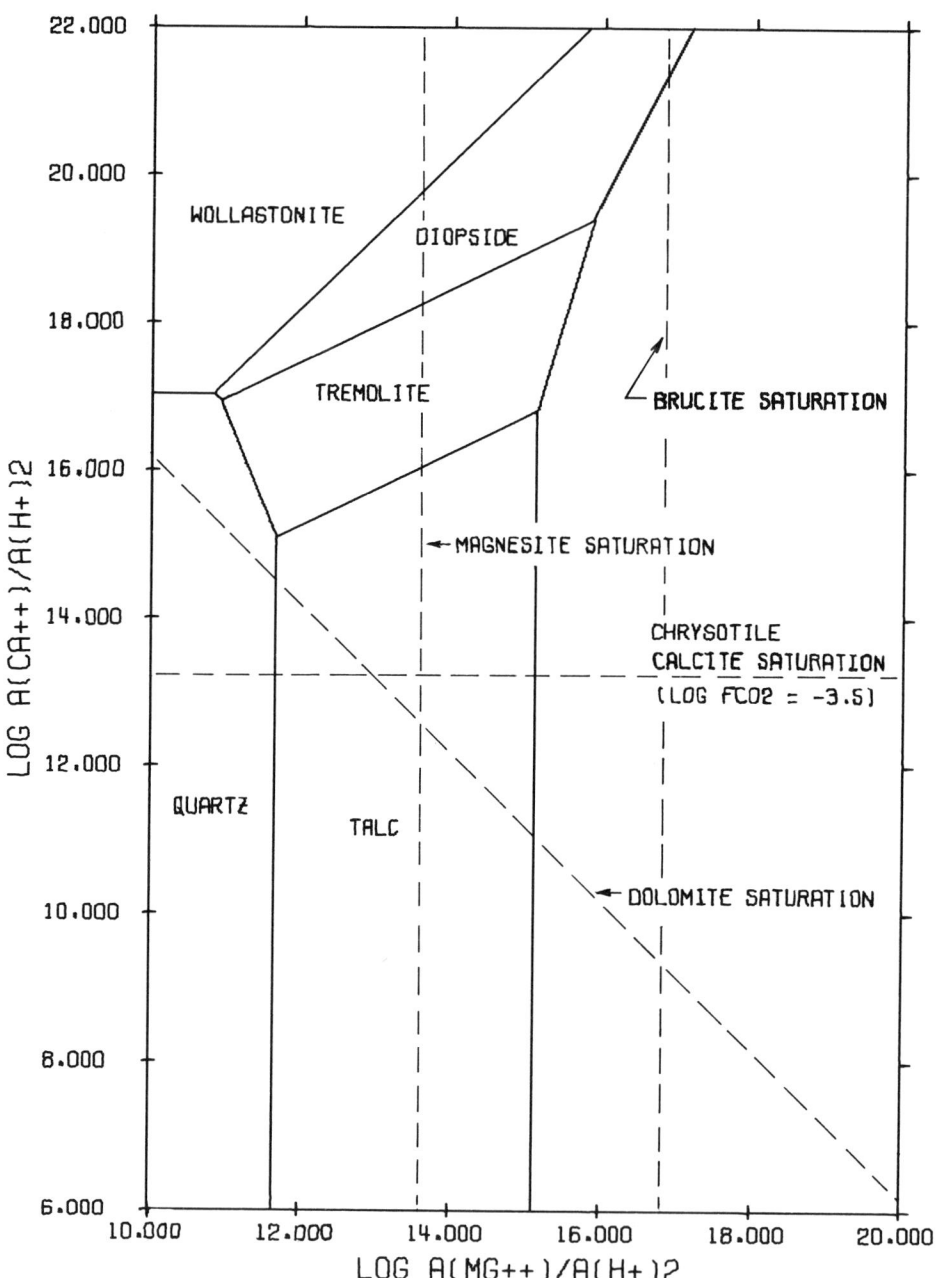

The System HCl—H$_2$O—CaO—CO$_2$—MgO—SiO$_2$ at 25°C.

Activity Diagrams

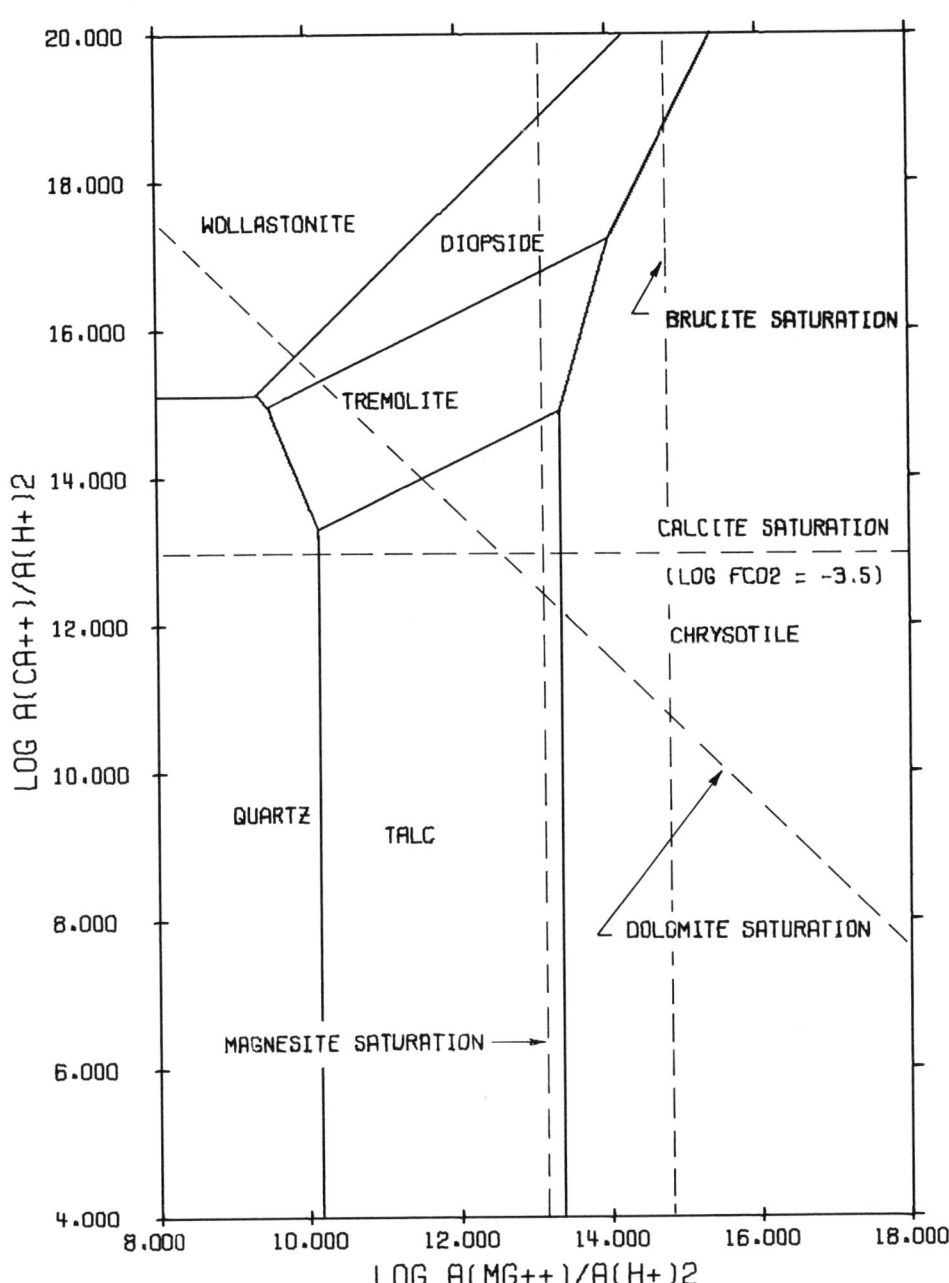

The System HCl—H$_2$O—CaO—CO$_2$—MgO—SiO$_2$ at 60°C.

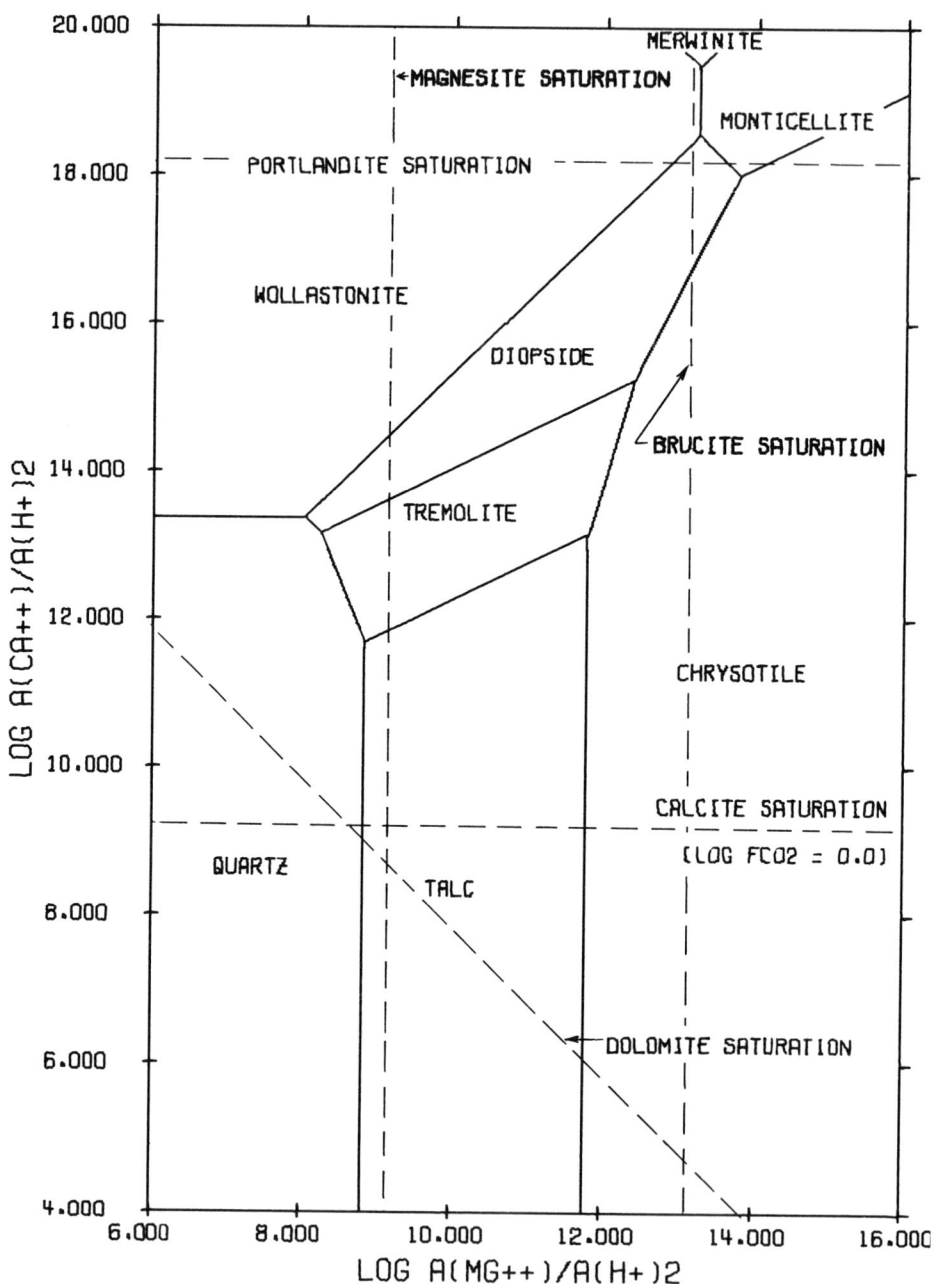

The System HCl—H$_2$O—CaO—CO$_2$—MgO—SiO$_2$ at 100°C.

Activity Diagrams

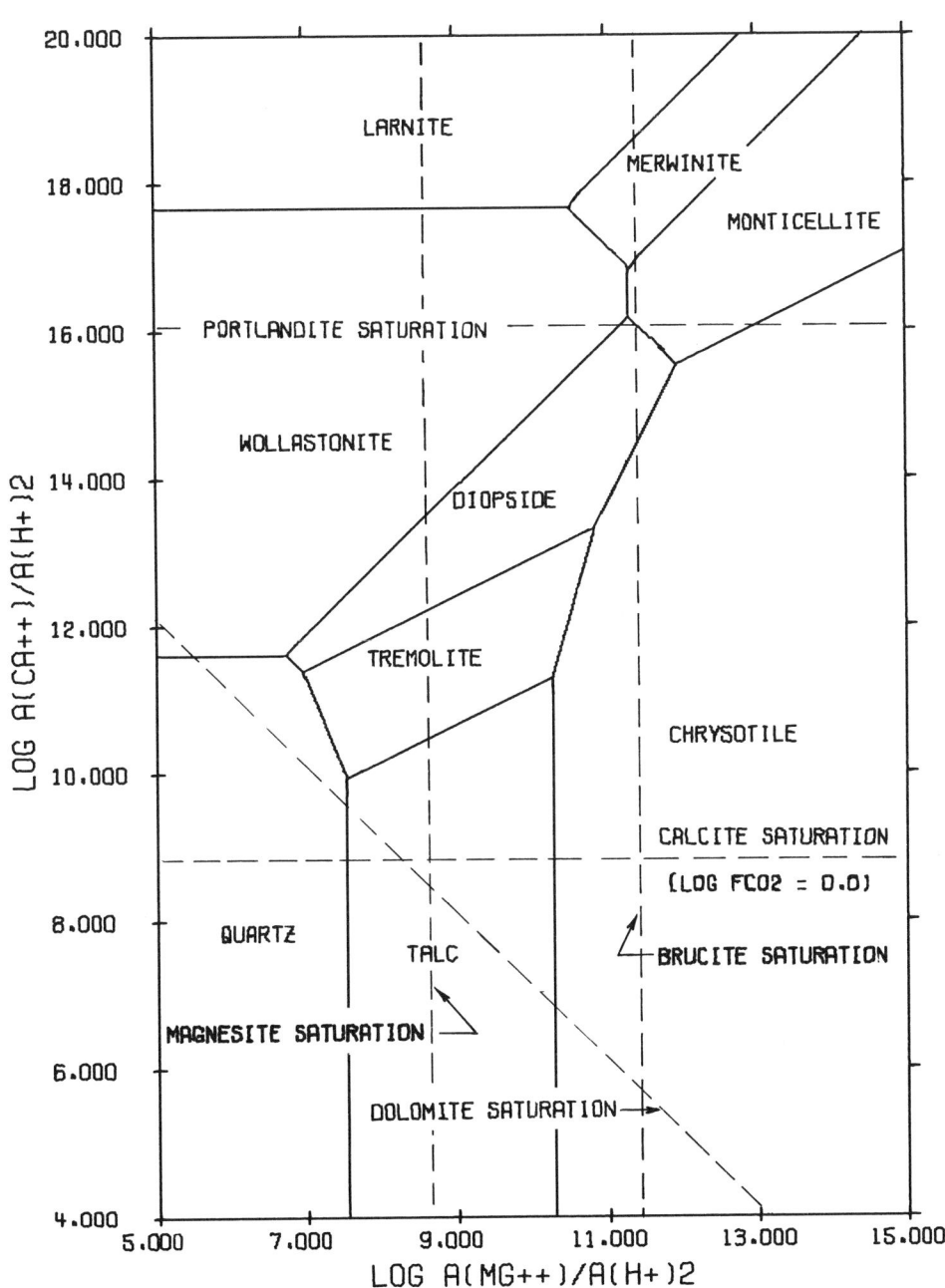

The System HCl—H₂O—CaO—CO₂—MgO—SiO₂ at 150°C.

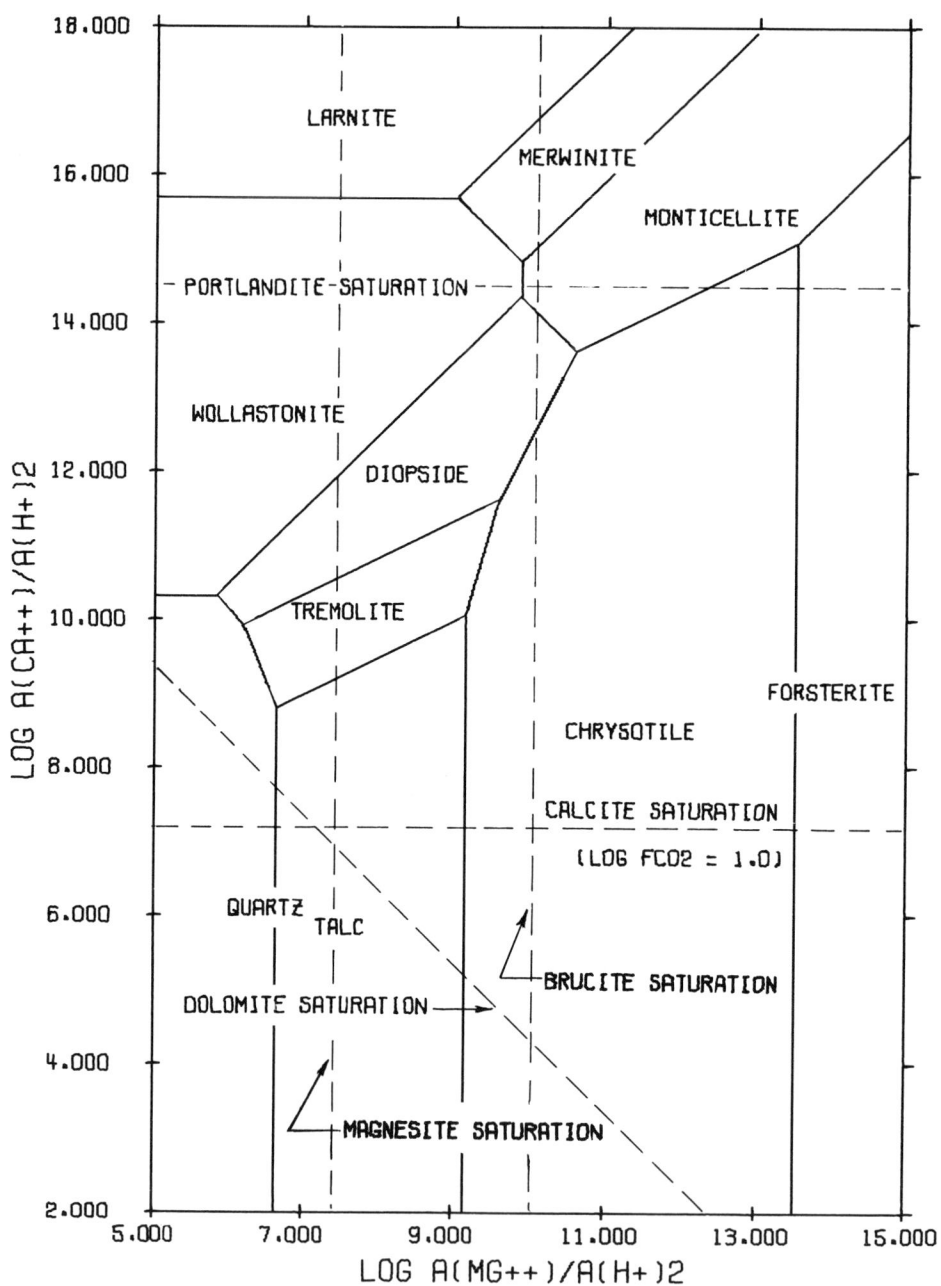

The System HCl—H$_2$O—CaO—CO$_2$—MgO—SiO$_2$ at 200°C.

Activity Diagrams

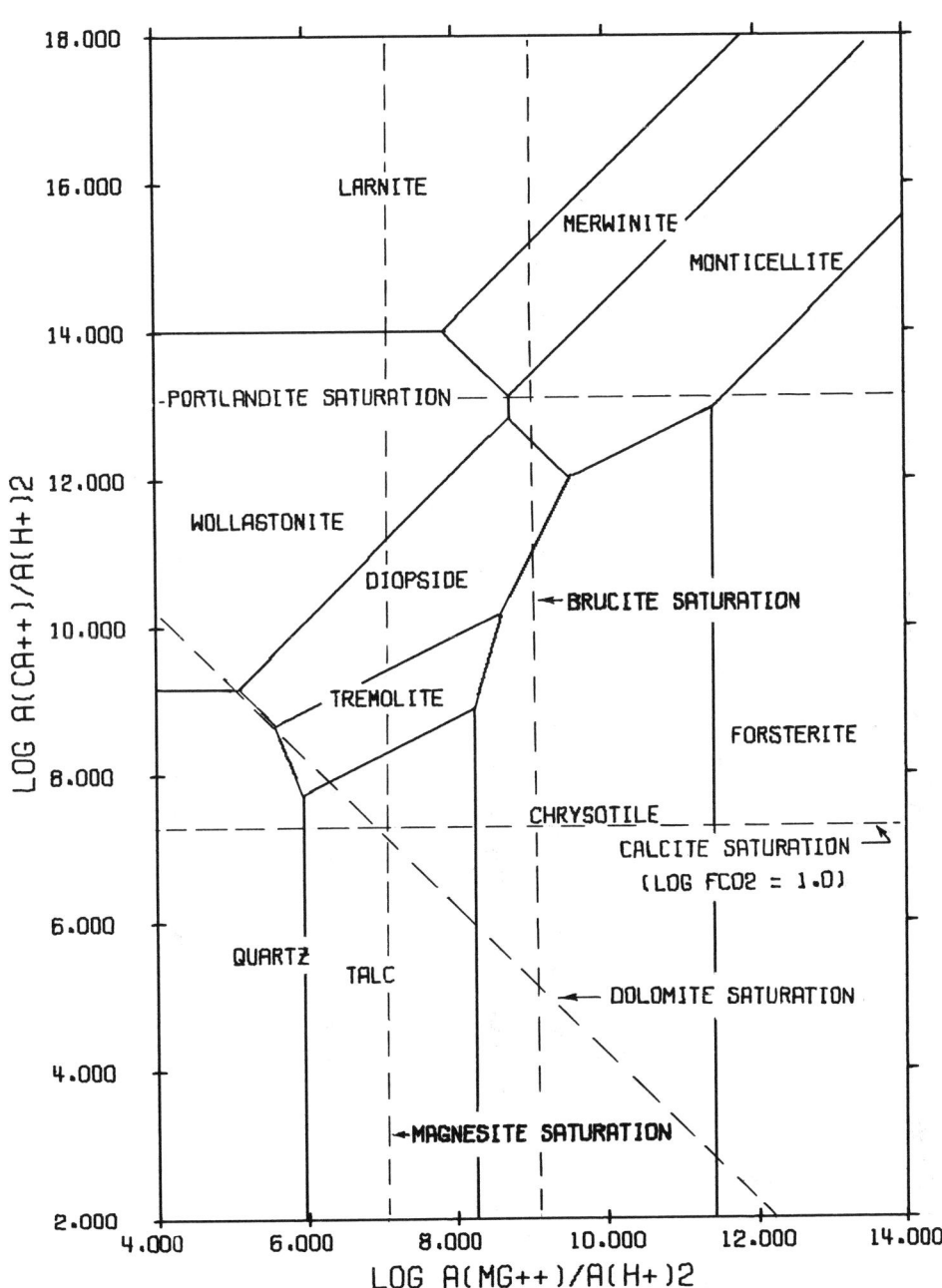

The System HCl—H₂O—CaO—CO₂—MgO—SiO₂ at 250°C.

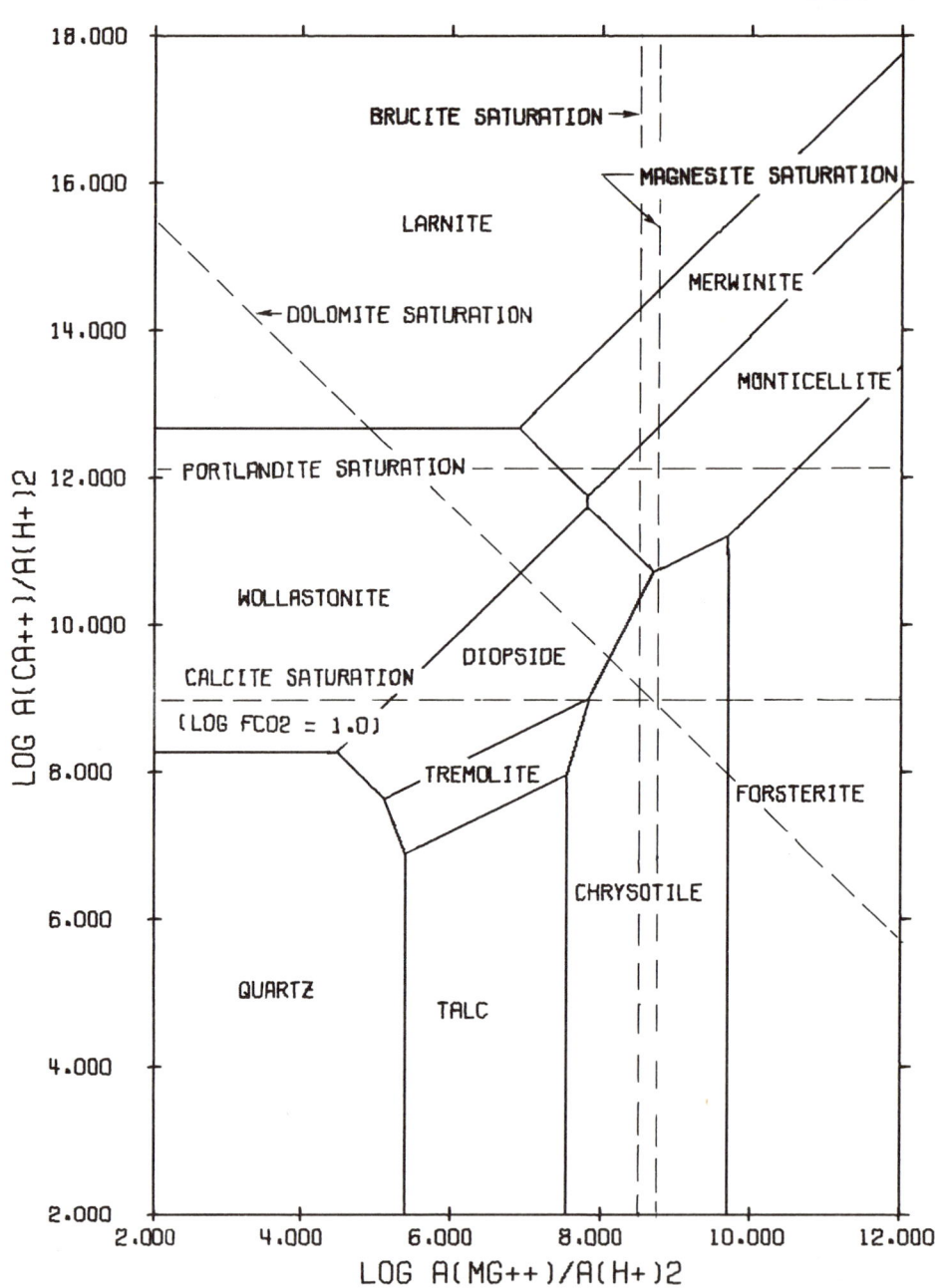

The System HCl—H₂O—CaO—CO₂—MgO—SiO₂ at 300°C.

Activity Diagrams

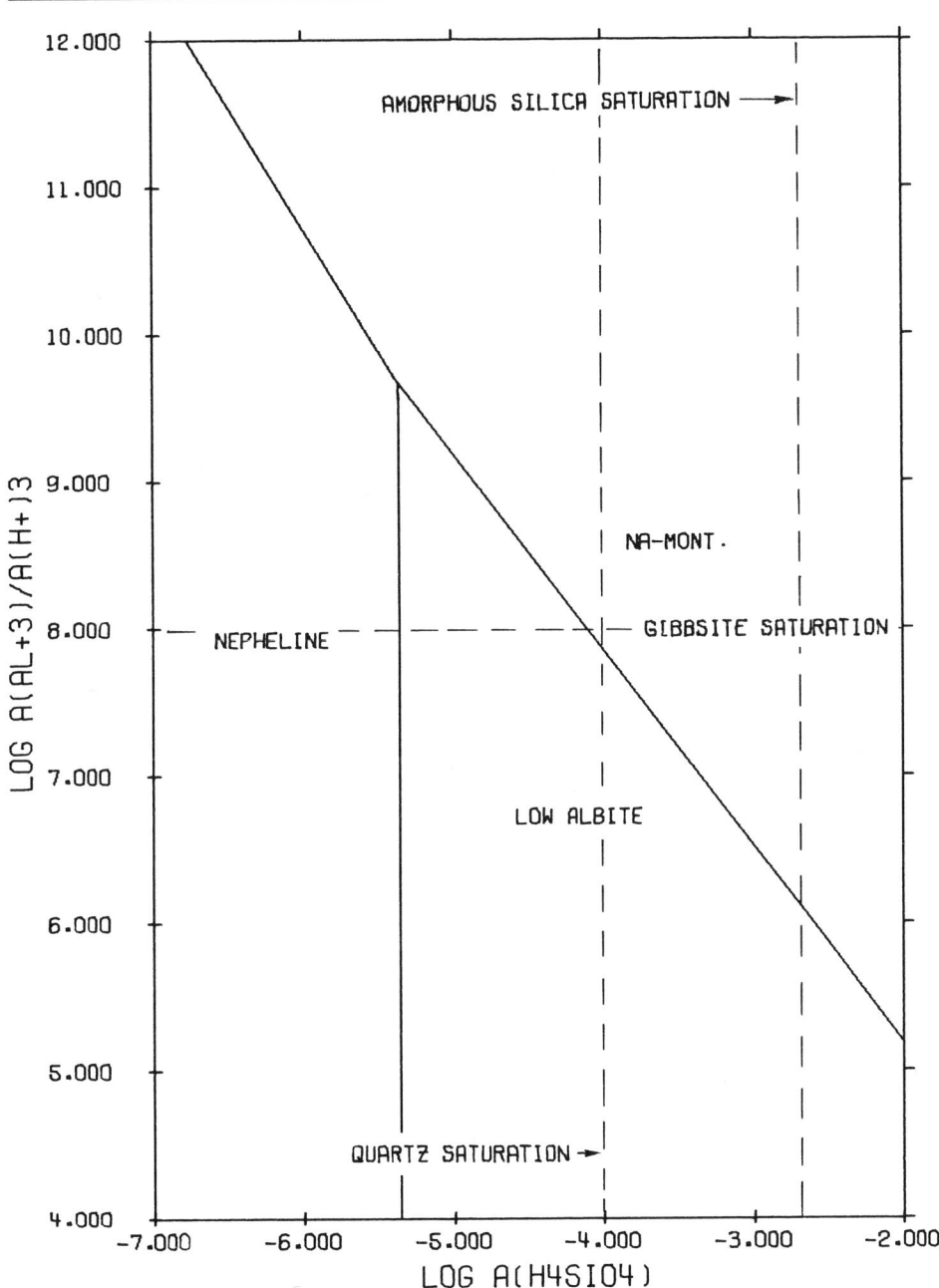

The System HCl—H$_2$O—Al$_2$O$_3$—Na$_2$O—SiO$_2$ at 25°C.

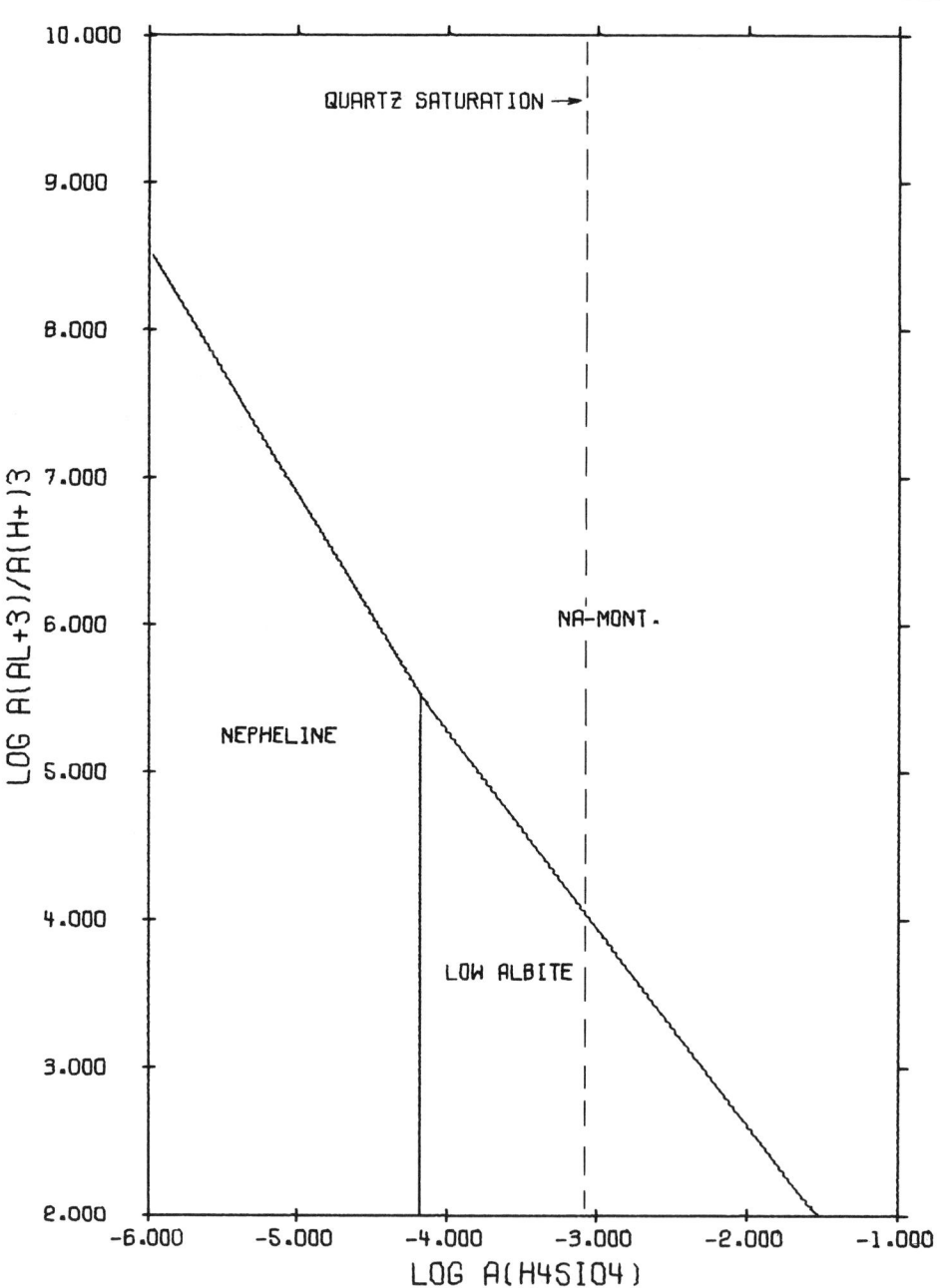

The System HCl—H$_2$O—Al$_2$O$_3$—Na$_2$O—SiO$_2$ at 100°C.

Activity Diagrams

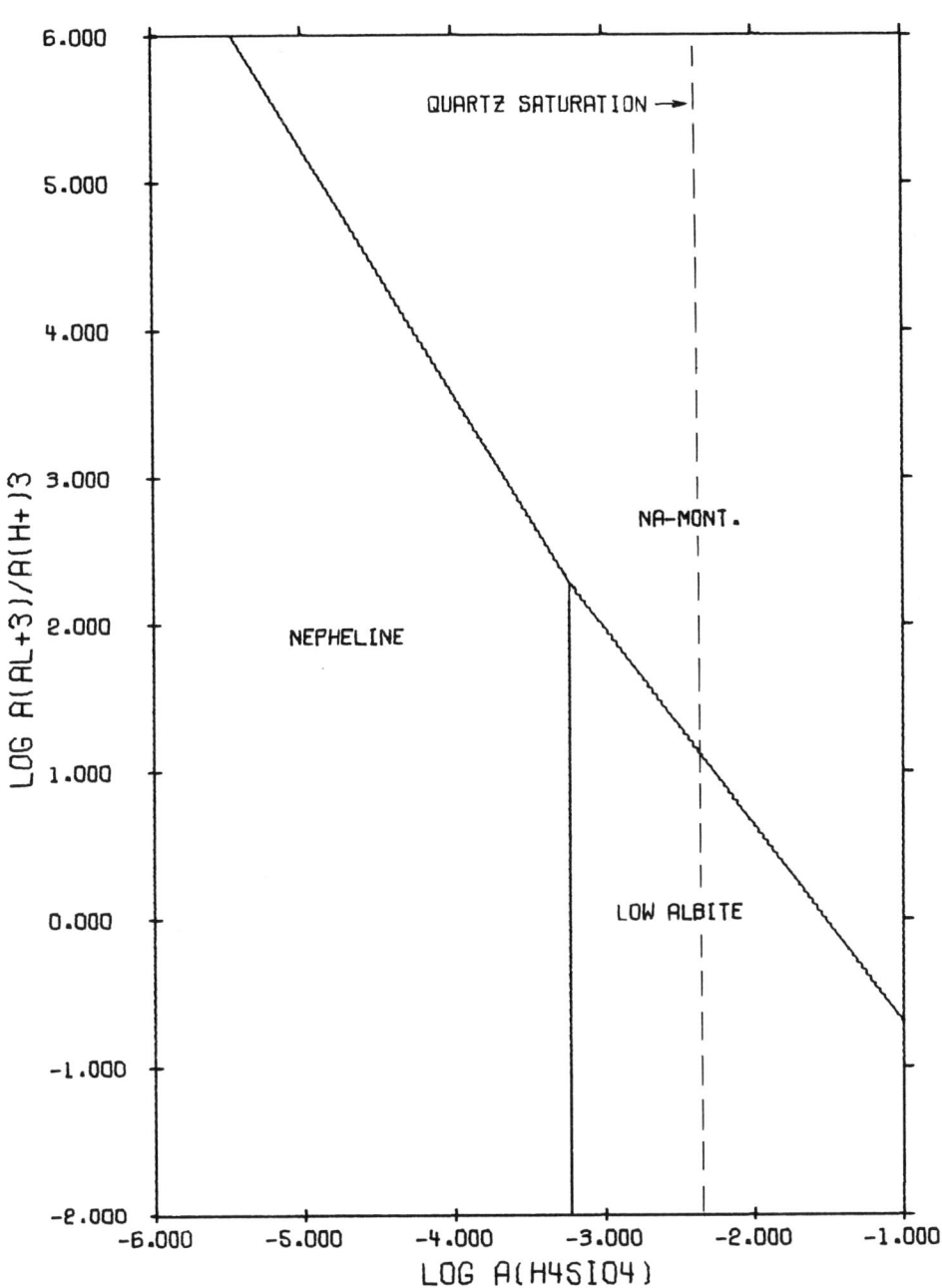

The System HCl—H$_2$O—Al$_2$O$_3$—Na$_2$O—SiO$_2$ at 200°C.

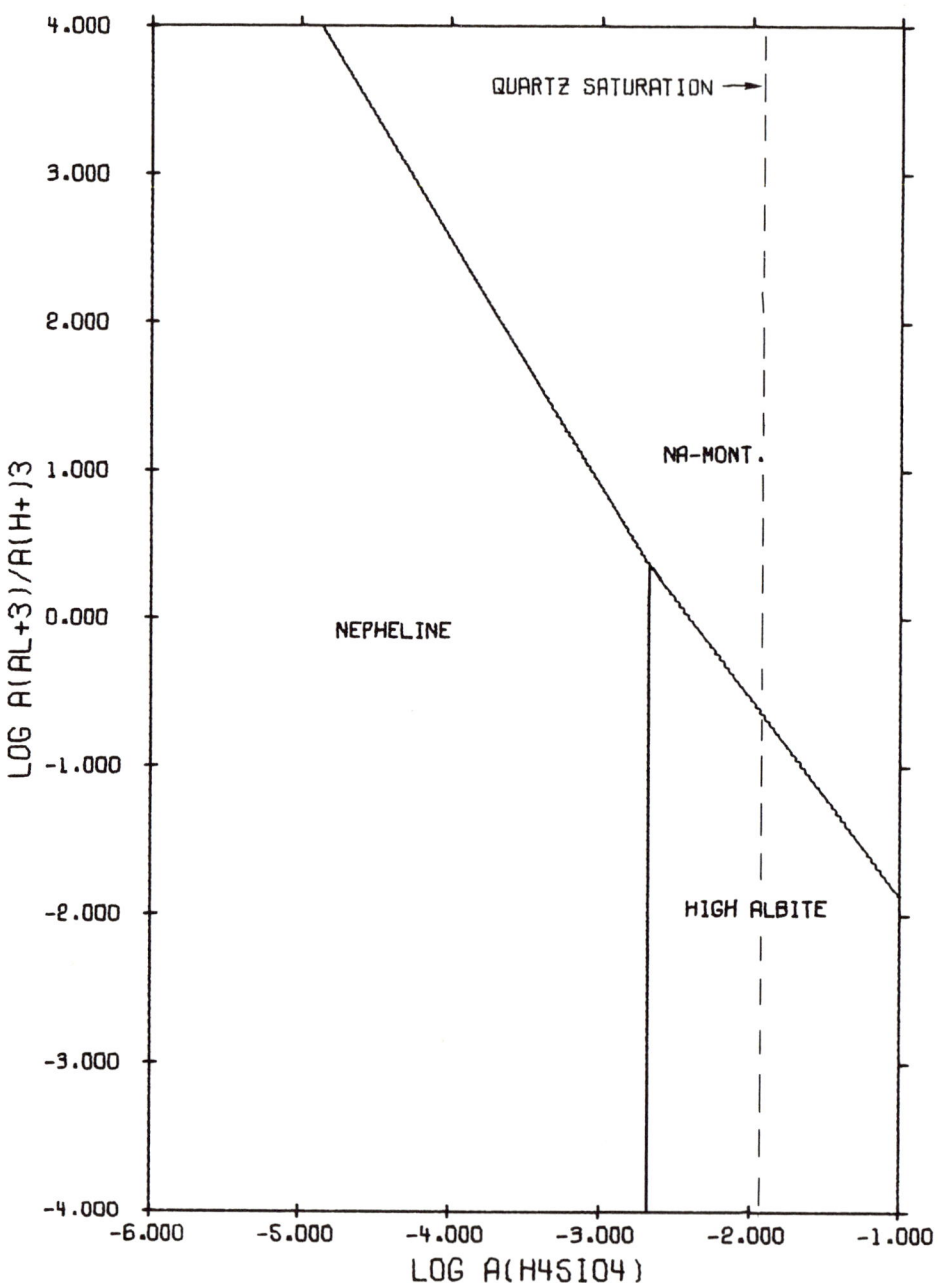

The System HCl—H₂O—Al₂O₃—Na₂O—SiO₂ at 300°C.

Activity Diagrams

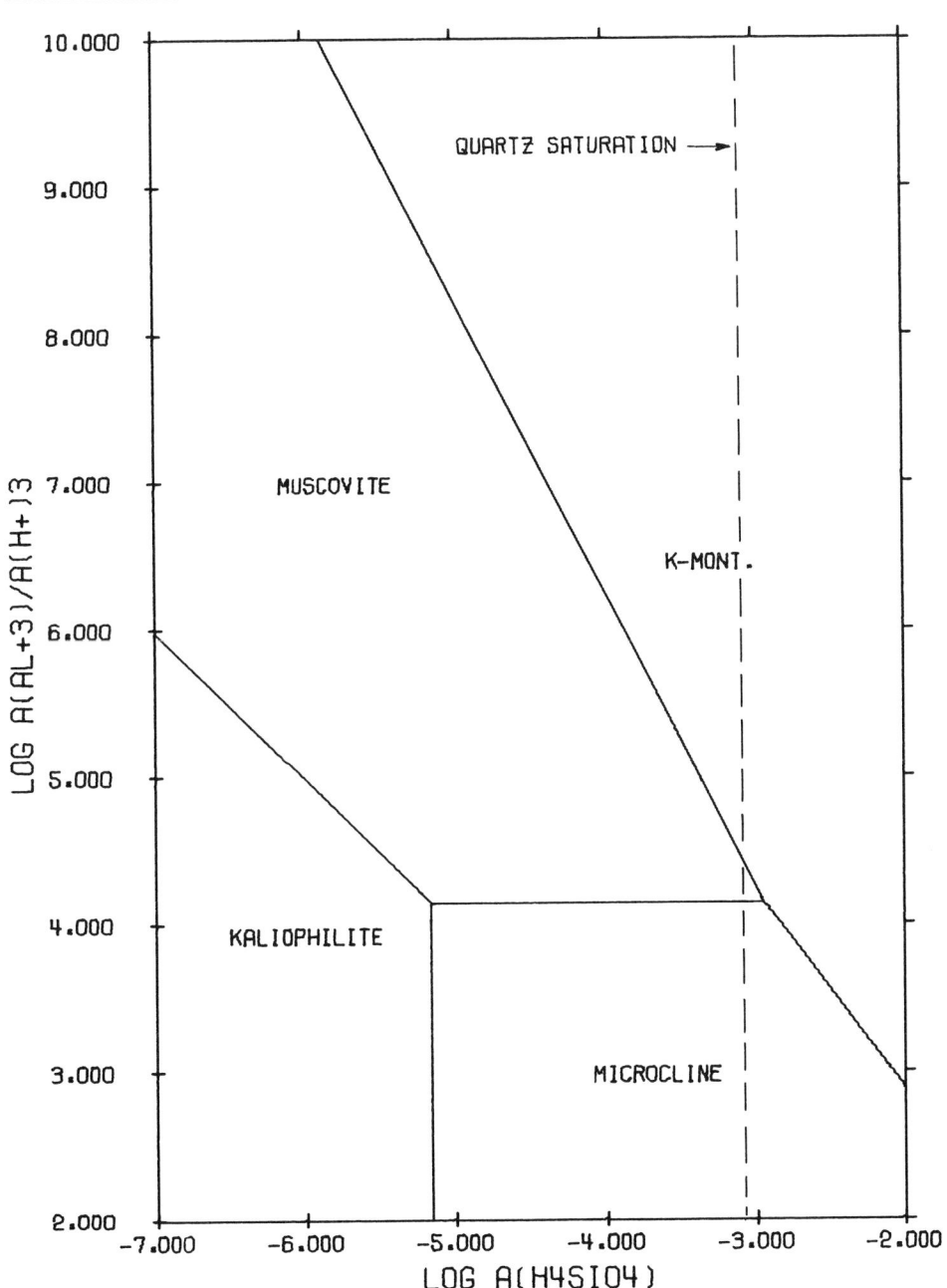

The System HCl—H$_2$O—Al$_2$O$_3$—K$_2$O—SiO$_2$ at 100°C.

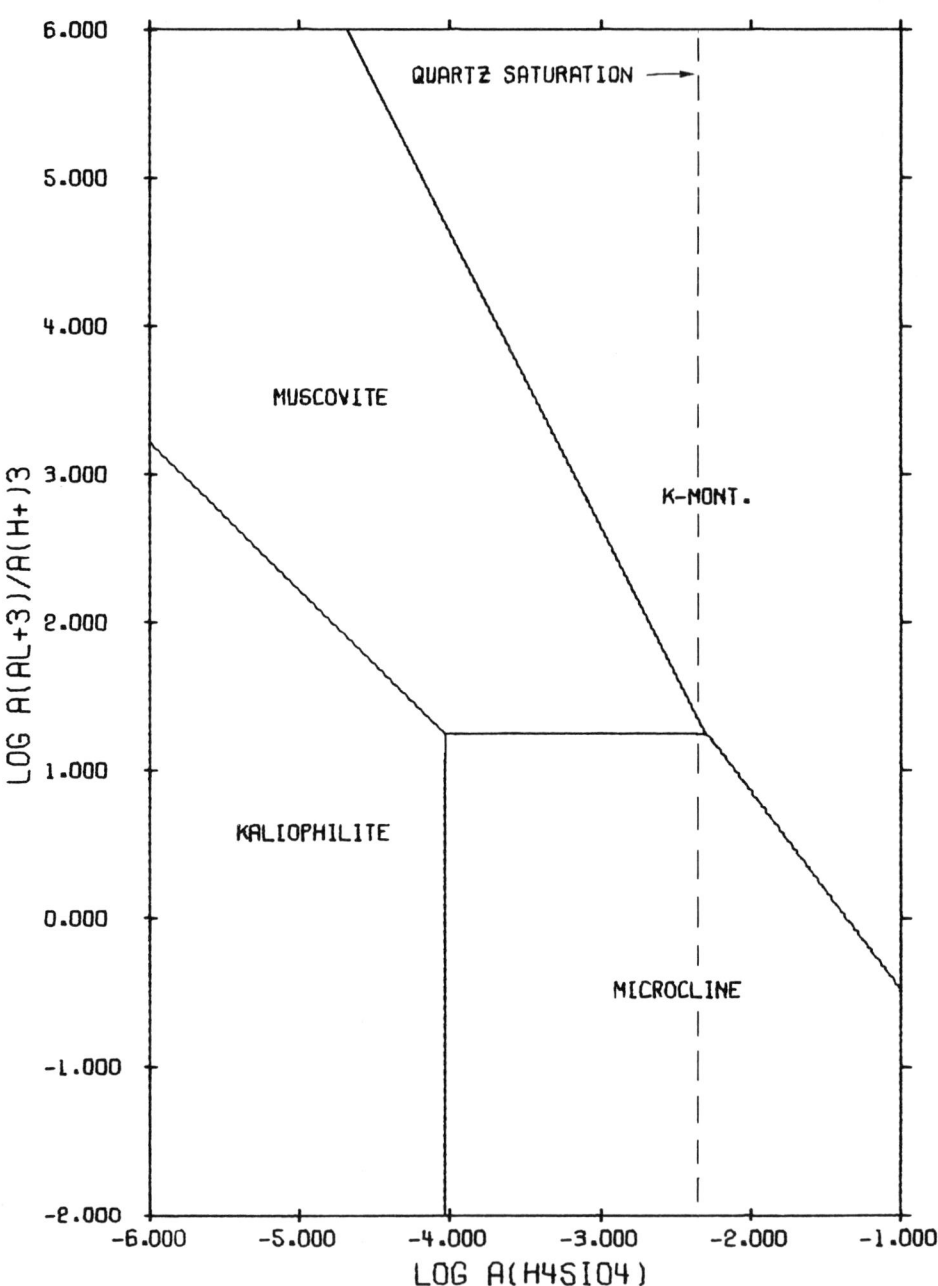

The System HCl—H$_2$O—Al$_2$O$_3$—K$_2$O—SiO$_2$ at 200°C.

Activity Diagrams

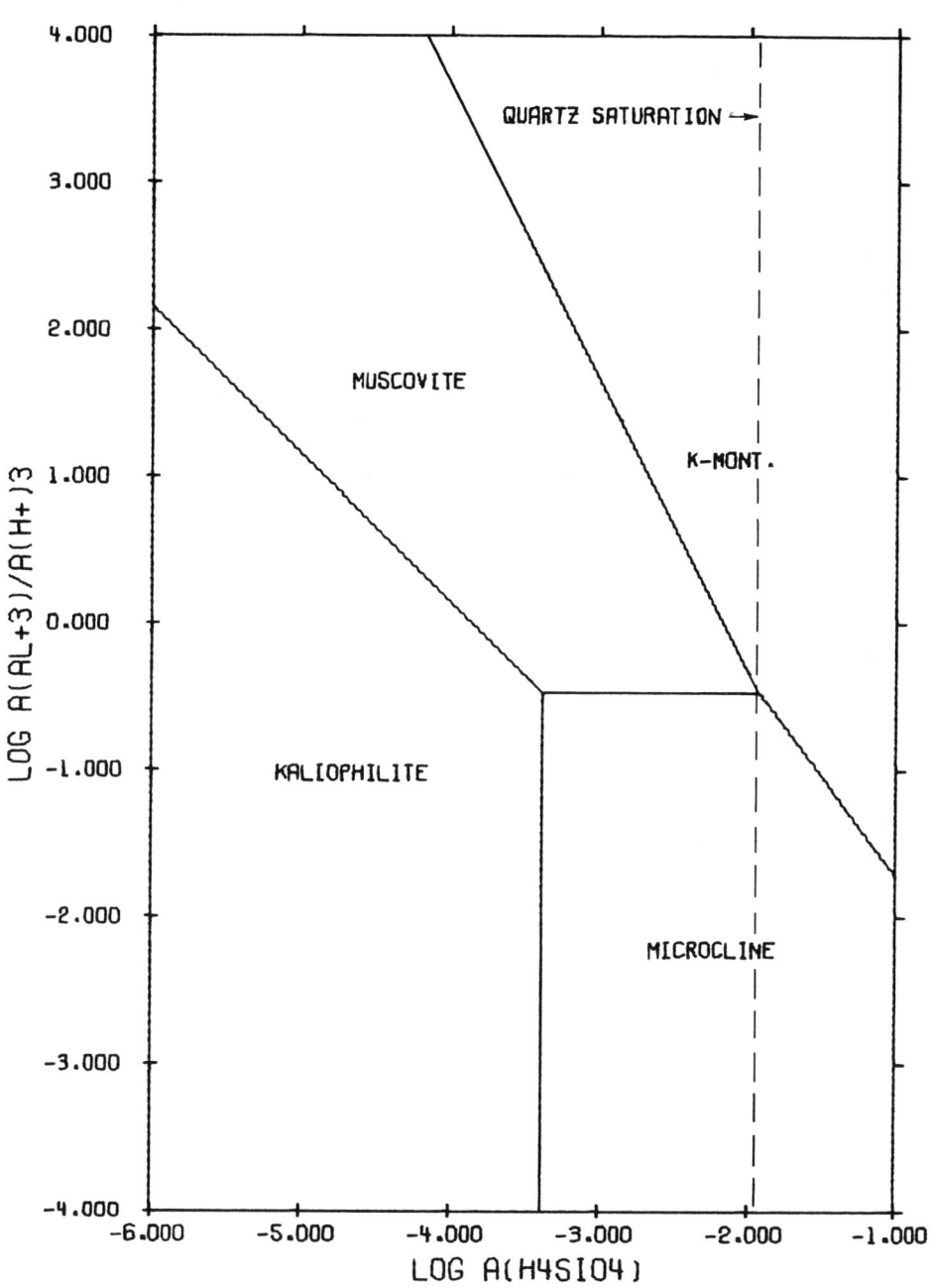

The System HCl—H$_2$O—Al$_2$O$_3$—K$_2$O—SiO$_2$ at 300°C.

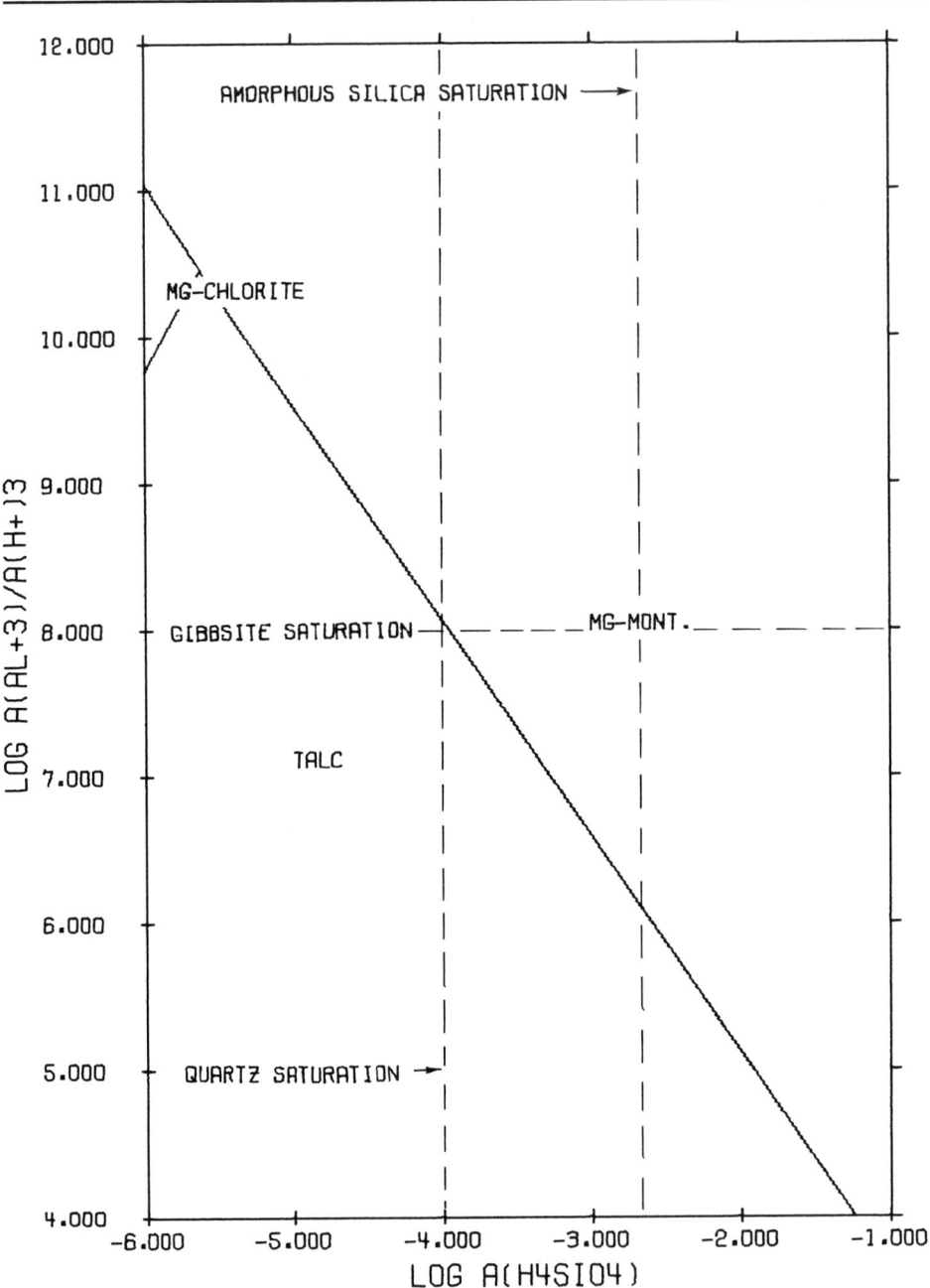

The System HCl—H_2O—Al_2O_3—MgO—SiO_2 at 25°C.

Activity Diagrams

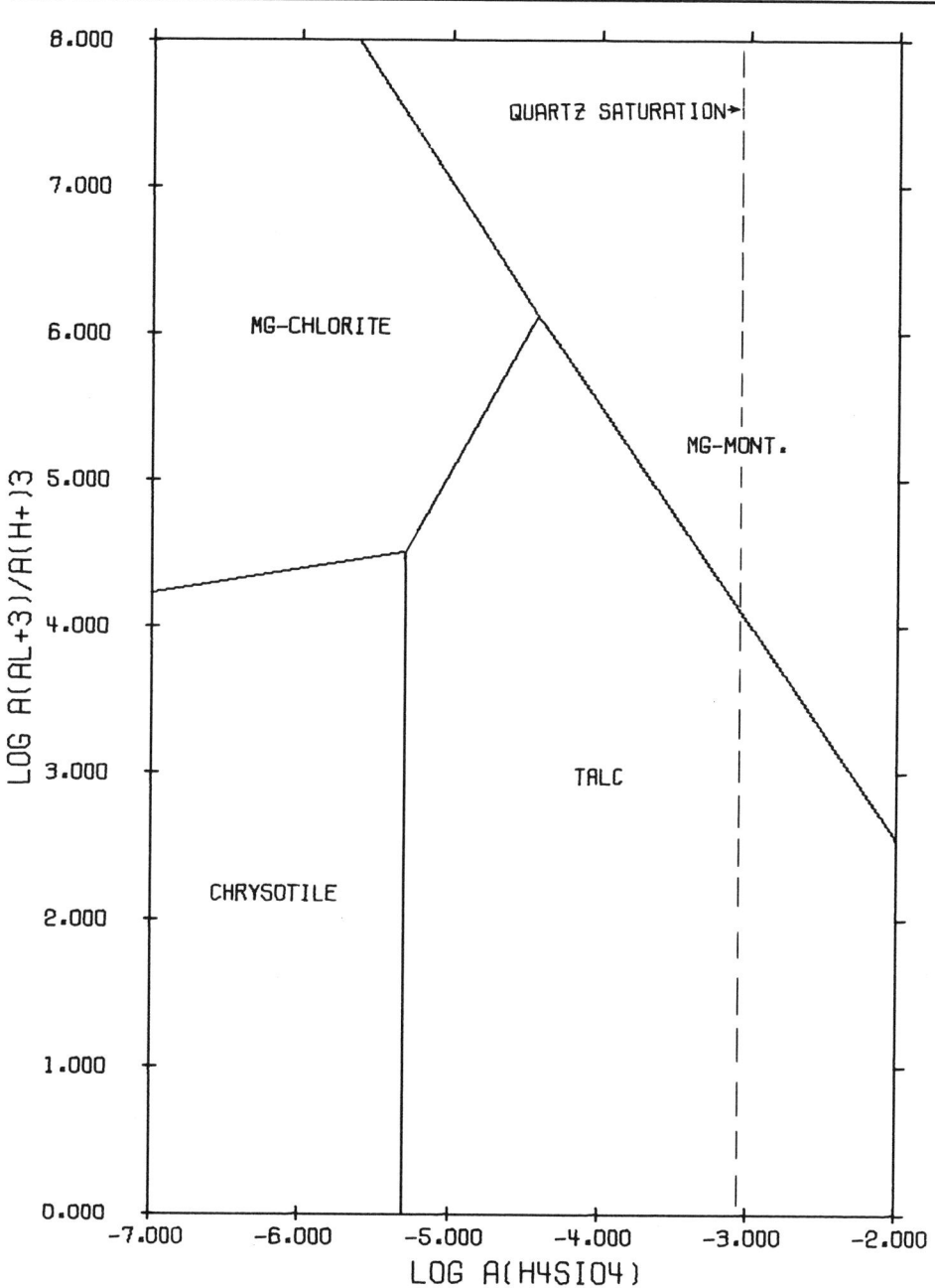

The System HCl—H$_2$O—Al$_2$O$_3$—MgO—SiO$_2$ at 100°C.

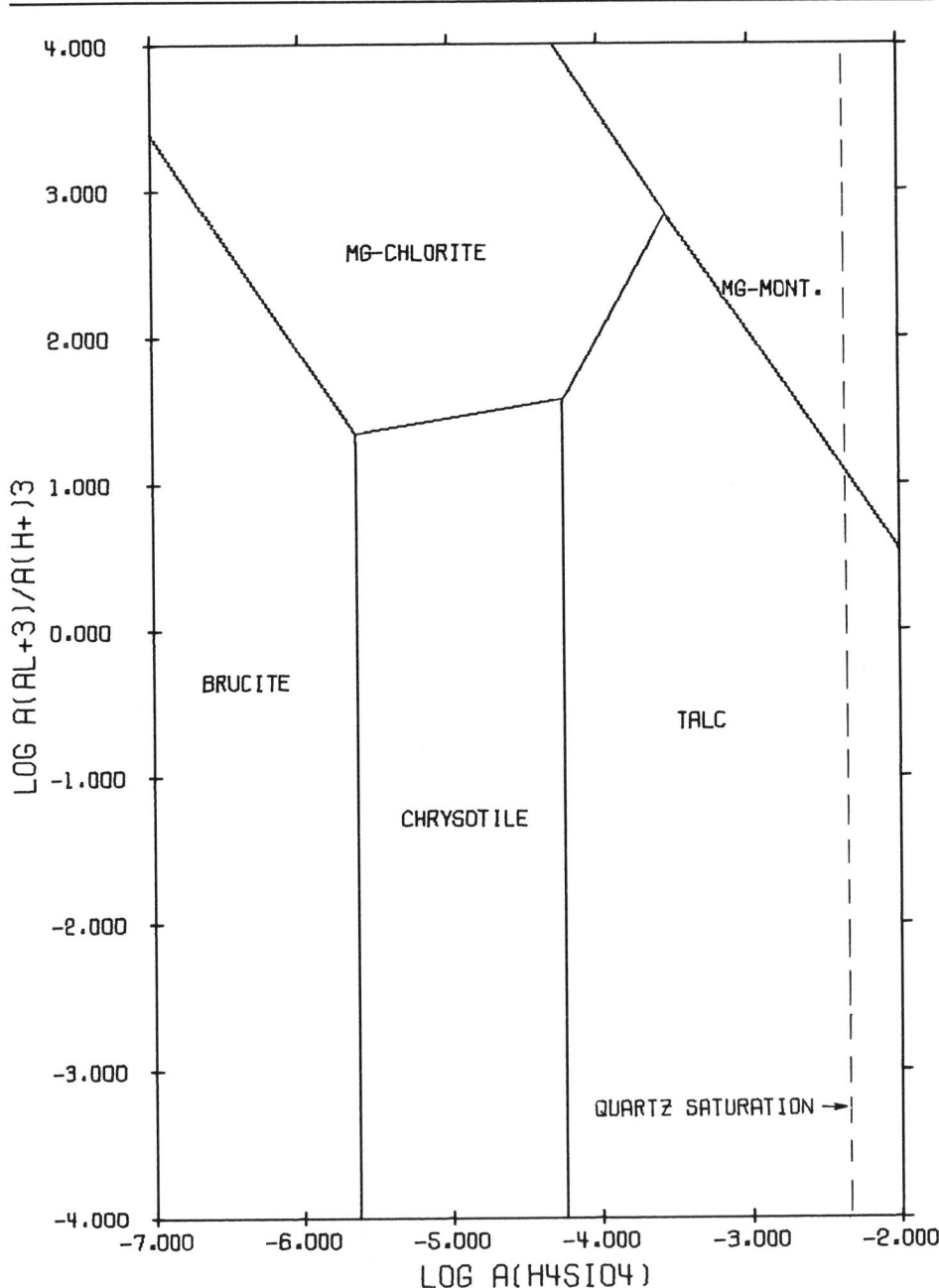

The System HCl—H₂O—Al₂O₃—MgO—SiO₂ at 200°C.

Activity Diagrams

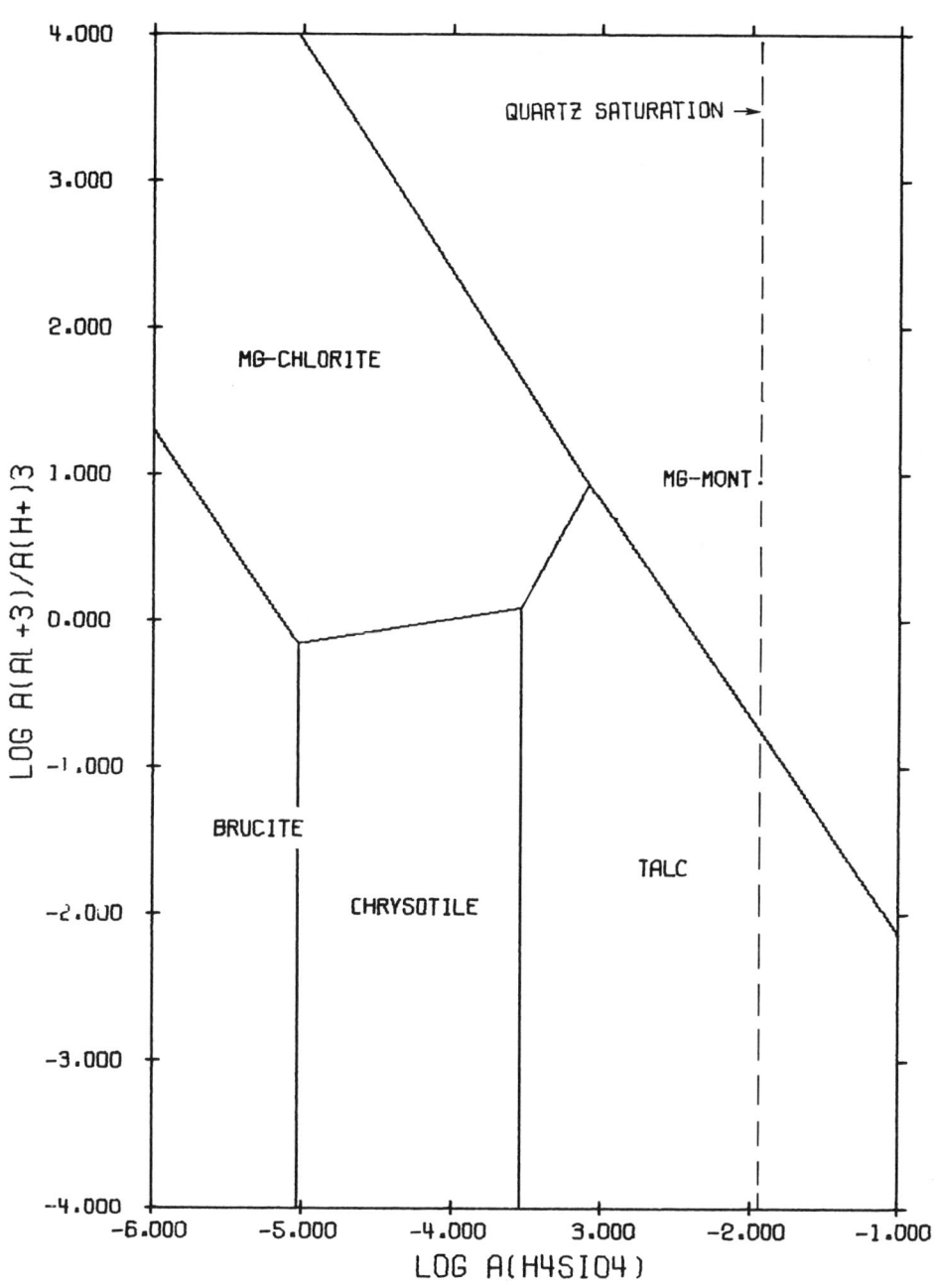

The System HCl—H_2O—Al_2O_3—MgO—SiO_2 at 300°C.

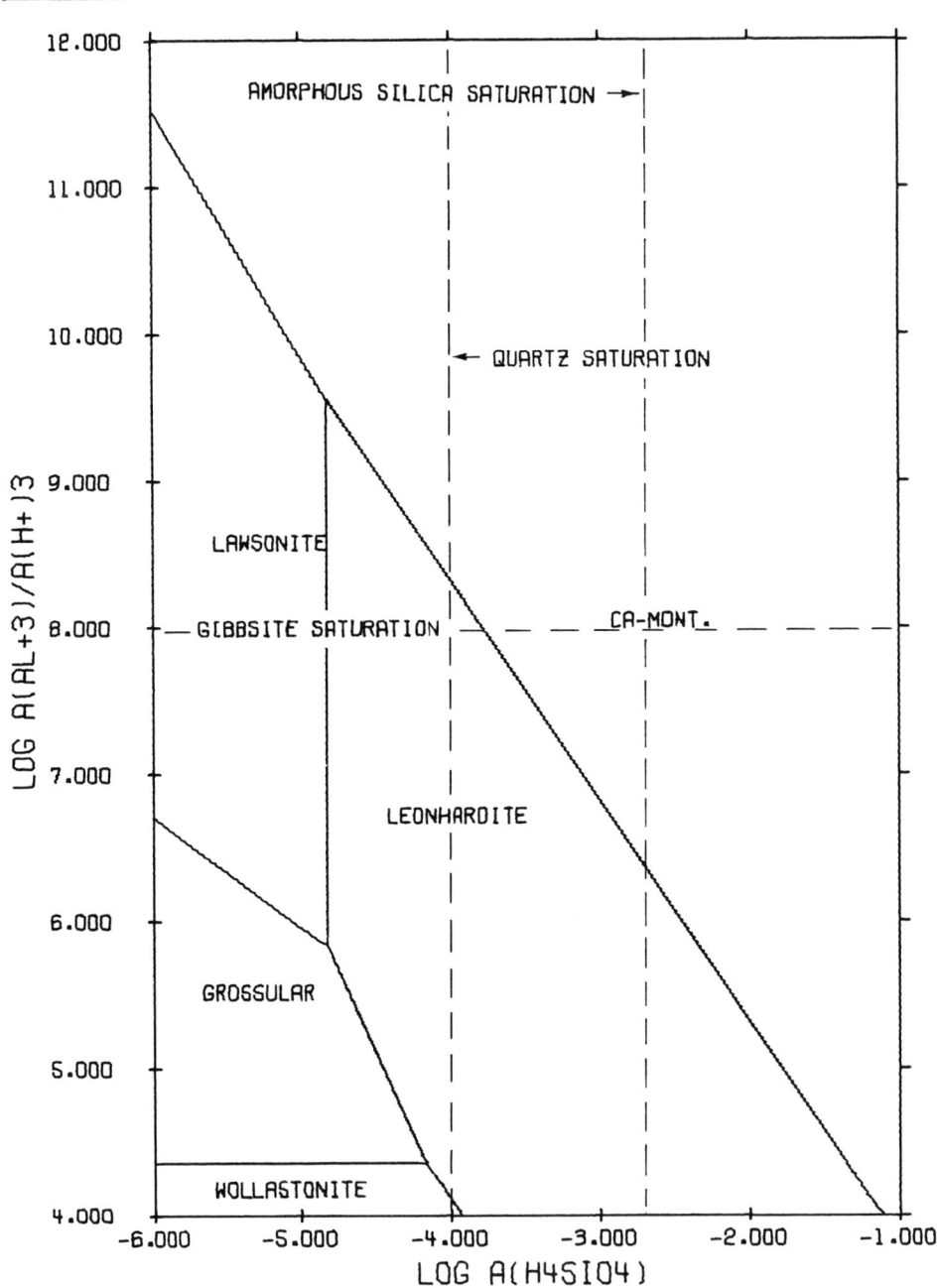

The System HCl—H$_2$O—Al$_2$O$_3$—CaO—SiO$_2$ at 25°C.

Activity Diagrams

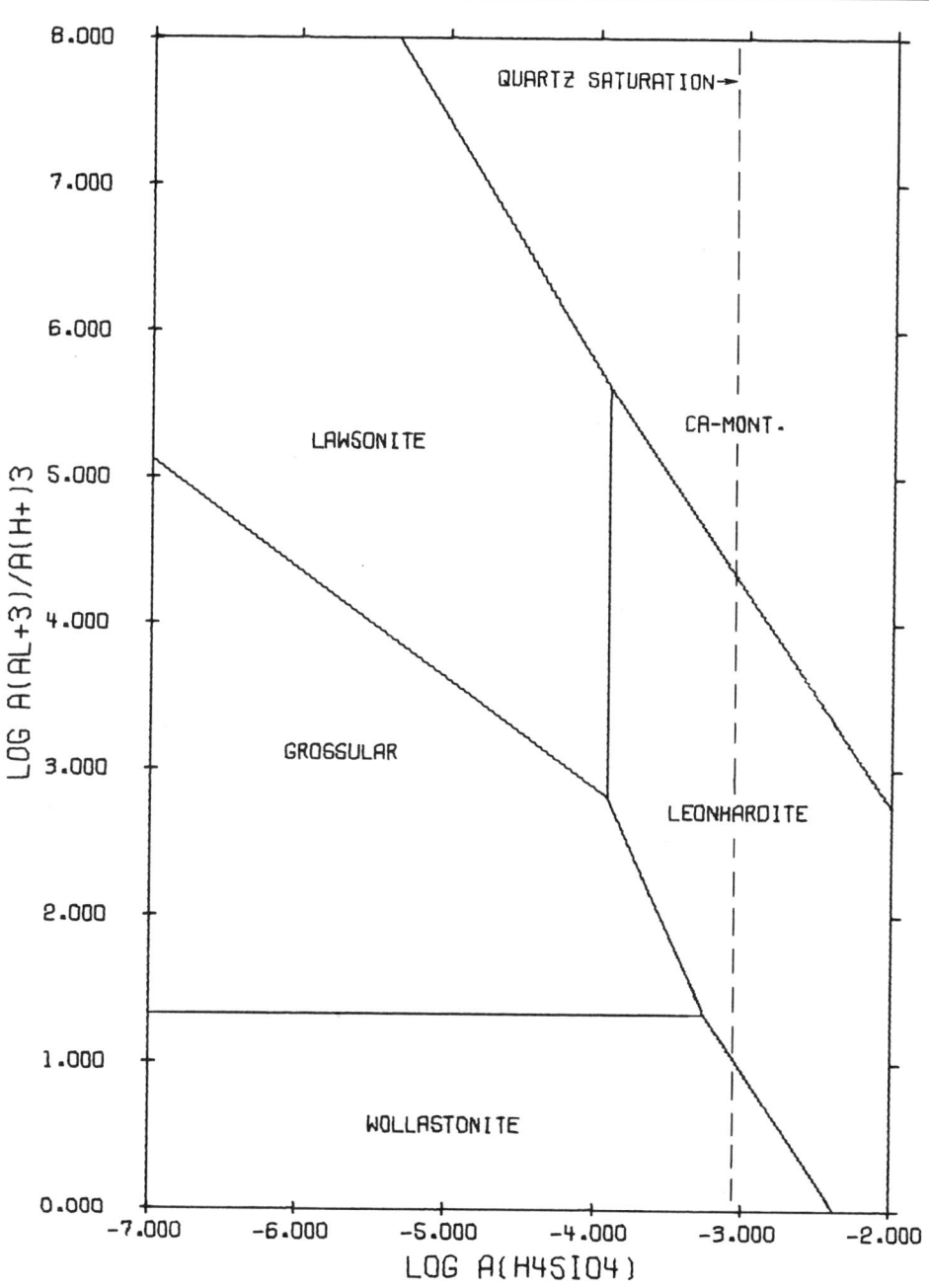

The System HCl—H$_2$O—Al$_2$O$_3$—CaO—SiO$_2$ at 100°C.

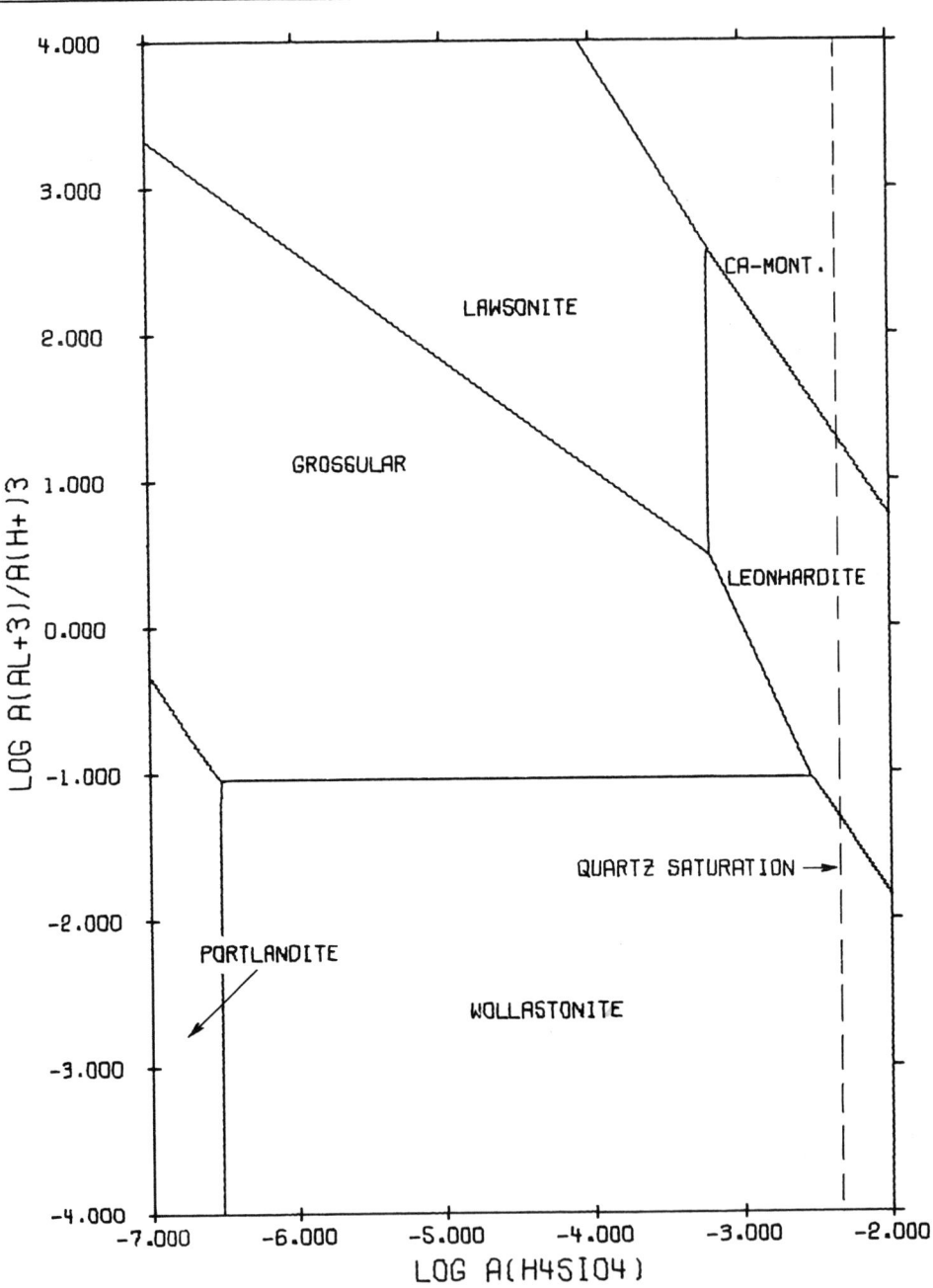

The System HCl—H$_2$O—Al$_2$O$_3$—CaO—SiO$_2$ at 200°C.

Activity Diagrams

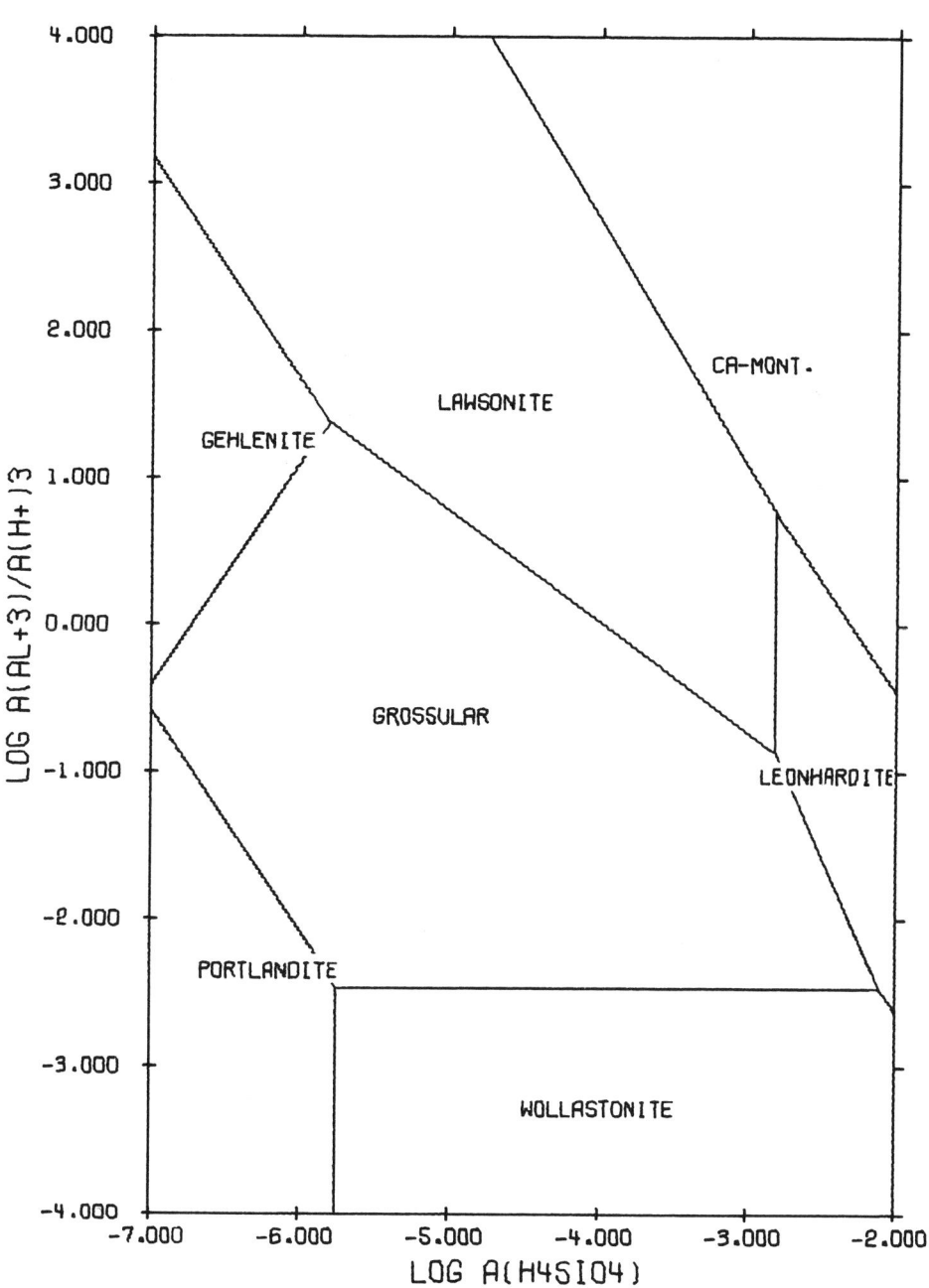

The System HCl—H$_2$O—Al$_2$O$_3$—CaO—SiO$_2$ at 300°C.

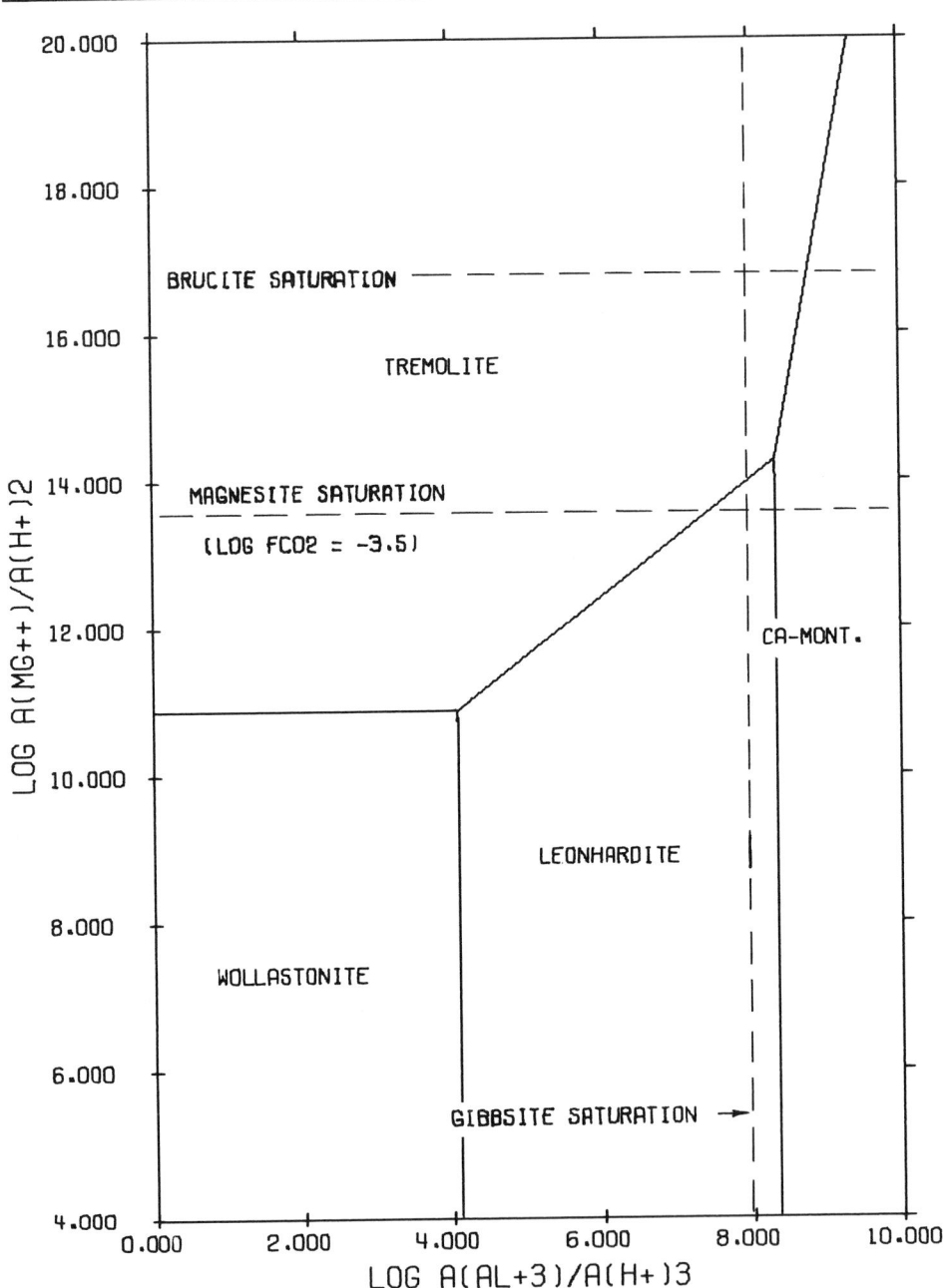

The System HCl—H₂O—Al₂O₃—CaO—CO₂—MgO—SiO₂ at 25°C; $\log a_{H_4SiO_4} = -4.00$ = quartz saturation.

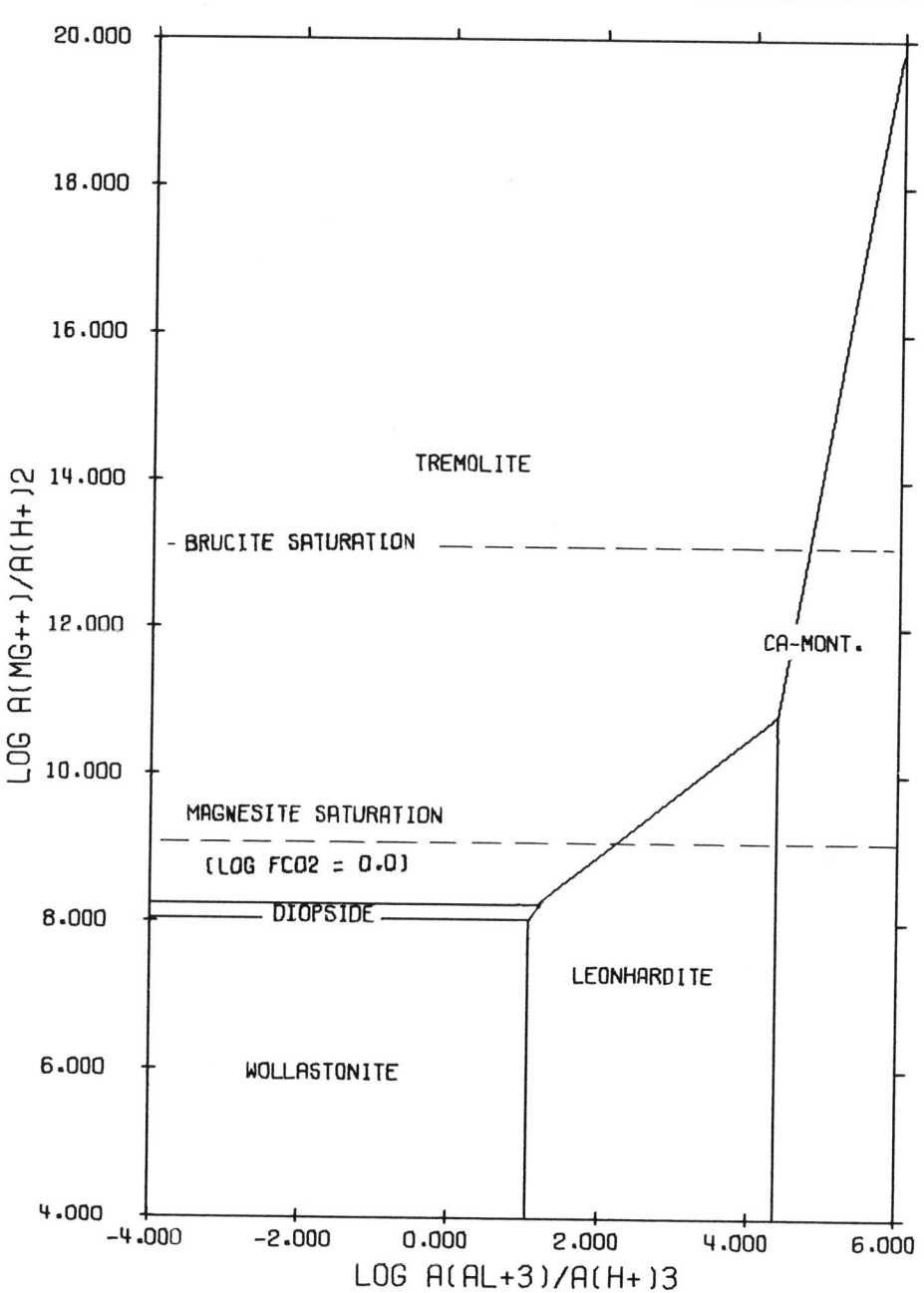

The System HCl—H$_2$O—Al$_2$O$_3$—CaO—CO$_2$—MgO—SiO$_2$ at 100°C; log $a_{H_4SiO_4}$ = −3.08 = quartz saturation.

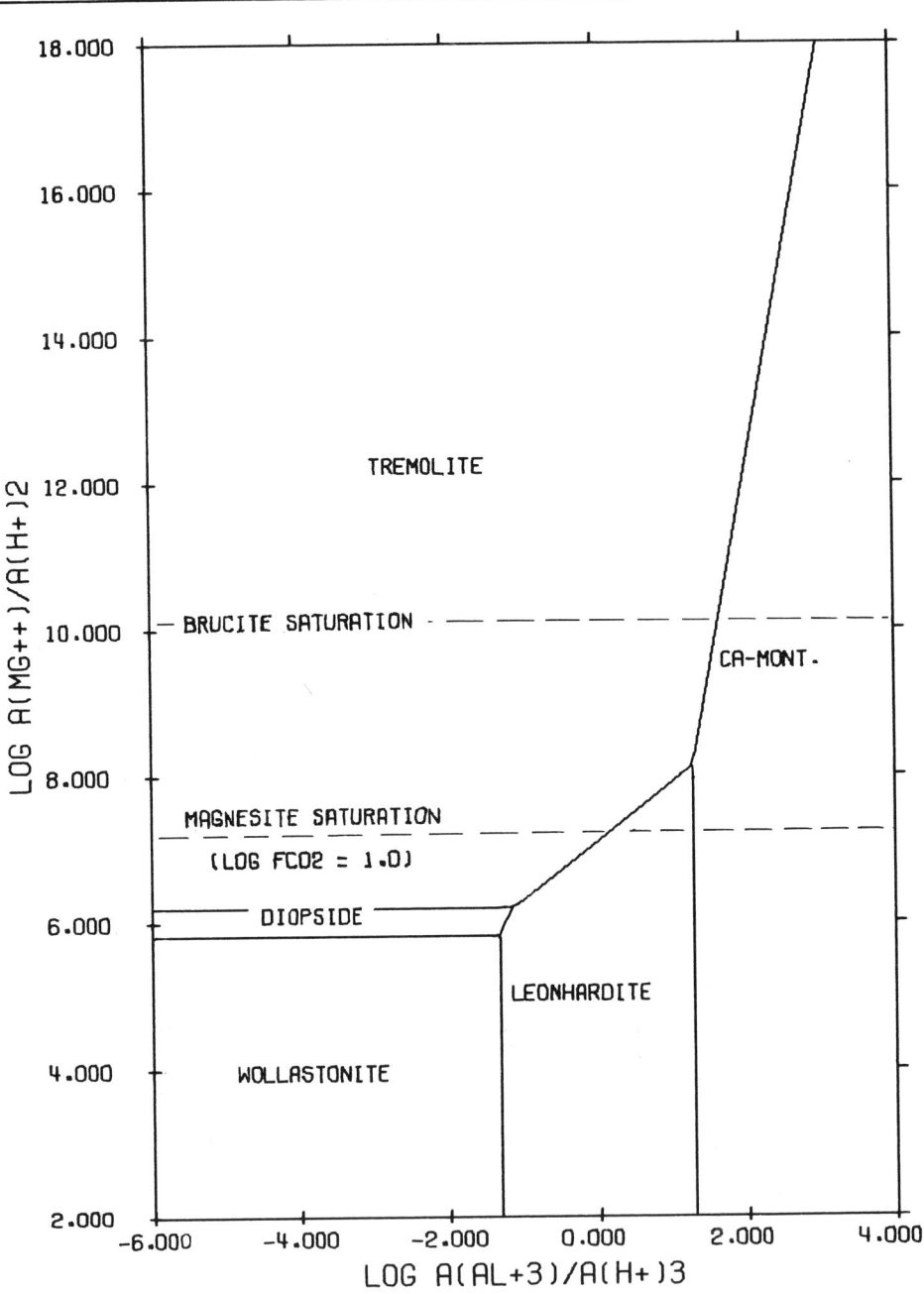

The System HCl—H$_2$O—Al$_2$O$_3$—CaO—CO$_2$—MgO—SiO$_2$ at 200°C; log $a_{H_4SiO_4}$ = −2.35 = quartz saturation.

Activity Diagrams

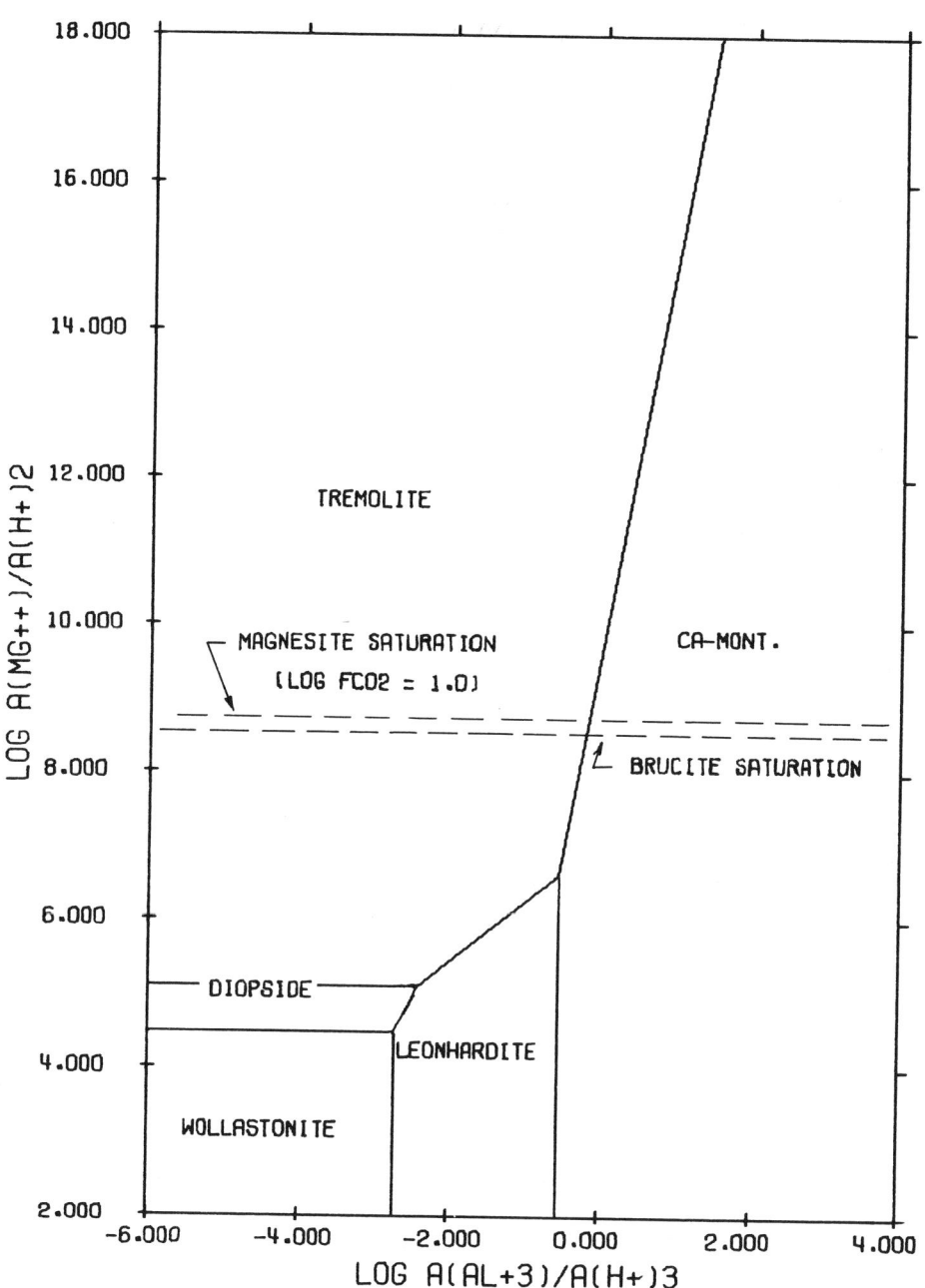

The System HCl—H$_2$O—Al$_2$O$_3$—CaO—CO$_2$—MgO—SiO$_2$ at 300°C; log $a_{H_4SiO_4}$ = −1.94 = quartz saturation.

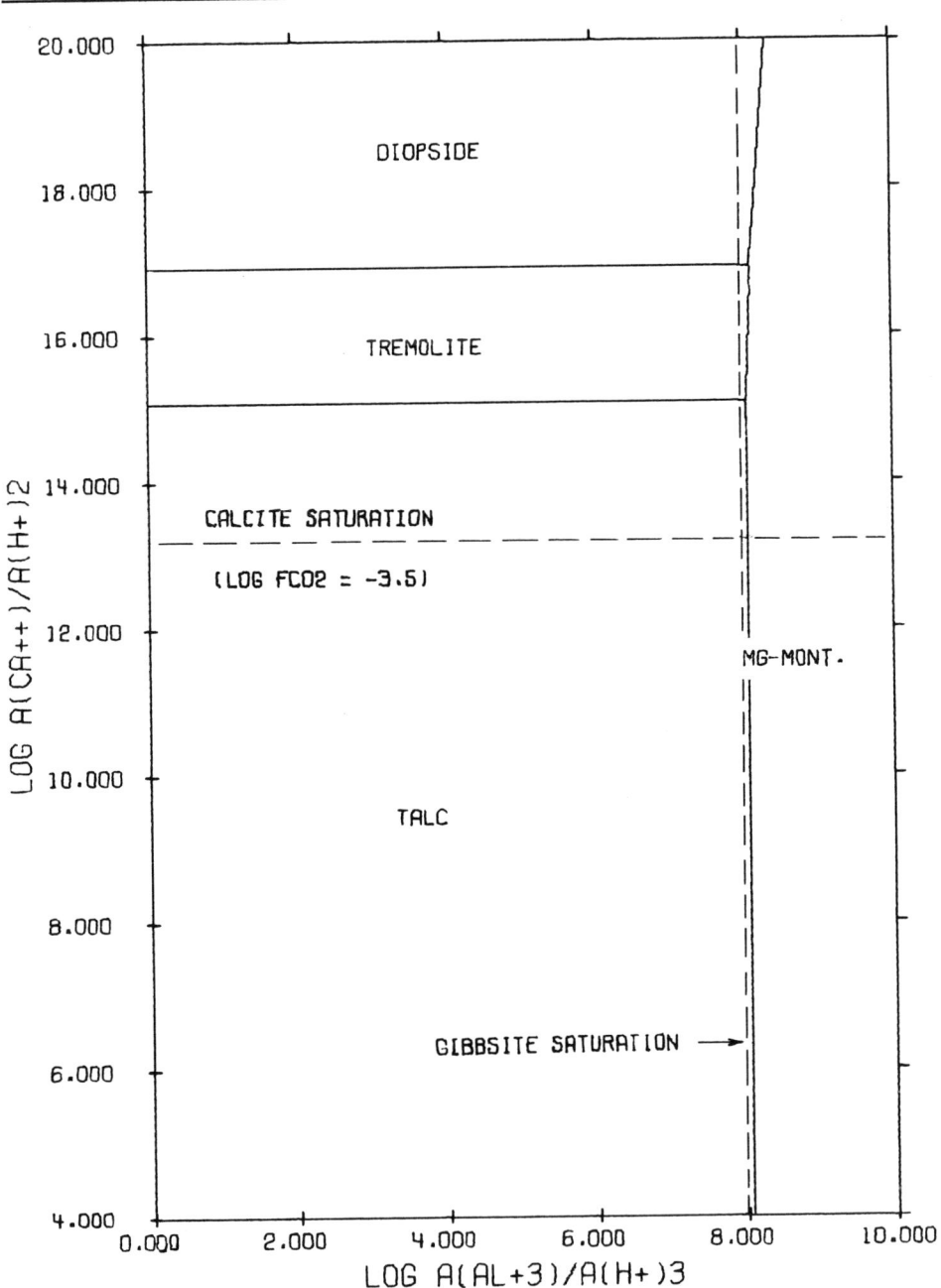

The System HCl—H_2O—Al_2O_3—CaO—CO_2—MgO—SiO_2 at 25°C; log $a_{H_4SiO_4}$ = −4.00 = quartz saturation.

Activity Diagrams

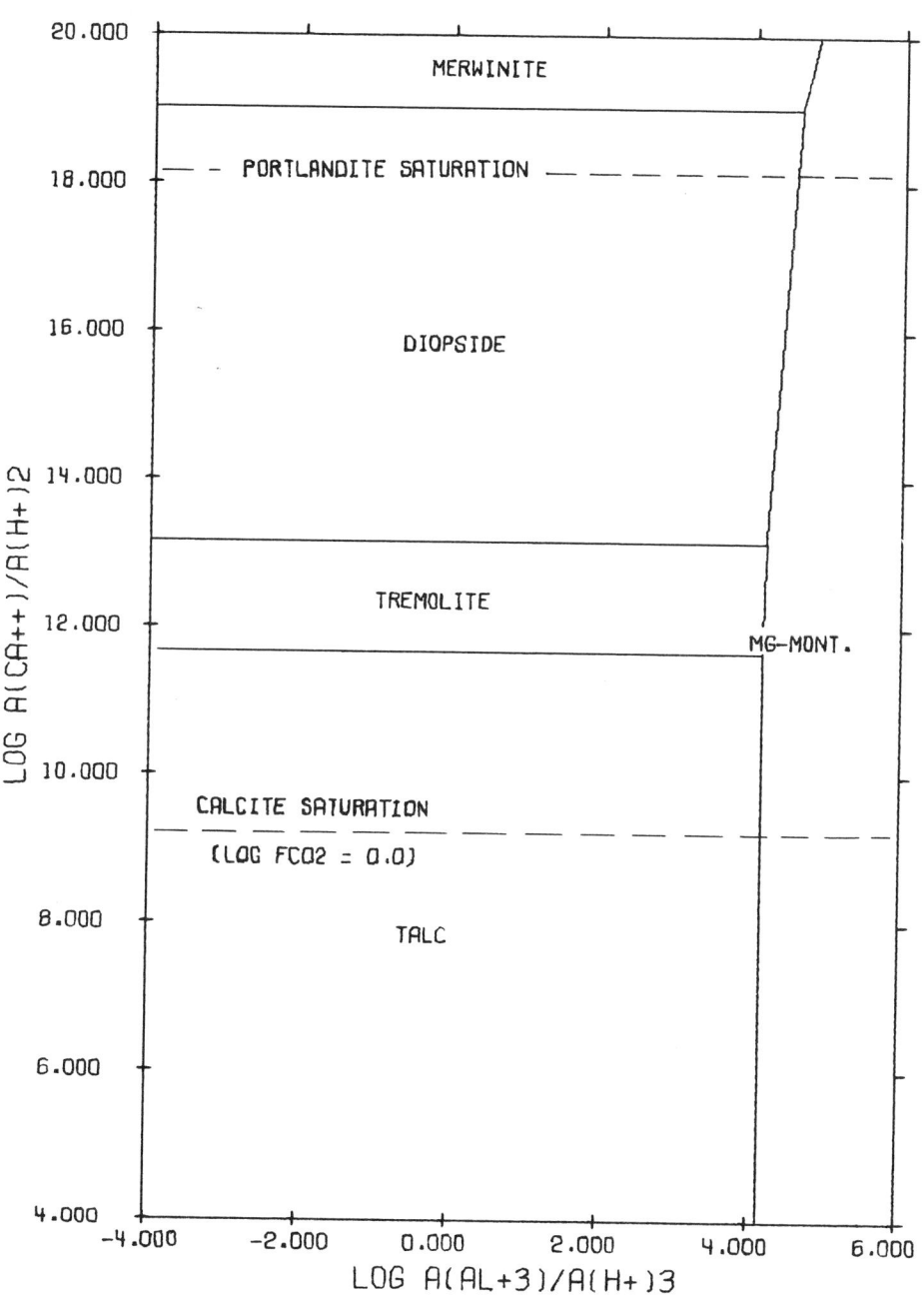

The System HCl—H_2O—Al_2O_3—CaO—CO_2—MgO—SiO_2 at 100°C; log $a_{H_4SiO_4}$ = 3.08 = quartz saturation.

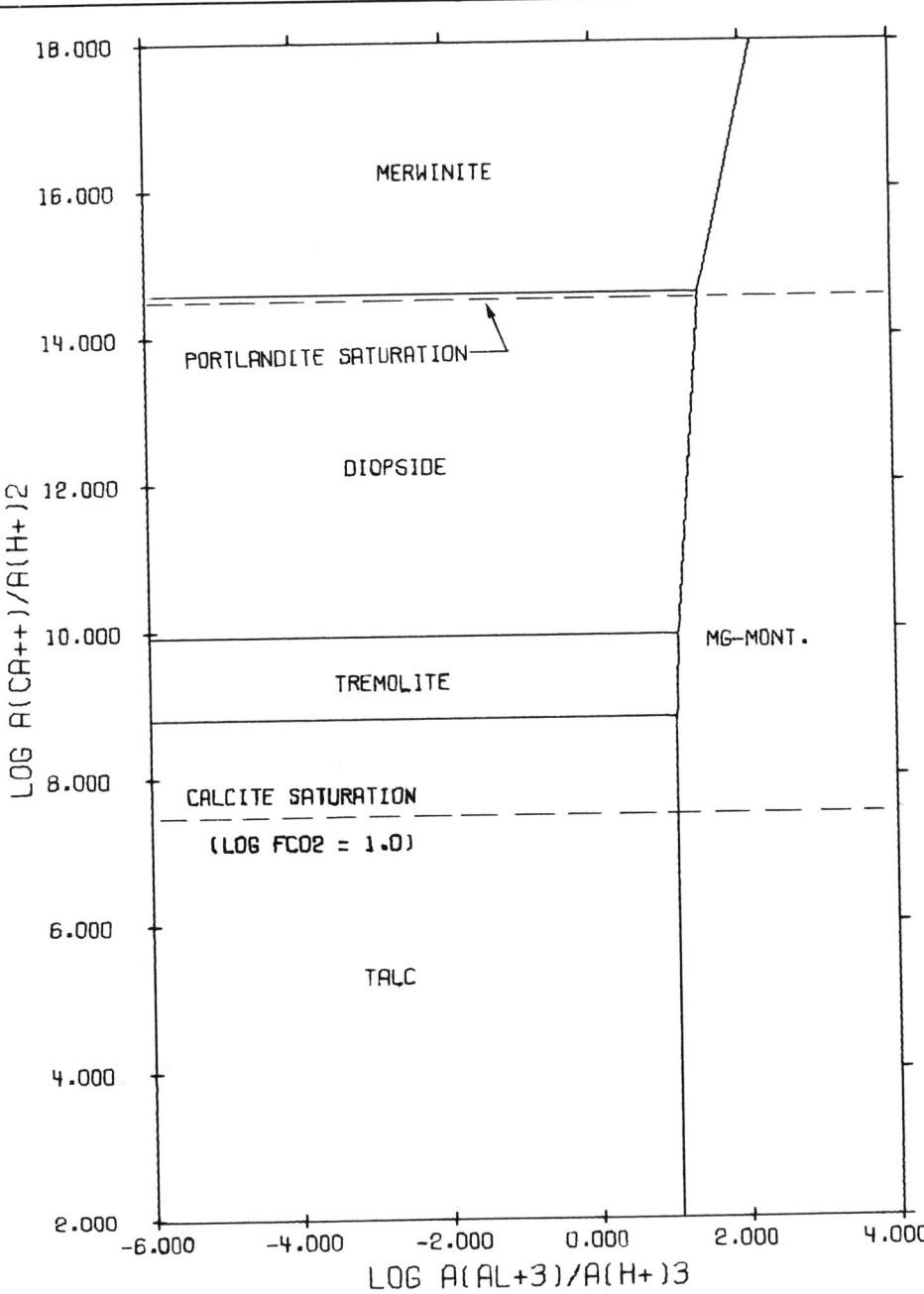

The System HCl—H_2O—Al_2O_3—CaO—CO_2—MgO—SiO_2 at 200°C; log $a_{H_4SiO_4}$ = −2.35 = quartz saturation.

Activity Diagrams

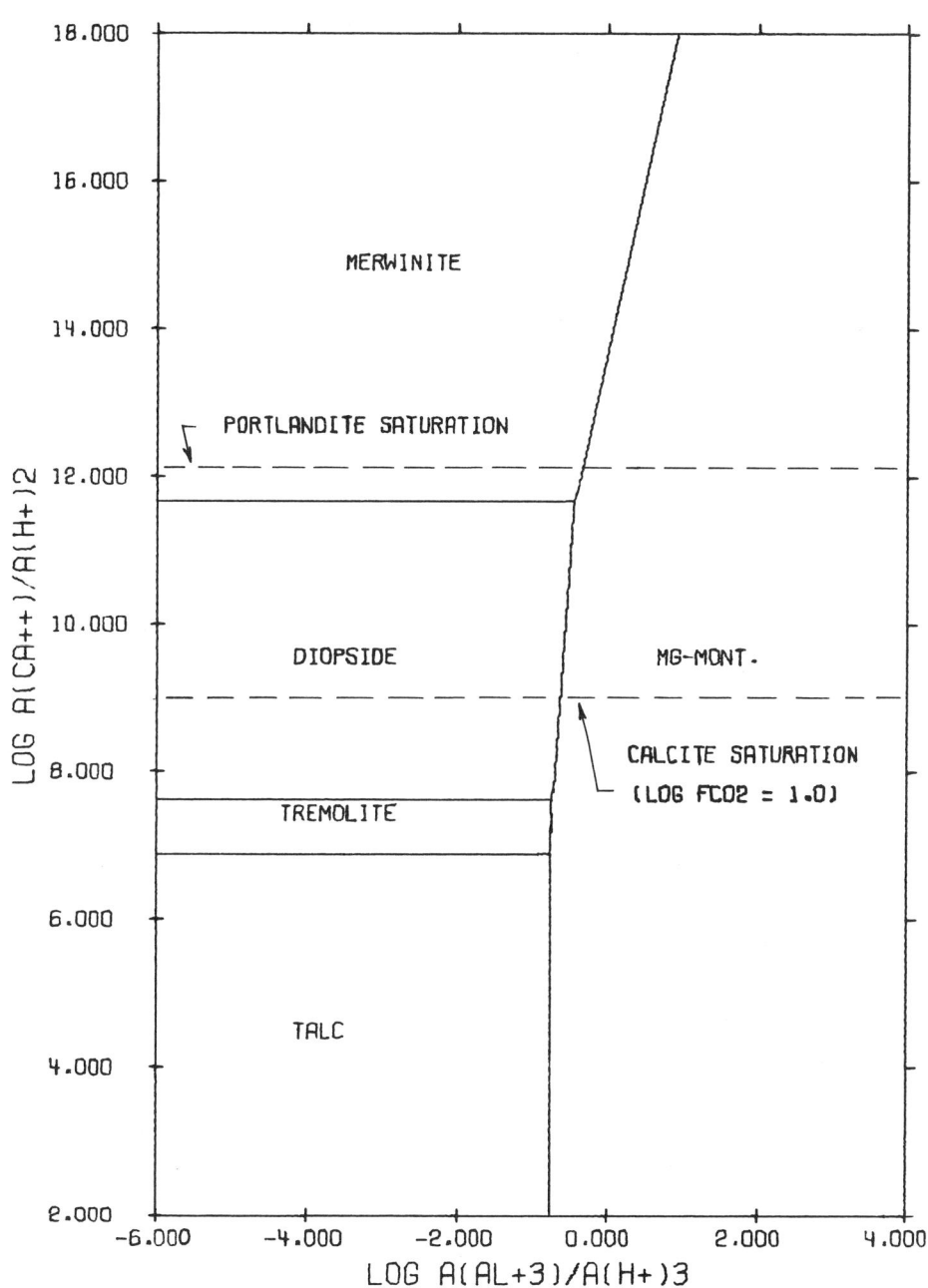

The System HCl—H_2O—Al_2O_3—CaO—CO_2—MgO—SiO_2 at 300°C; log $a_{H_4SiO_4}$ = -1.94 = quartz saturation.

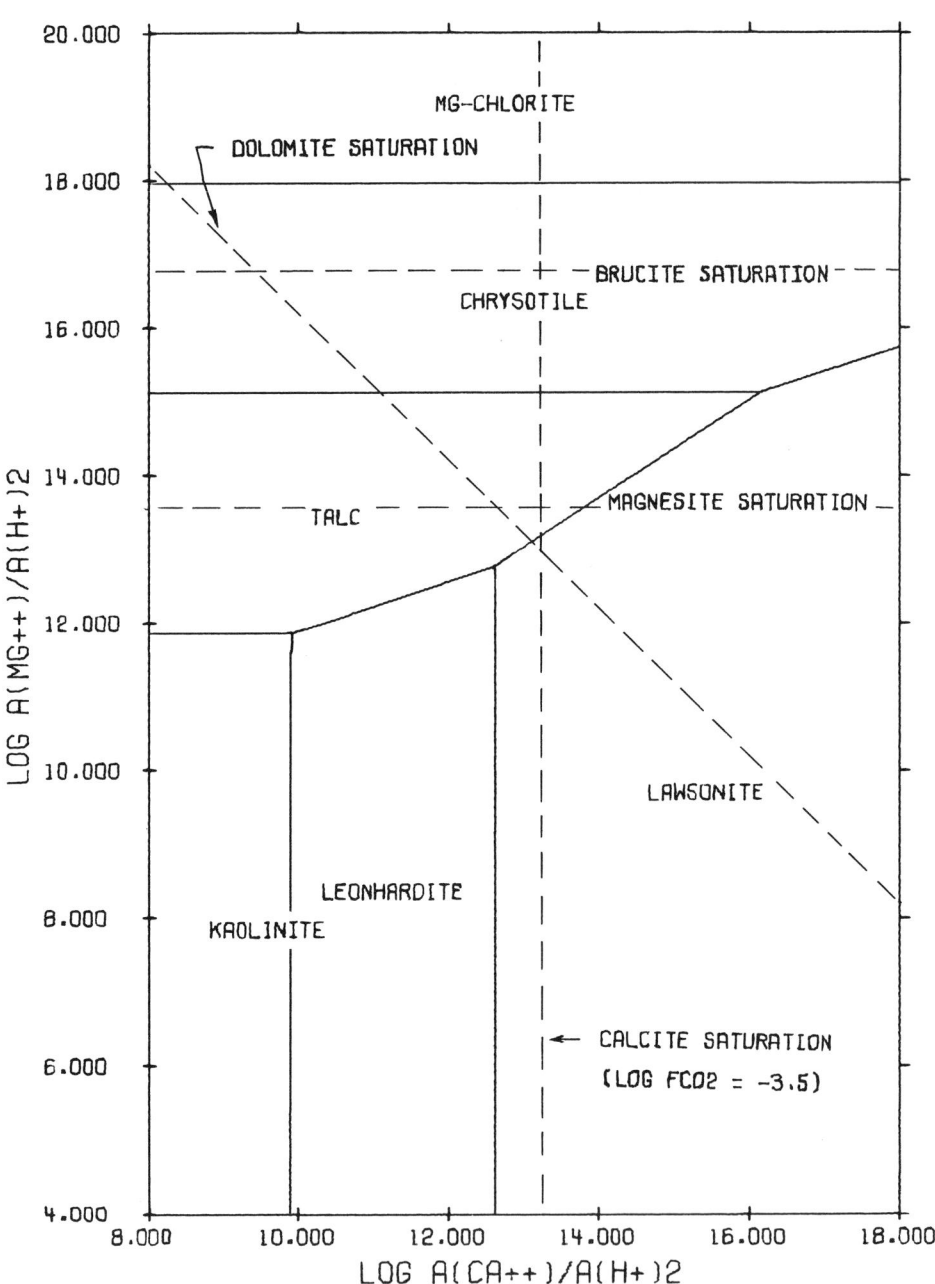

The System HCl—H$_2$O—Al$_2$O$_3$—CaO—CO$_2$—MgO—SiO$_2$ at 25°C; log $a_{Al^{+3}}/a_{H^{+3}}$ = 7.96 = gibbsite saturation.

Activity Diagrams

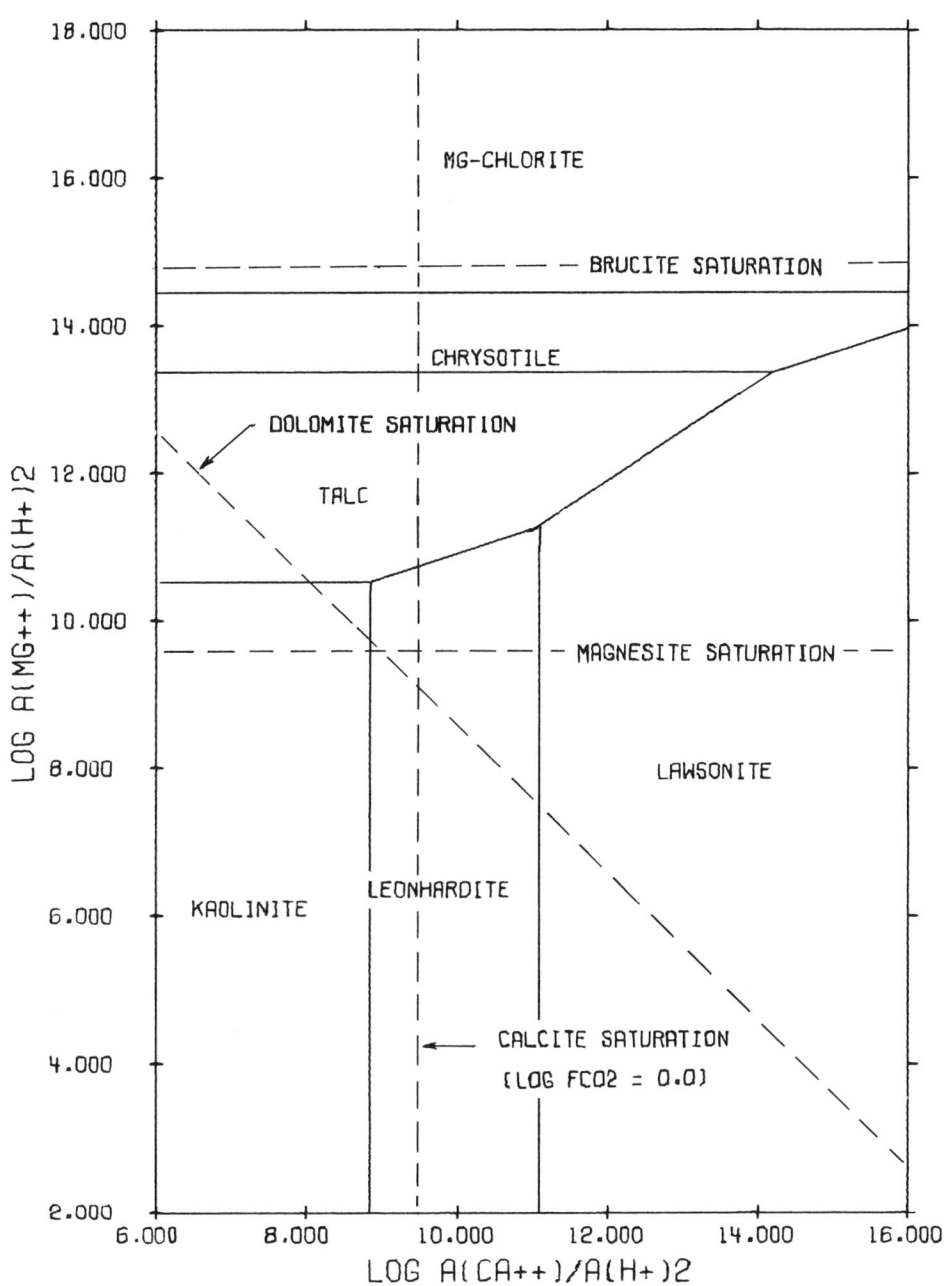

The System HCl—H$_2$O—Al$_2$O$_3$—CaO—CO$_2$—MgO—SiO$_2$ at 60°C; log $a_{Al^{+3}}/a_{H^{+3}}$ = 6.17 = gibbsite saturation.

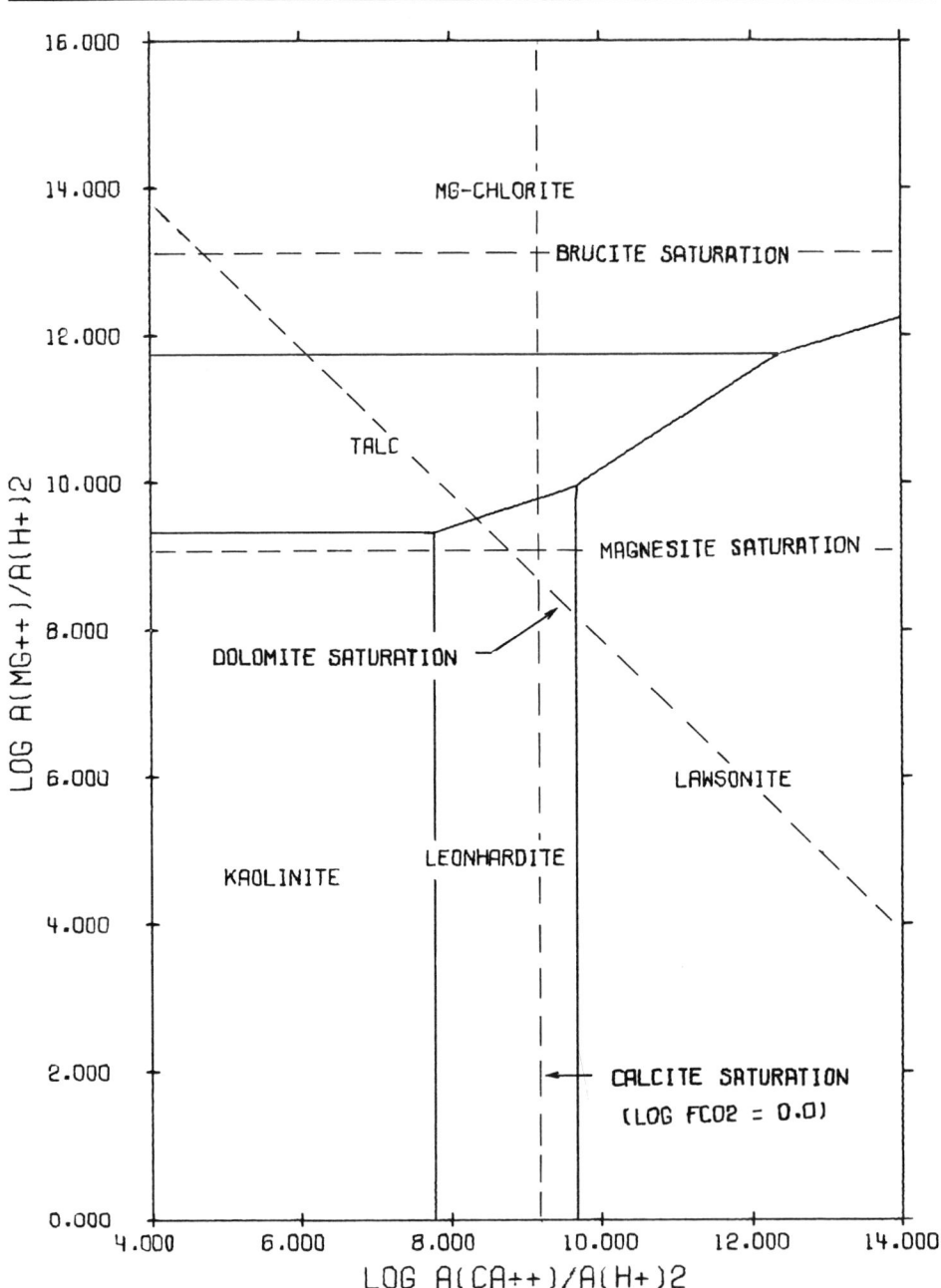

The System HCl—H$_2$O—Al$_2$O$_3$—CaO—CO$_2$—MgO—SiO$_2$ at 100°C; log $a_{Al^{+3}}/a_{H^{+3}}$ = 4.58 = corundum saturation.

Activity Diagrams

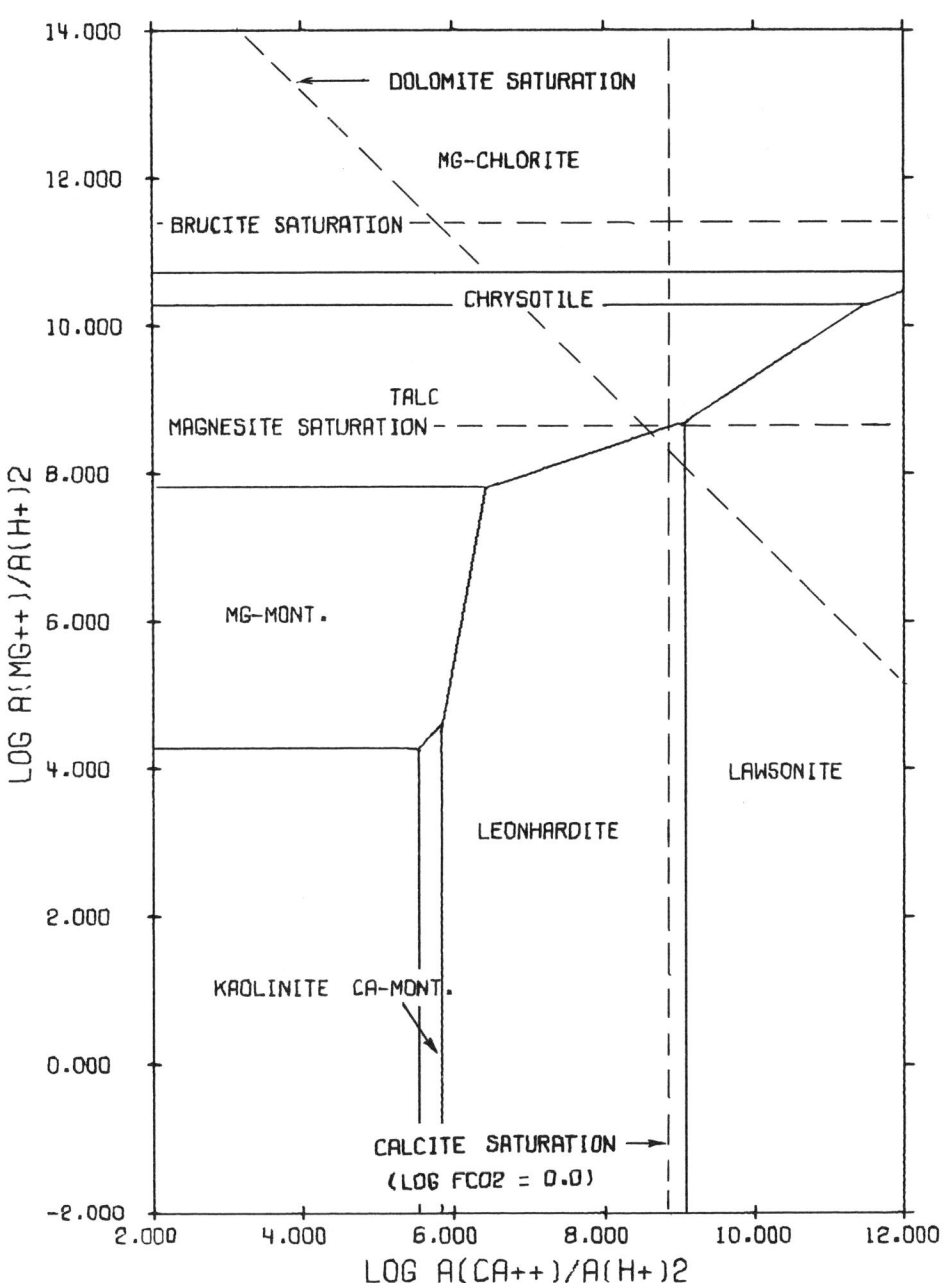

The System HCl—H_2O—Al_2O_3—CaO—CO_2—MgO—SiO_2 at 150°C; $\log a_{Al^{+3}}/a_{H^{+3}} = 2.65$ = corundum saturation.

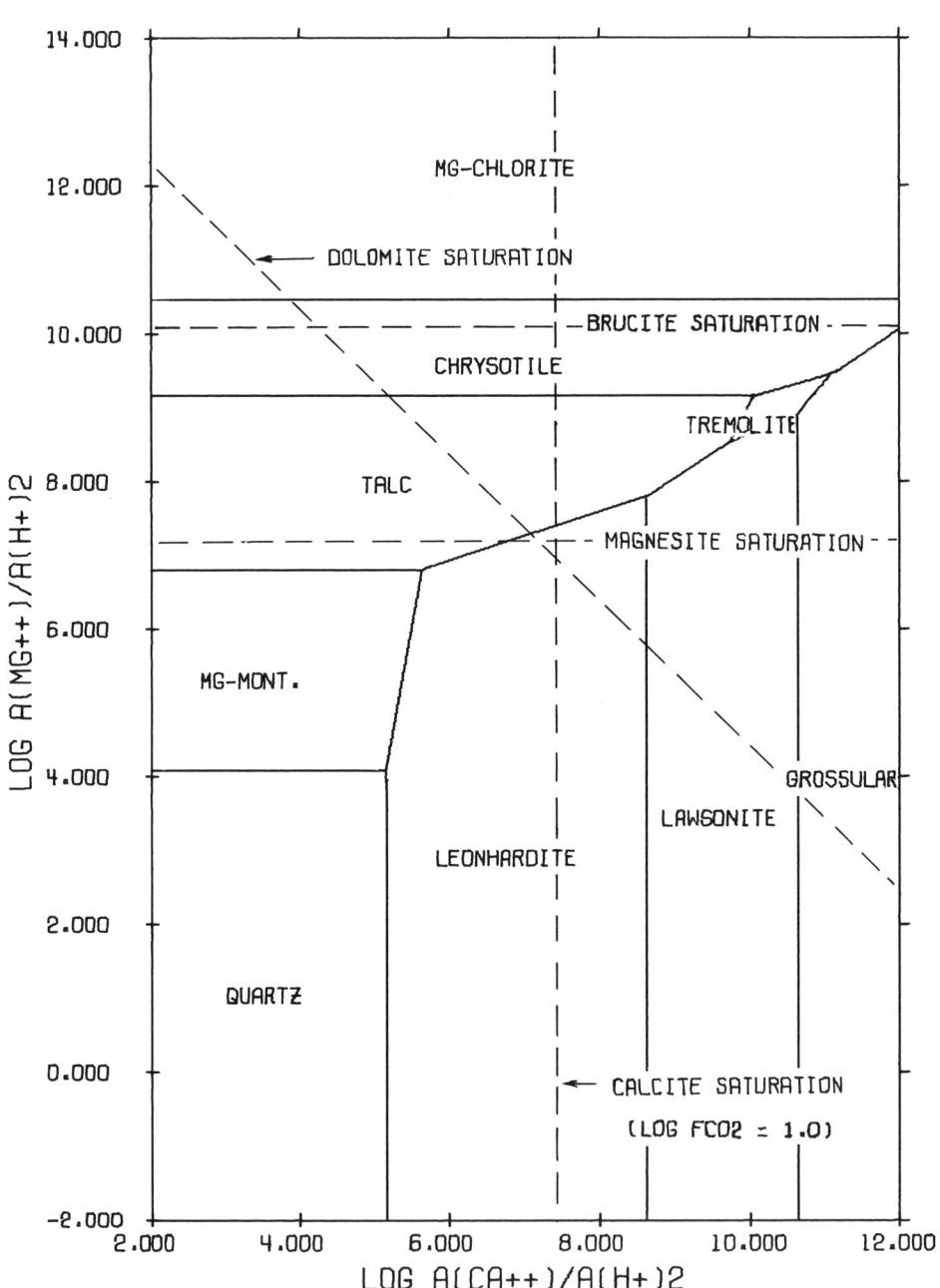

The System HCl—H$_2$O—Al$_2$O$_3$—CaO—CO$_2$—MgO—SiO$_2$ at 200°C; log $a_{Al^{+3}}/a_{H^{+3}}$ = 1.25 = corundum saturation.

Activity Diagrams

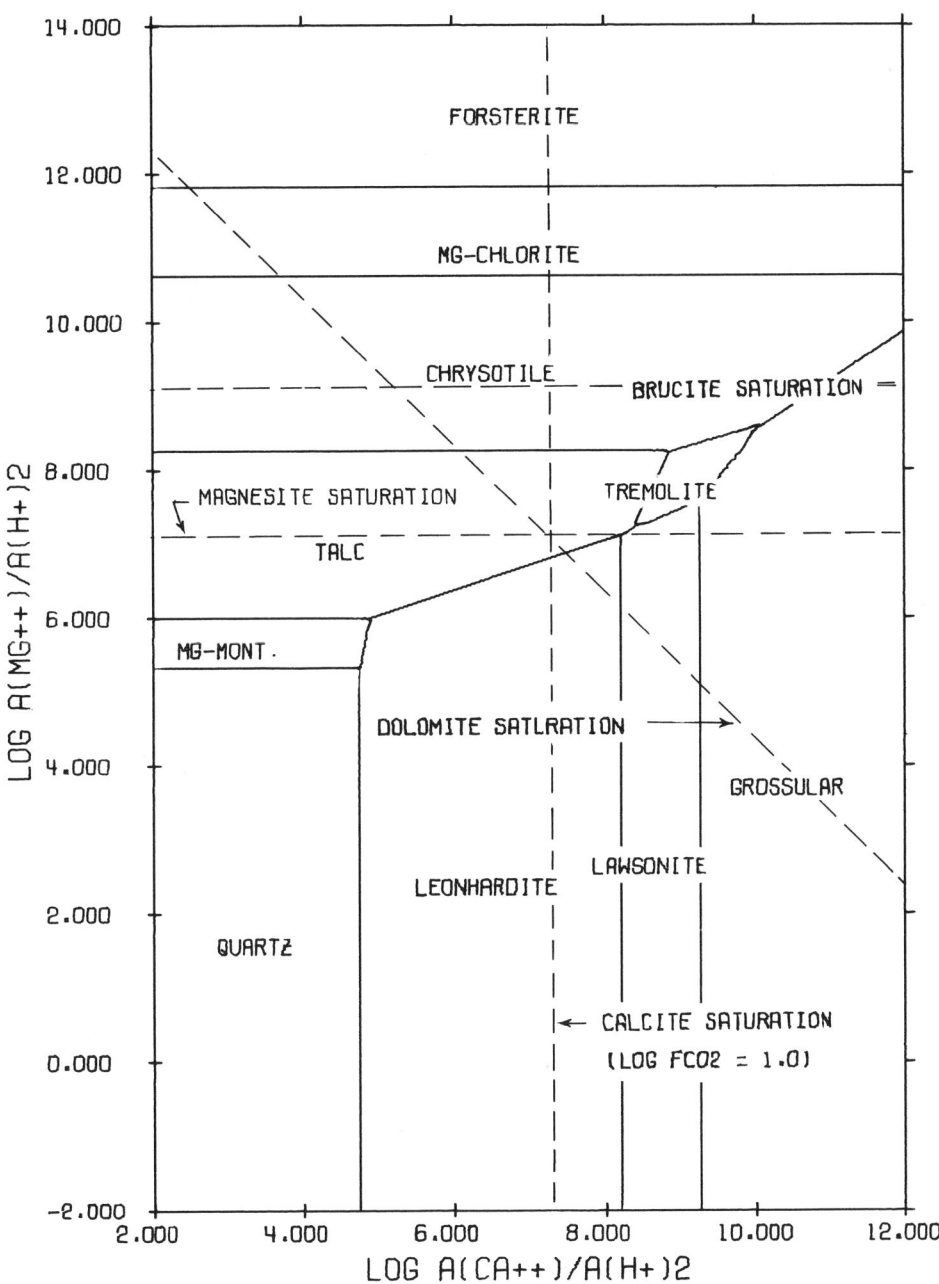

The System HCl—H$_2$O—Al$_2$O$_3$—CaO—CO$_2$—MgO—SiO$_2$ at 250°C; log $a_{Al^{+3}}/a_{H^{+3}} = 0.12$ = corundum saturation.

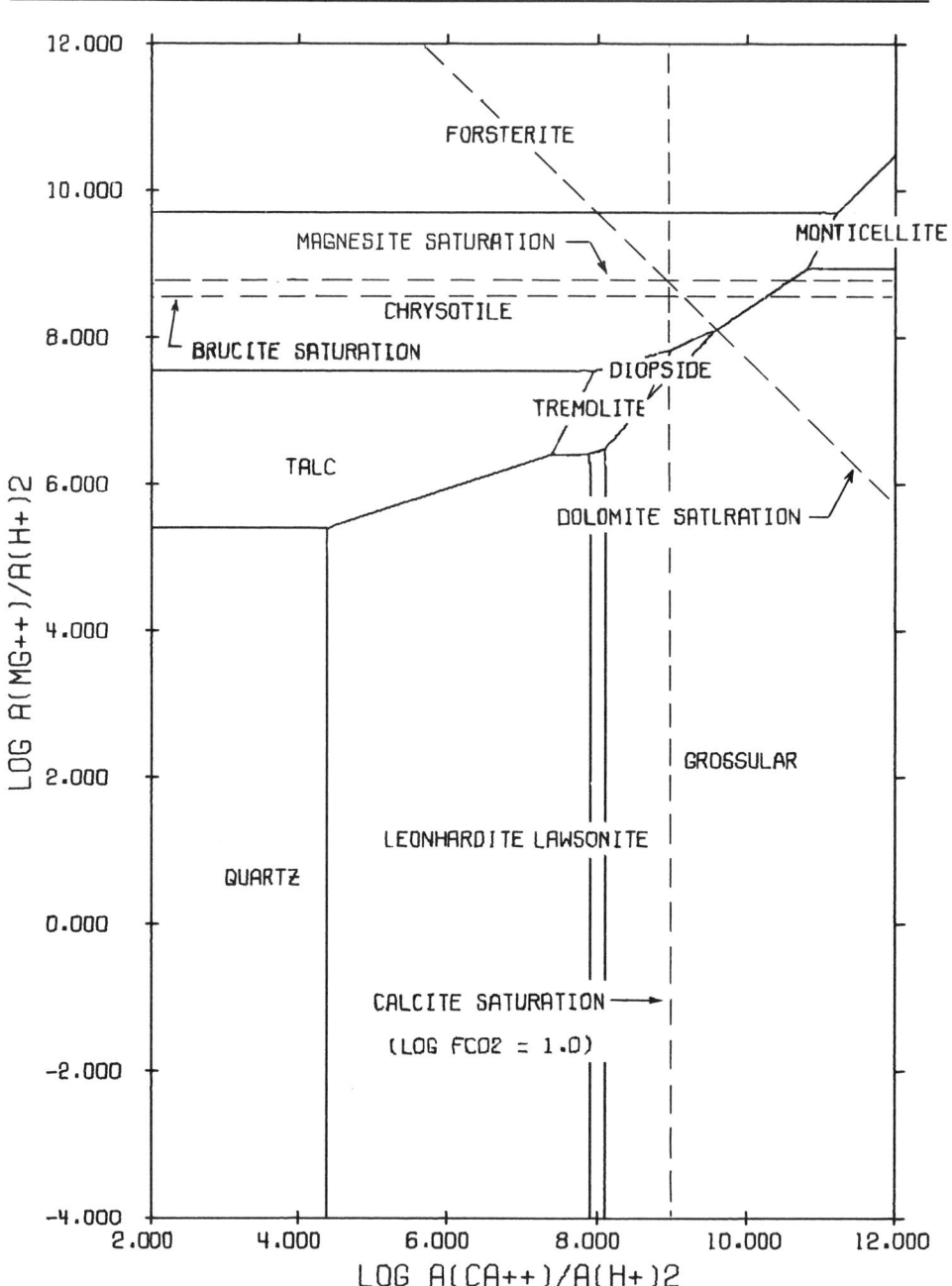

The System HCl—H$_2$O—Al$_2$O$_3$—CaO—CO$_2$—MgO—SiO$_2$ at 300°C; log $a_{Al^{+3}}/a_{H^{+3}} = -0.78$ = corundum saturation.

Activity Diagrams

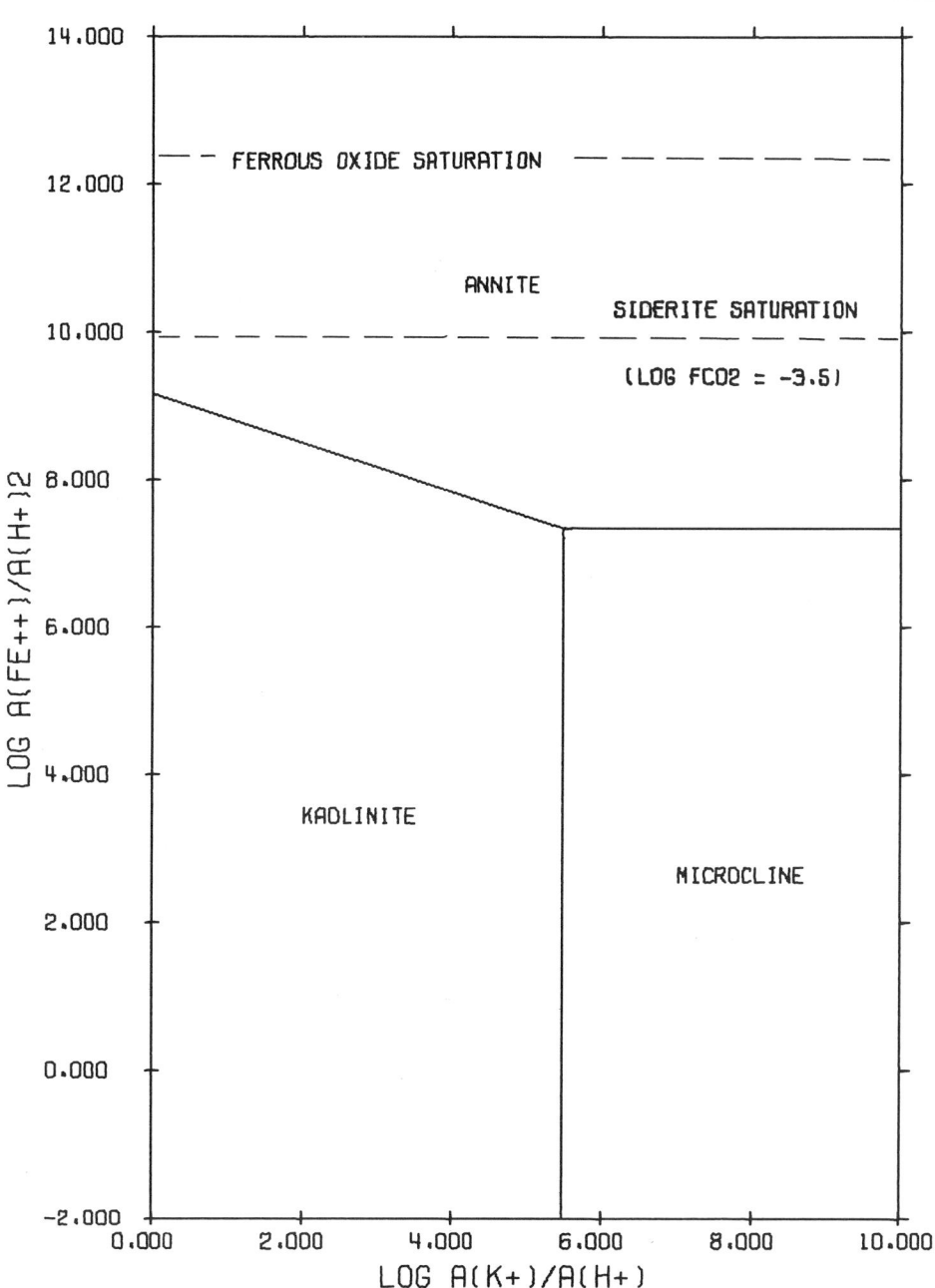

The System HCl—H$_2$O—Al$_2$O$_3$—CO$_2$—FeO—K$_2$O—SiO$_2$ at 25°C; log $a_{H_4SiO_4}$ = -4.00 = quartz saturation.

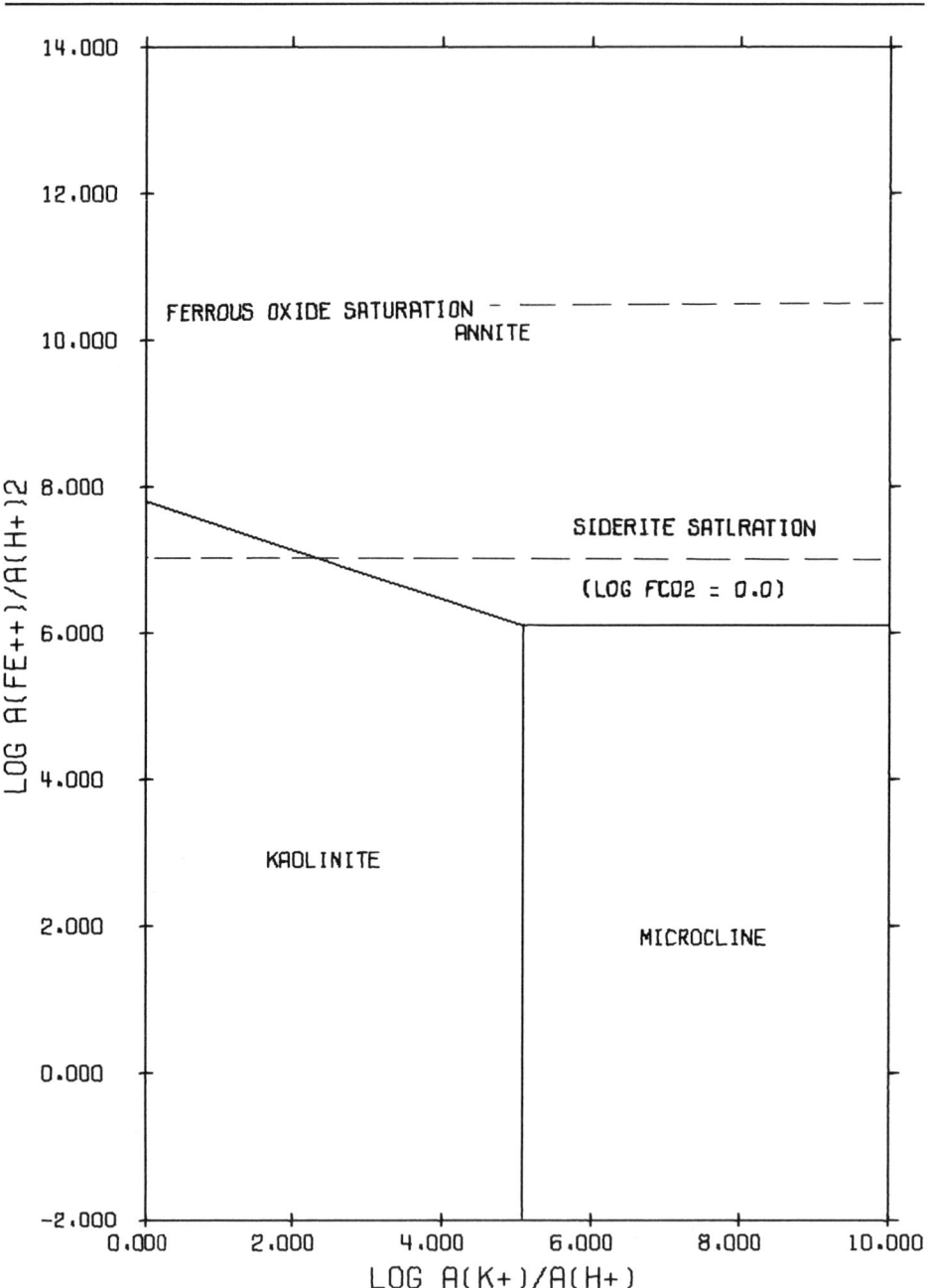

The System HCl—H$_2$O—Al$_2$O$_3$—CO$_2$—FeO—K$_2$O—SiO$_2$ at 60°C; log $a_{\text{H}_4\text{SiO}_4}$ = −3.52 = quartz saturation.

Activity Diagrams

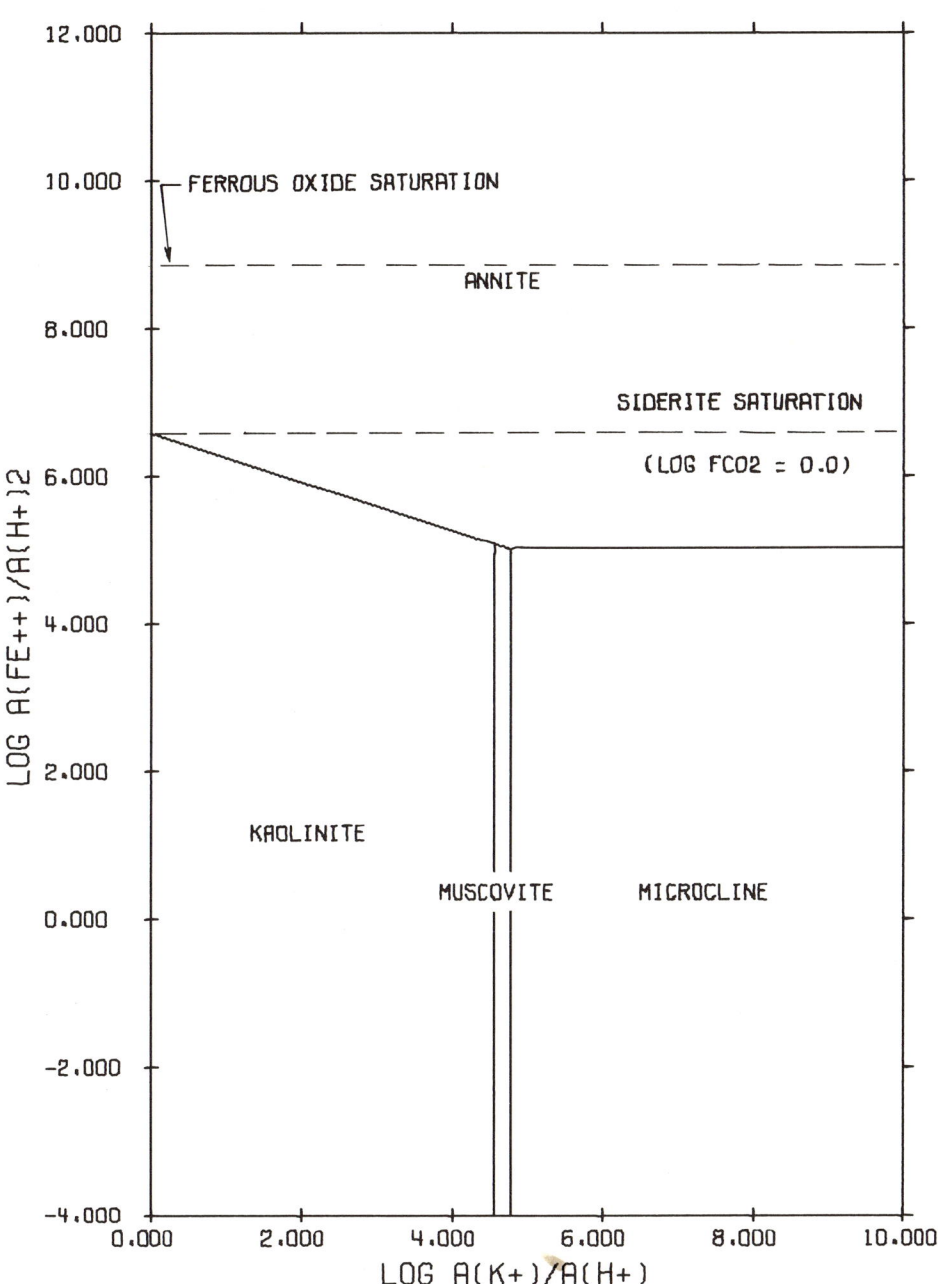

The System HCl—H$_2$O—Al$_2$O$_3$—CO$_2$—FeO—K$_2$O—SiO$_2$ at 100°C; log $a_{H_4SiO_4}$ = −3.08 = quartz saturation.

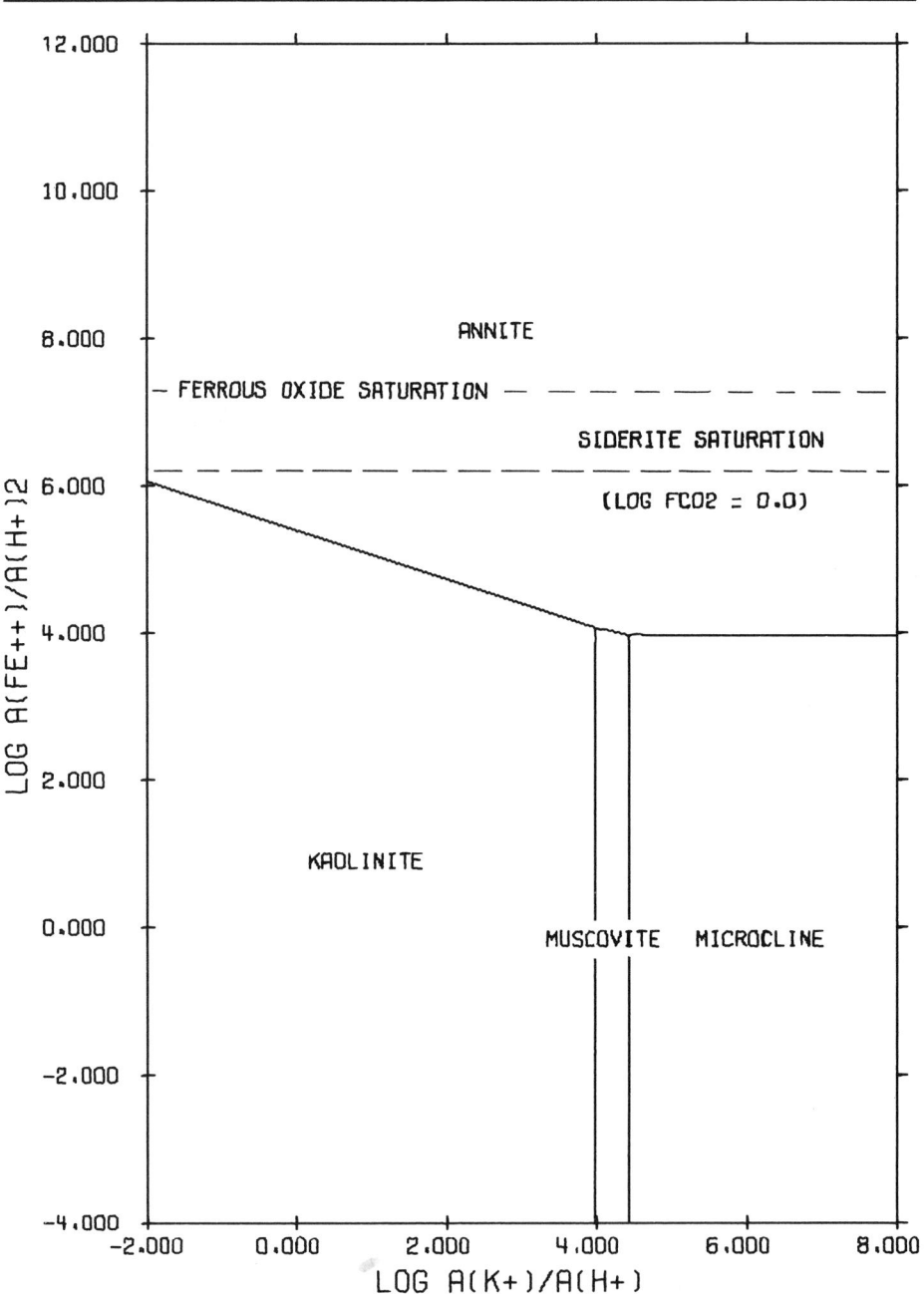

The System HCl—H$_2$O—Al$_2$O$_3$—CO$_2$—FeO—K$_2$O—SiO$_2$ at 150°C; log $a_{H_4SiO_4}$ = −2.67 = quartz saturation.

Activity Diagrams

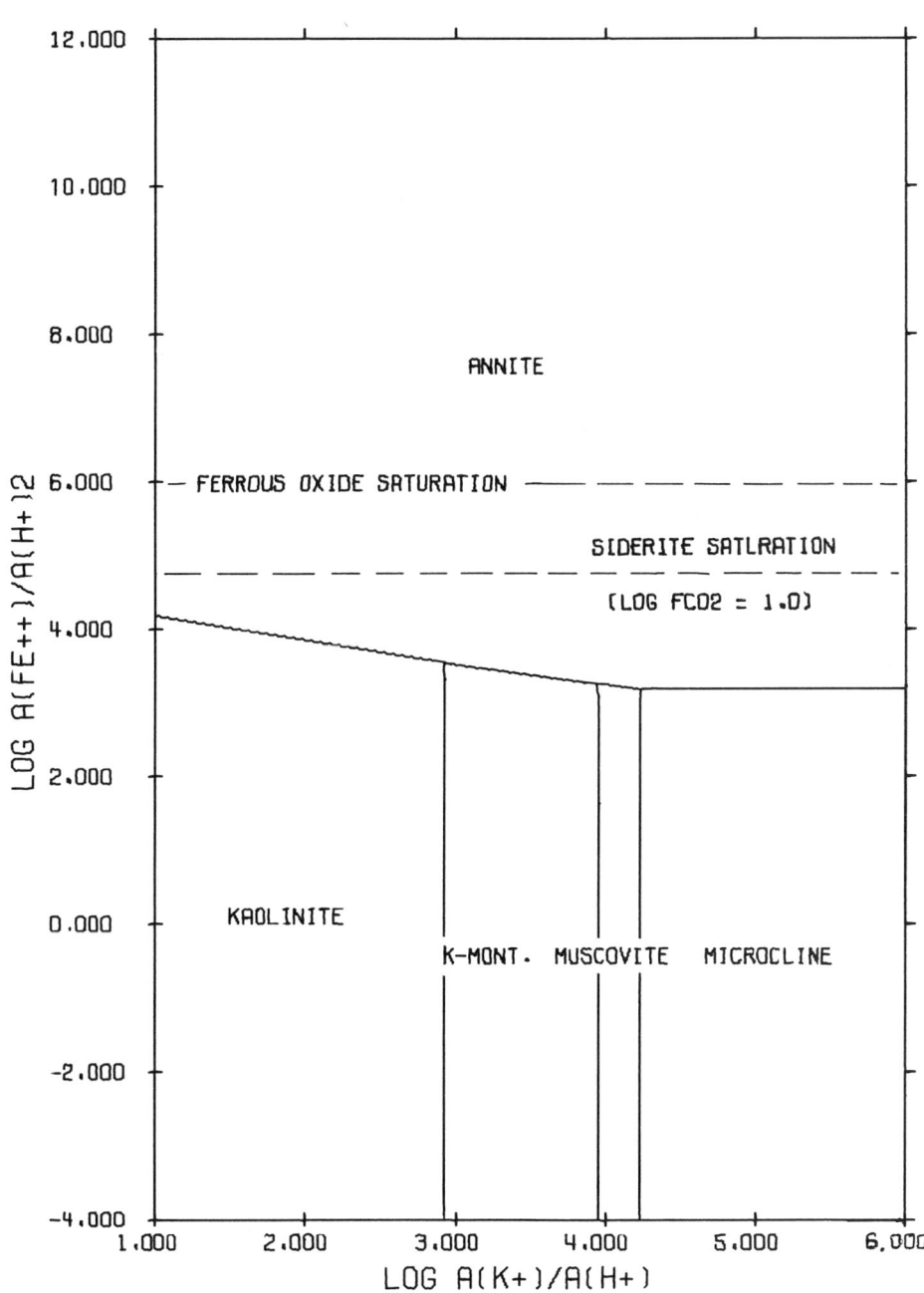

The System HCl—H_2O—Al_2O_3—CO_2—FeO—K_2O—SiO_2 at 200°C; log $a_{H_4SiO_4}$ = -2.35 = quartz saturation.

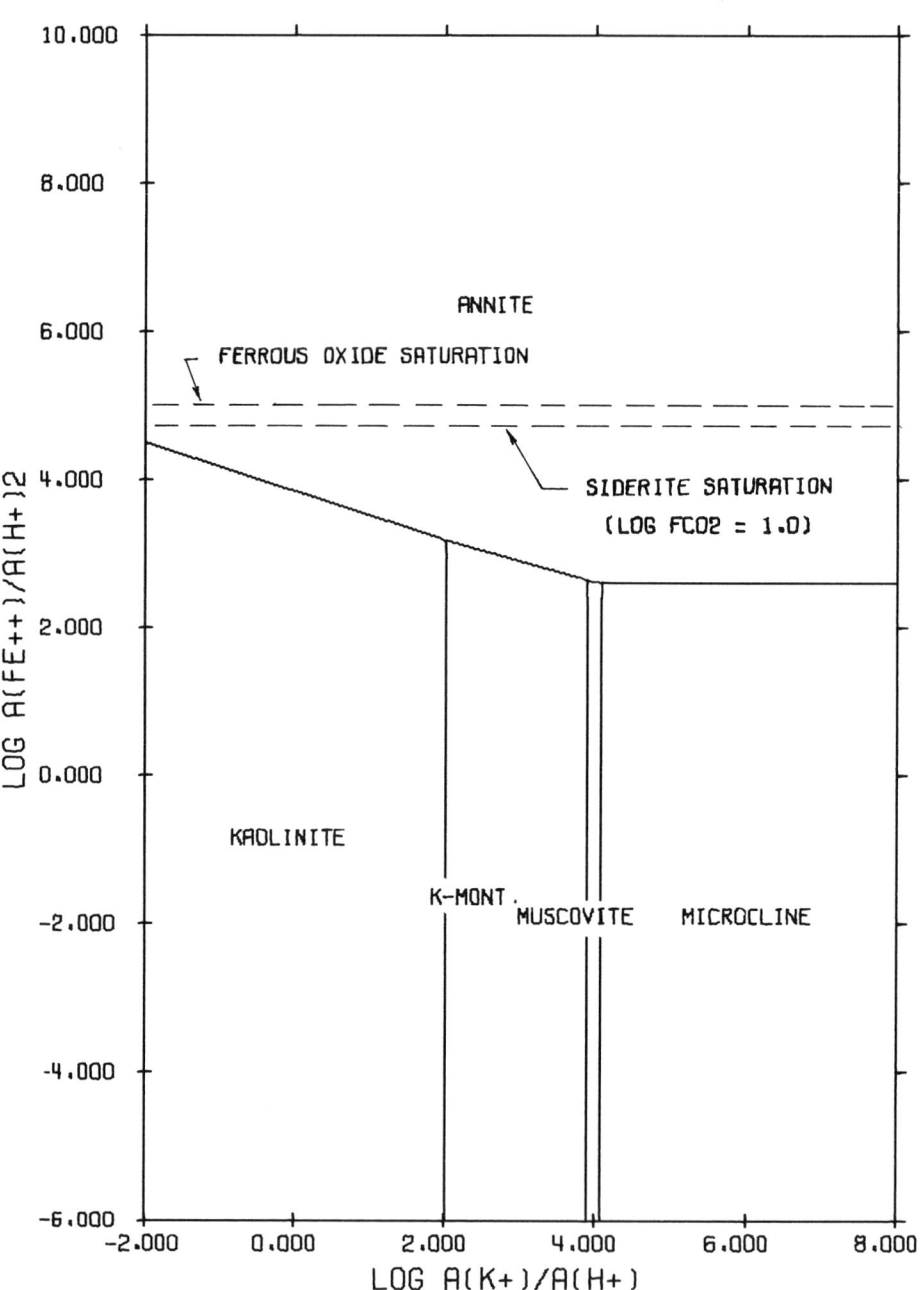

The System HCl—H$_2$O—Al$_2$O$_3$—CO$_2$—FeO—K$_2$O—SiO$_2$ at 250°C; log $a_{H_4SiO_4}$ = −2.11 = quartz saturation.

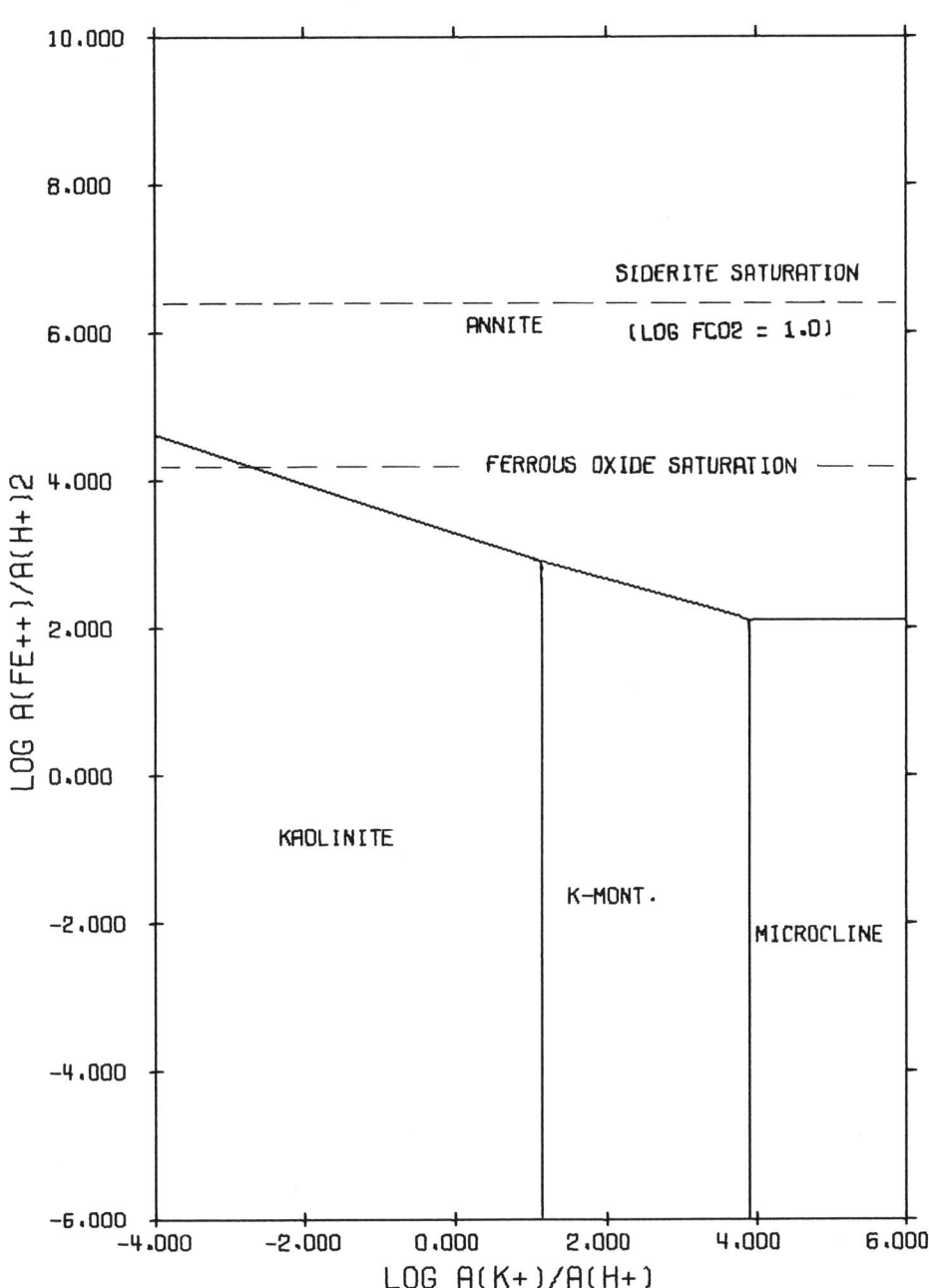

The System HCl—H$_2$O—Al$_2$O$_3$—CO$_2$—FeO—K$_2$O—SiO$_2$ at 300°C; log $a_{H_4SiO_4}$ = −1.94 = quartz saturation.

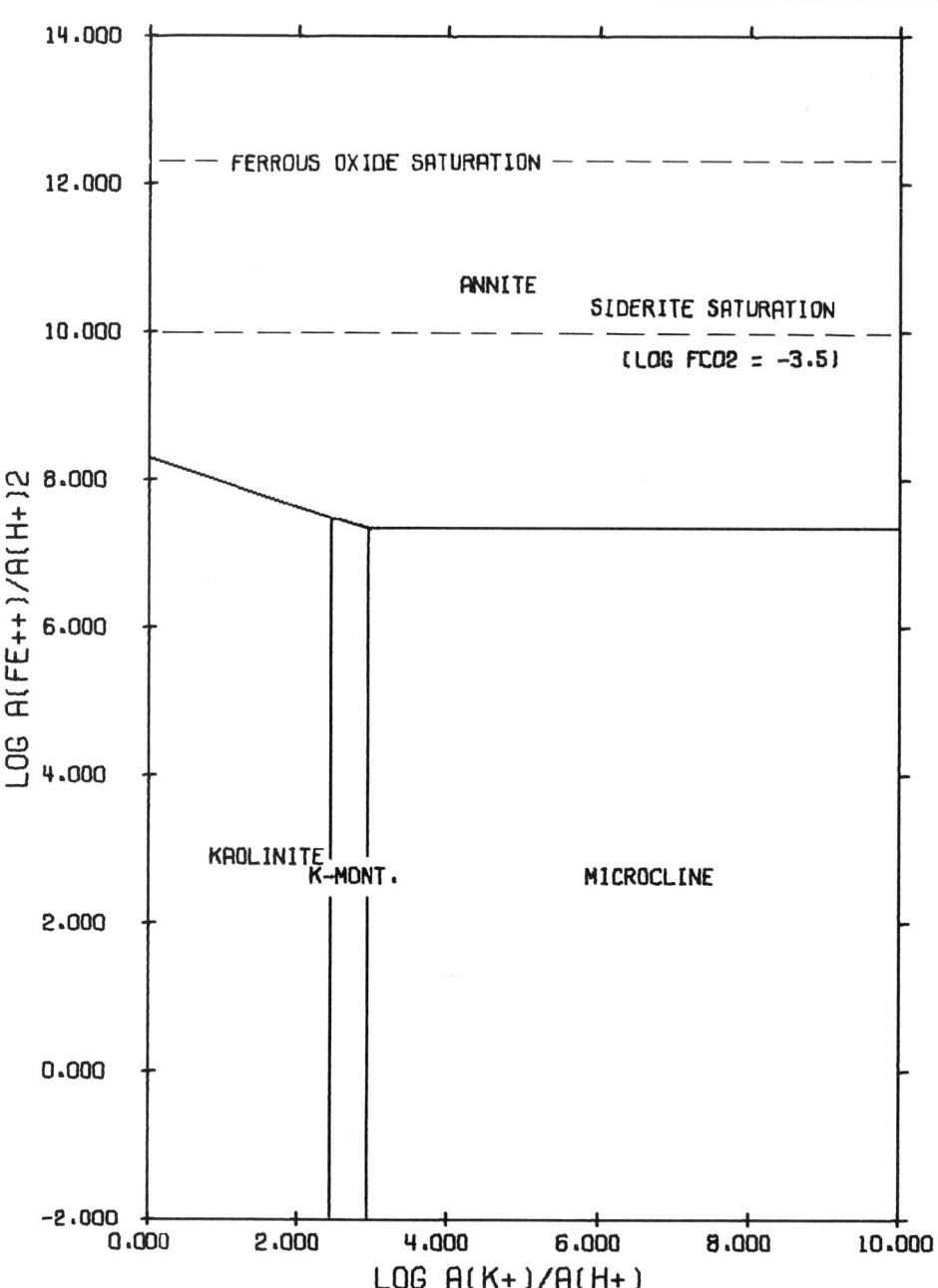

The System HCl—H$_2$O—Al$_2$O$_3$—CO$_2$—FeO—K$_2$O—SiO$_2$ at 25°C; log $a_{H_4SiO_4}$ = −2.70 = amorphous silica saturation.

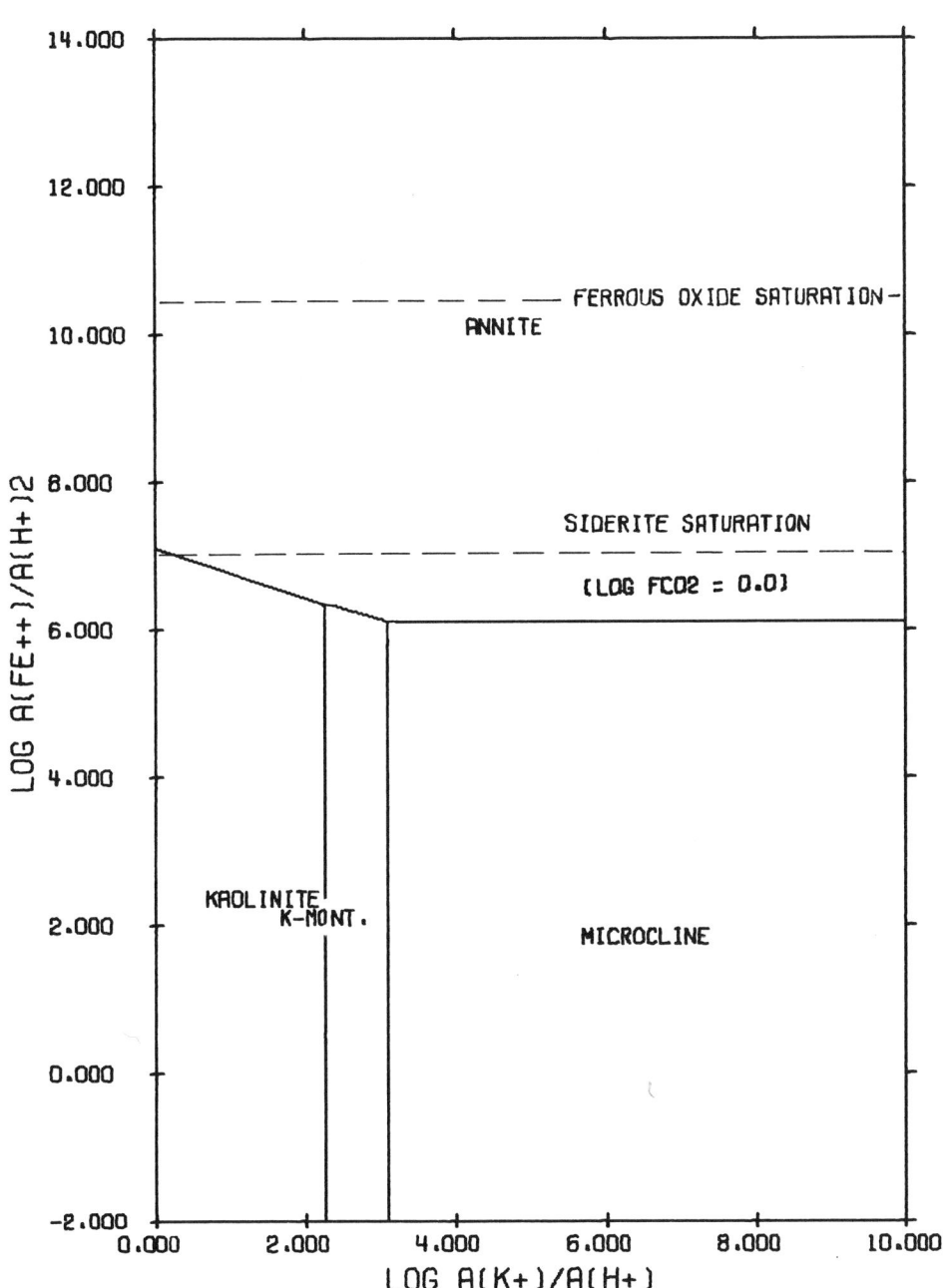

The System HCl—H_2O—Al_2O_3—CO_2—FeO—K_2O—SiO_2 at 60°C; $\log a_{H_4SiO_4} = -2.47$ = amorphous silica saturation.

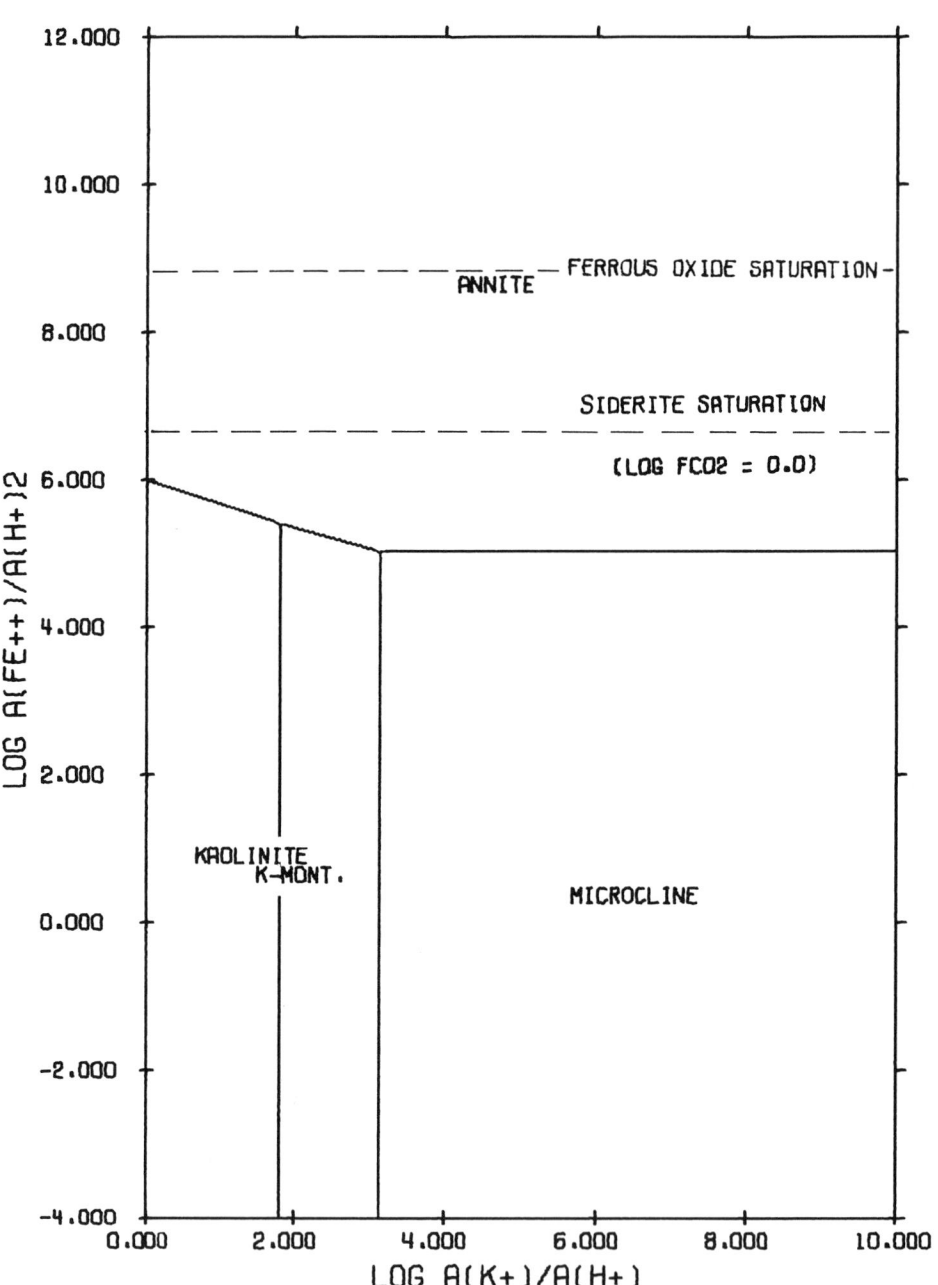

The System HCl—H$_2$O—Al$_2$O$_3$—CO$_2$—FeO—K$_2$O—SiO$_2$ at 100°C; log $a_{H_4SiO_4}$ = −2.20 = amorphous silica saturation.

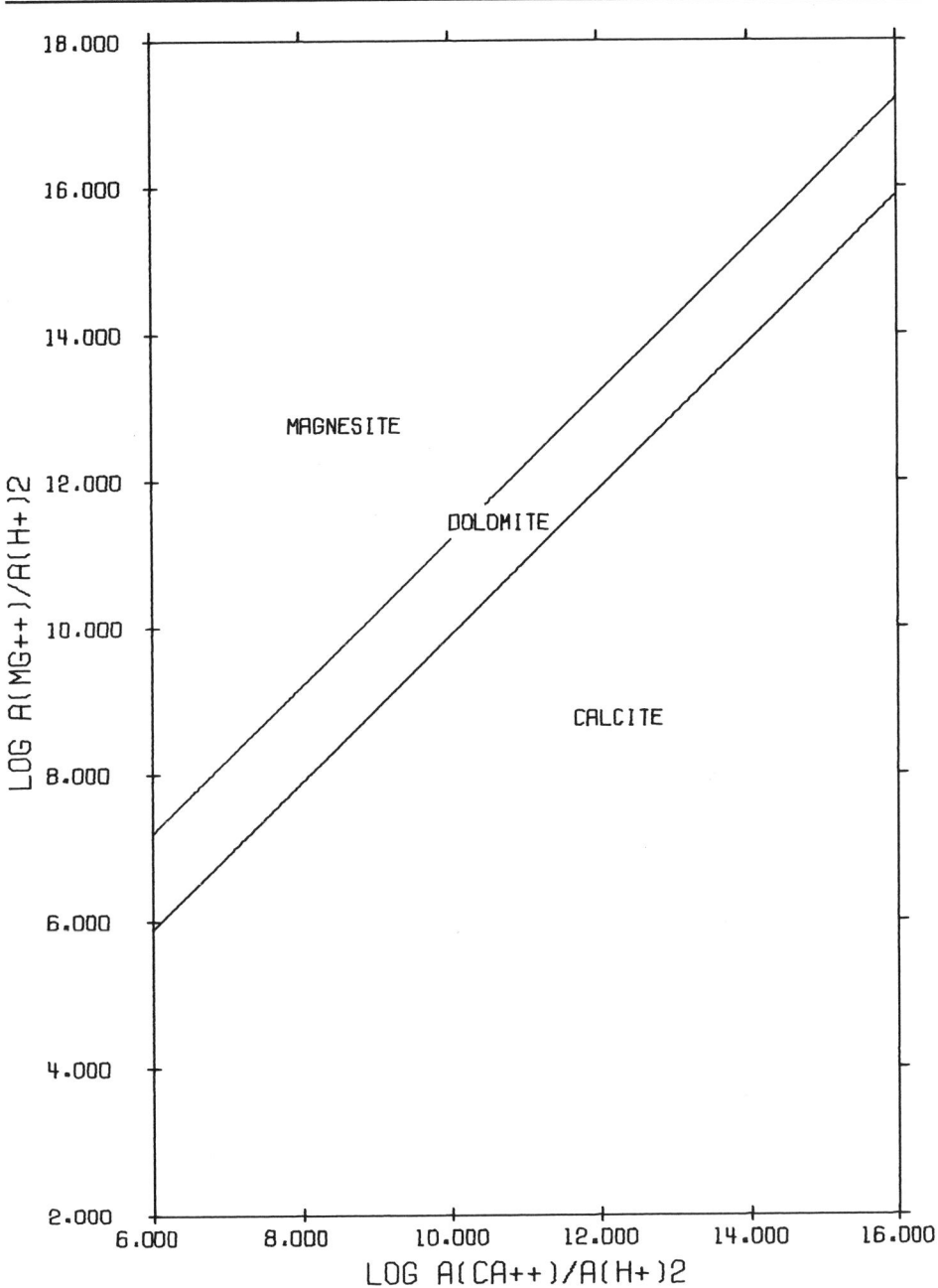

The System HCl—H$_2$O—CaO—CO$_2$—MgO at 0°C.

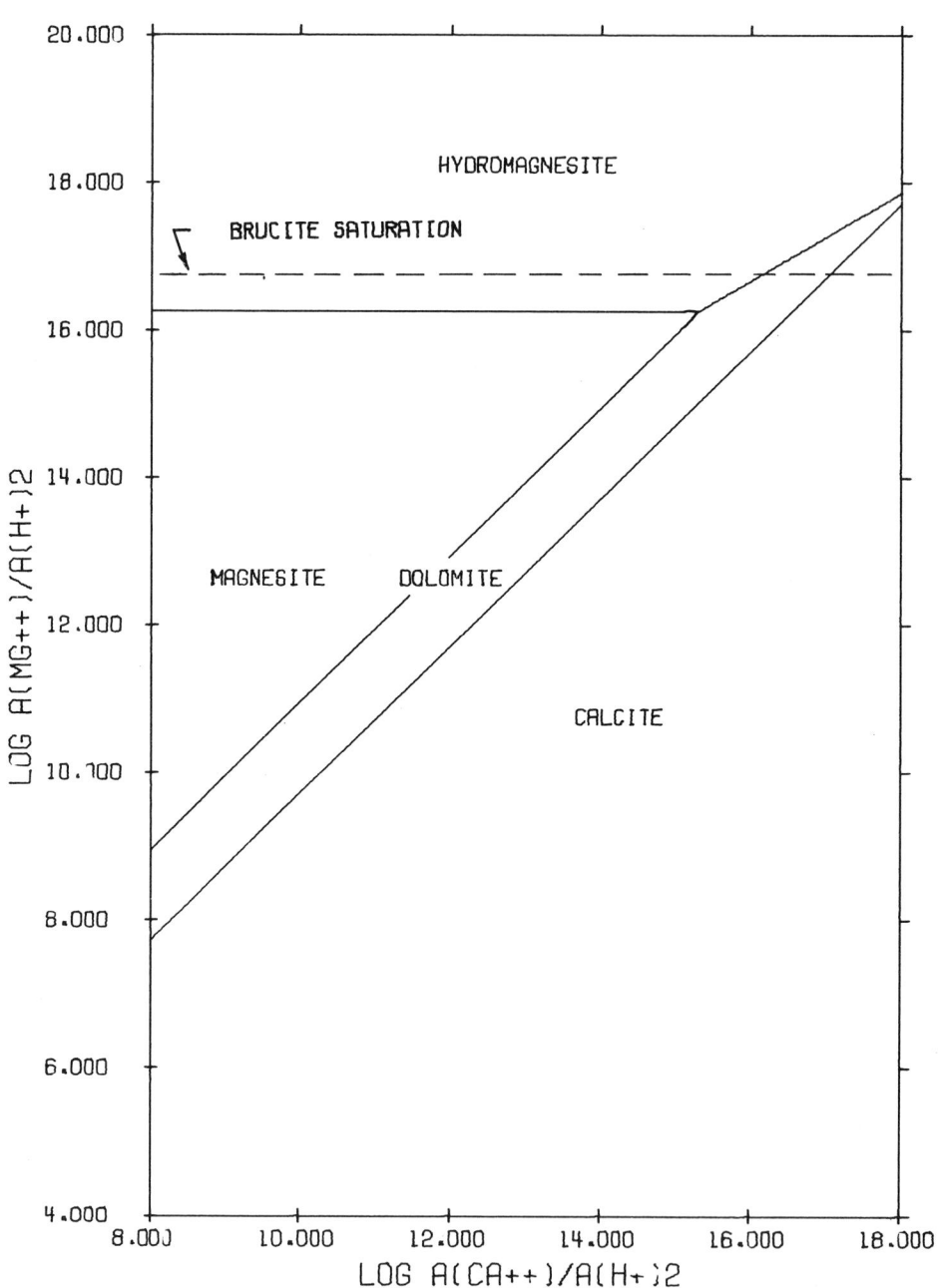

The System HCl—H$_2$O—CaO—CO$_2$—MgO at 25°C.

Activity Diagrams

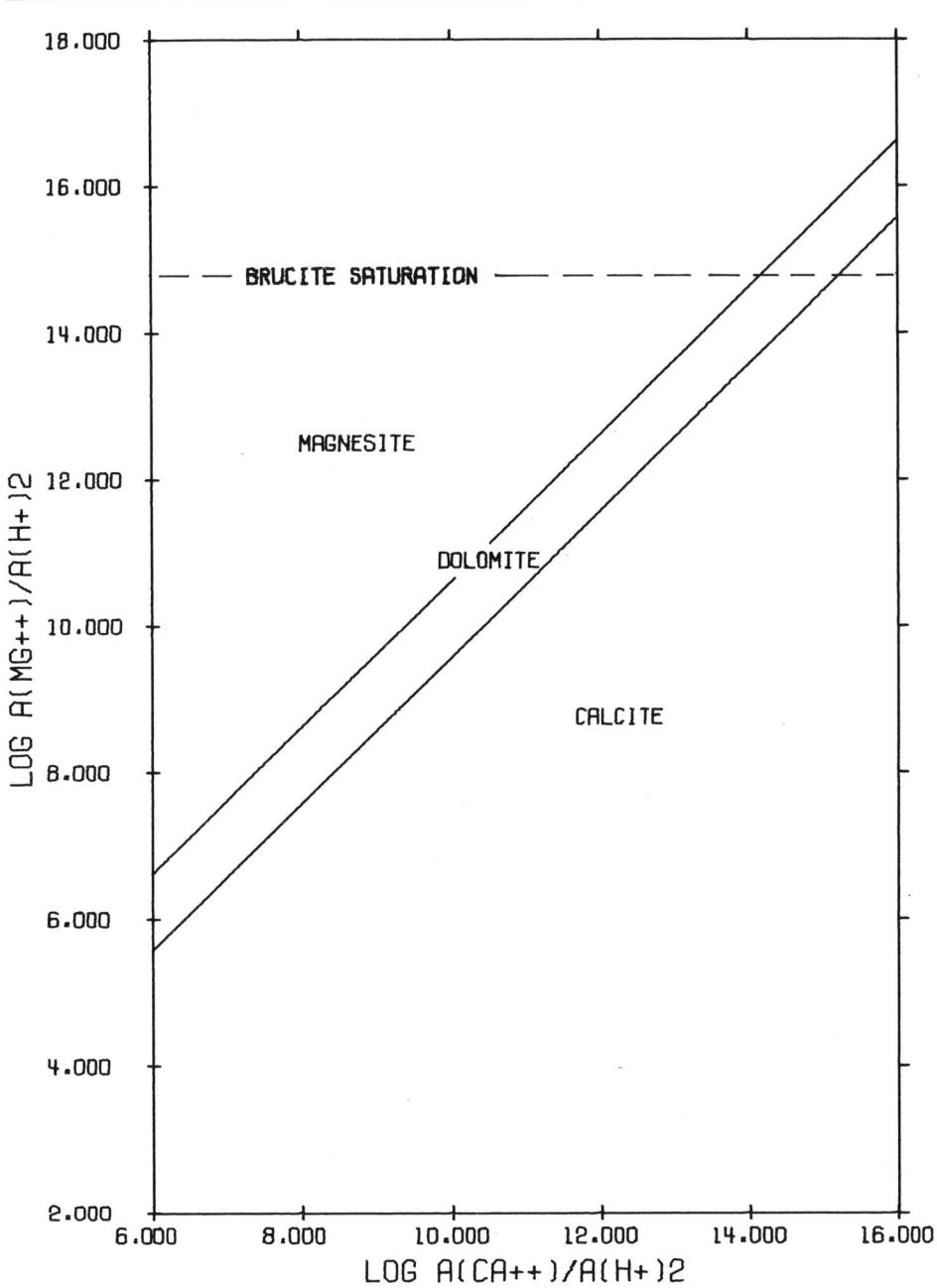

The System HCl—H$_2$O—CaO—CO$_2$—MgO at 60°C.

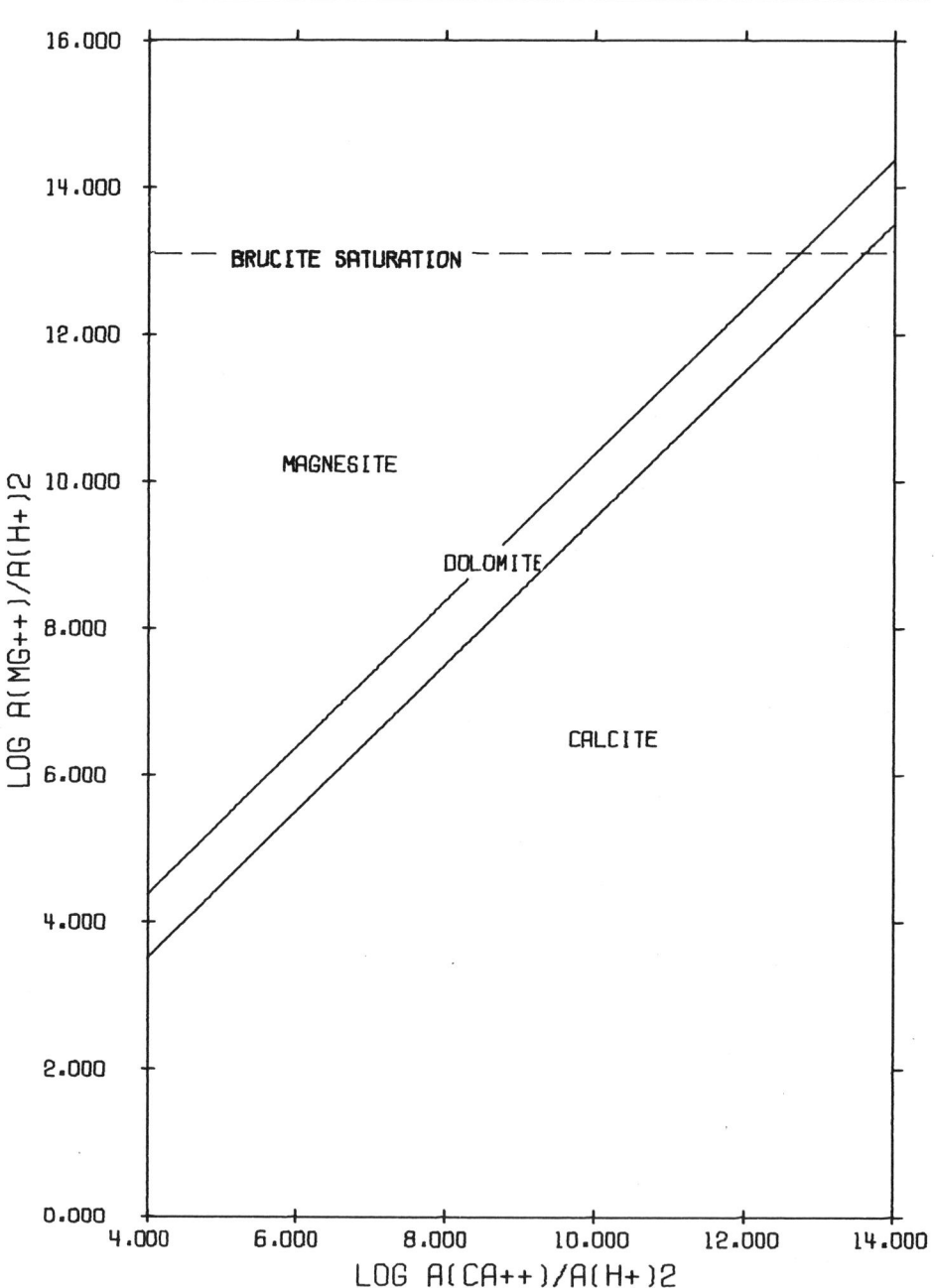

The System HCl—H₂O—CaO—CO₂--MgO at 100°C.

Activity Diagrams

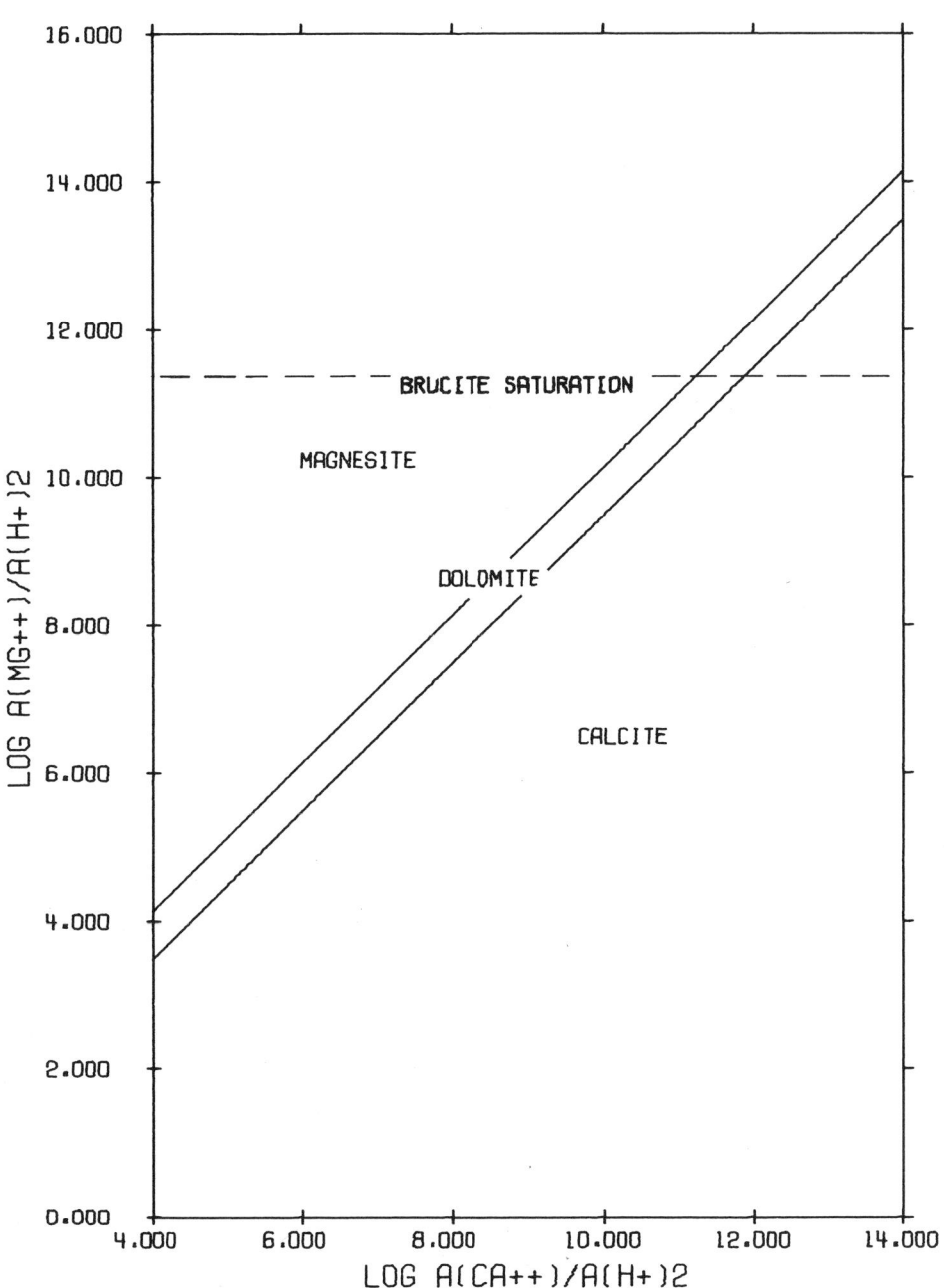

The System HCl—H$_2$O—CaO—CO$_2$—MgO at 150°C.

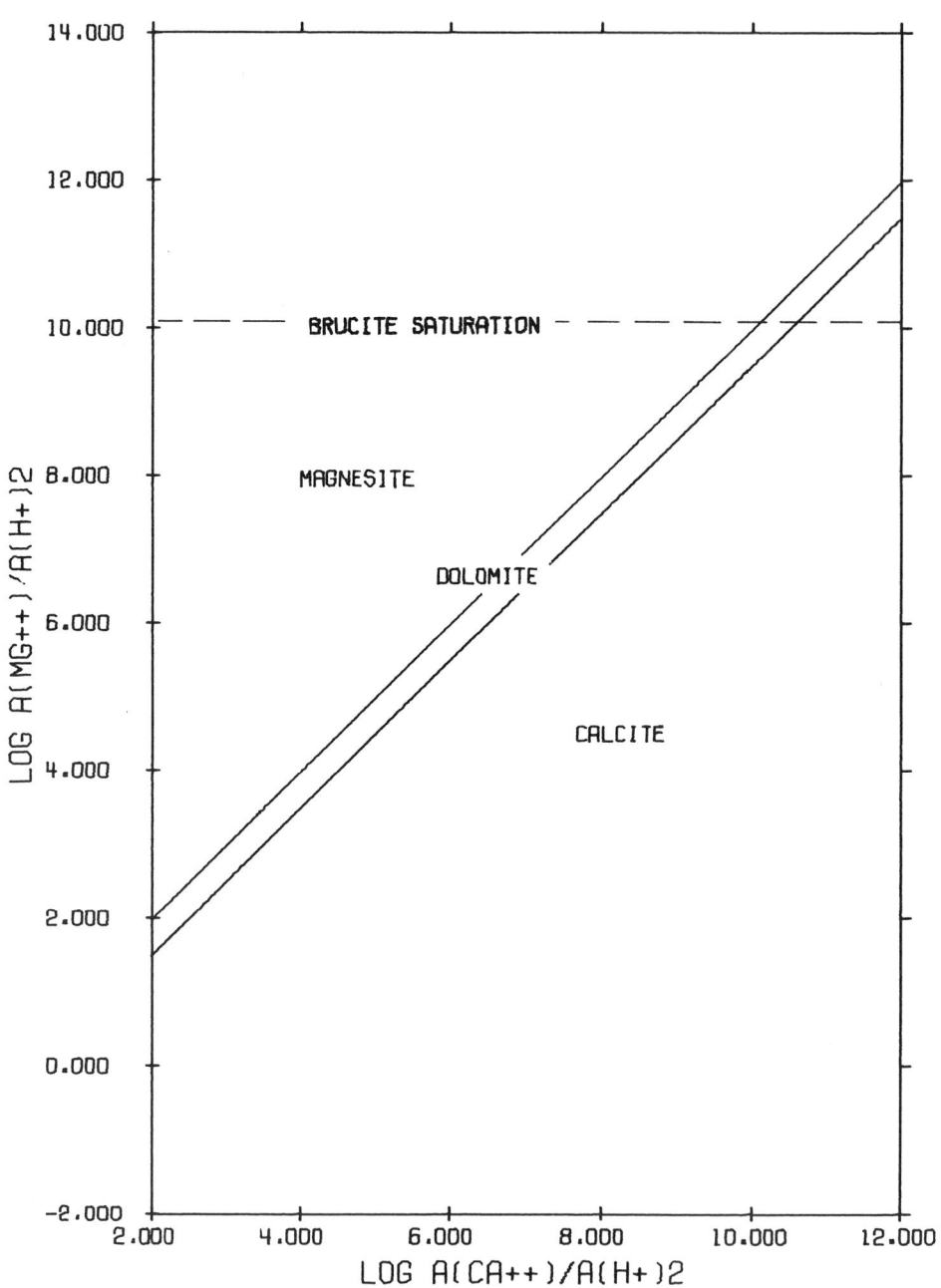

The System HCl—H$_2$O—CaO—CO$_2$—MgO at 200°C.

Activity Diagrams

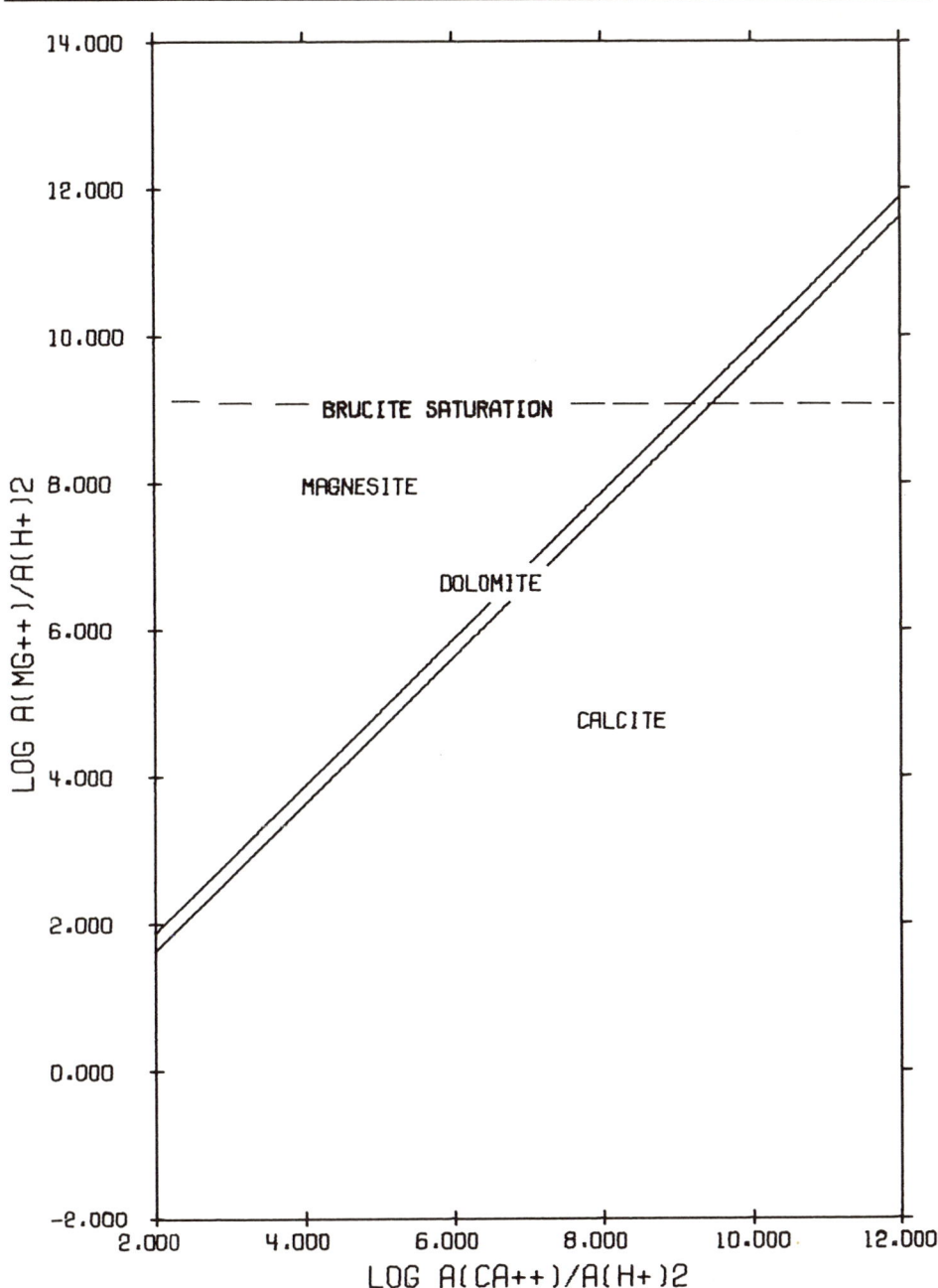

The System HCl—H₂O—CaO—CO₂—MgO at 250°C.

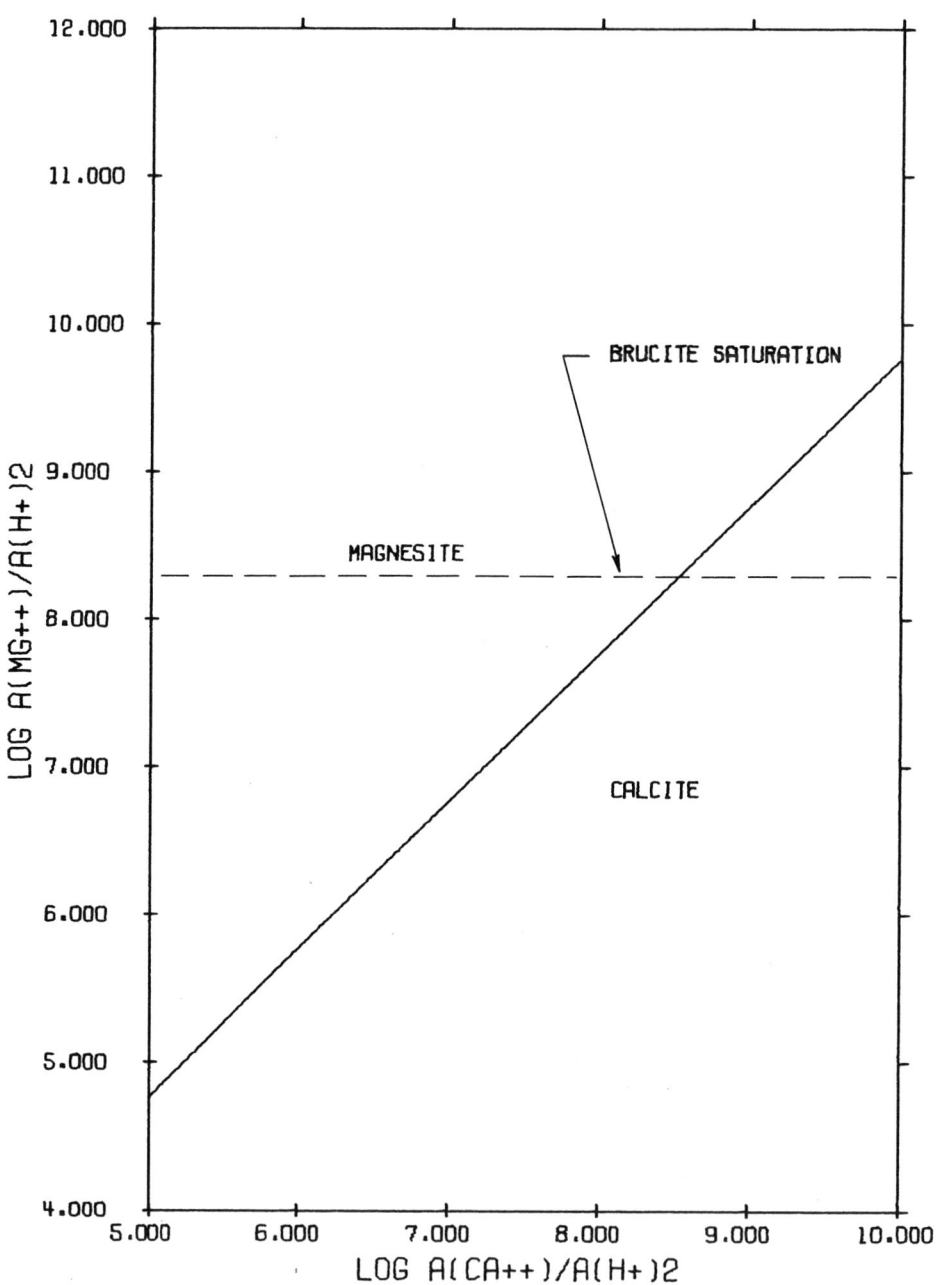

The System HCl—H₂O—CaO—CO₂—MgO at 300°C.

Activity Diagrams

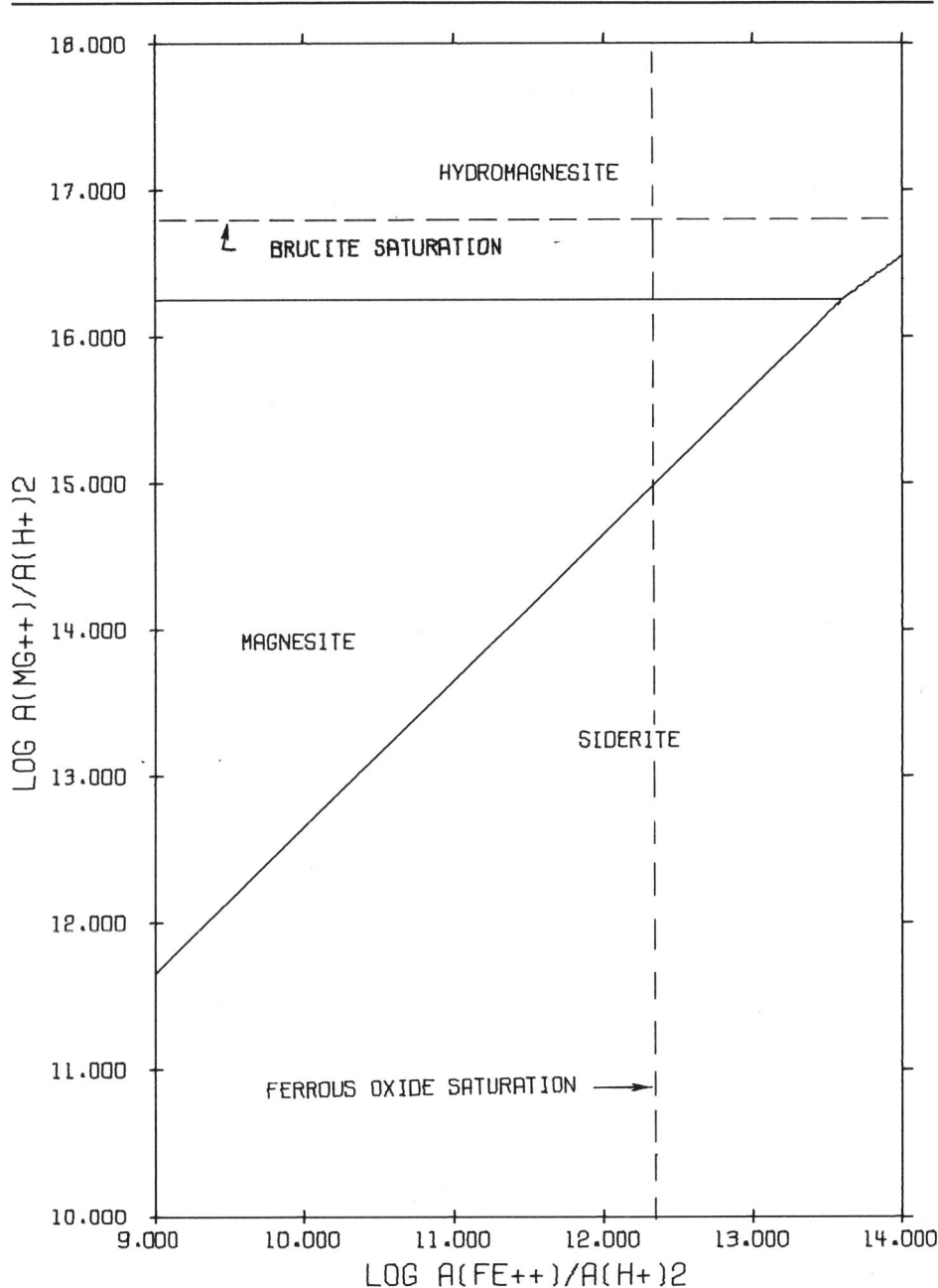

The System HCl—H$_2$O—CO$_2$—FeO—MgO at 25°C.

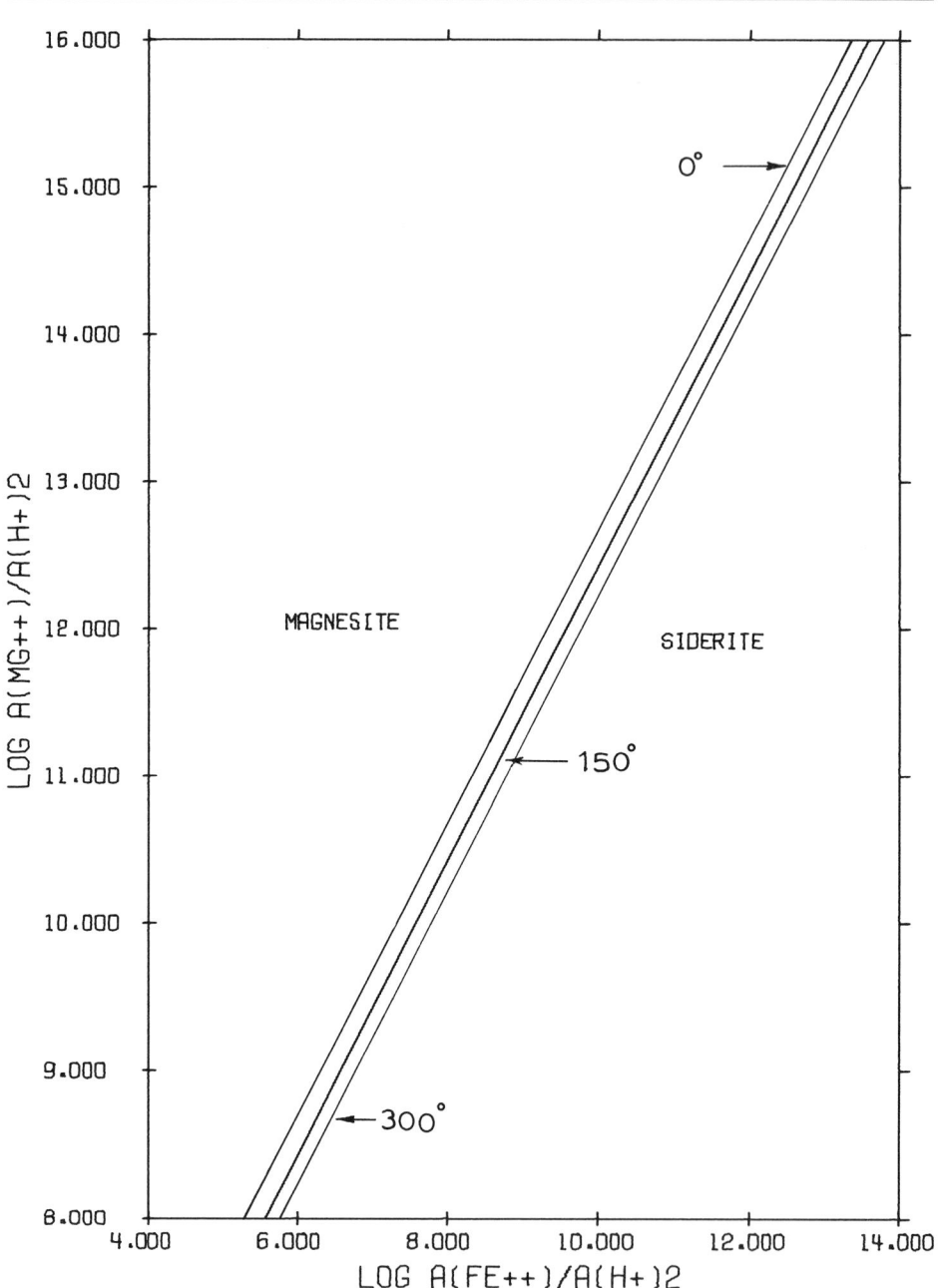

The System HCl—H$_2$O—CO$_2$—FeO—MgO at 0°, 150°, and 300°C.

Activity Diagrams

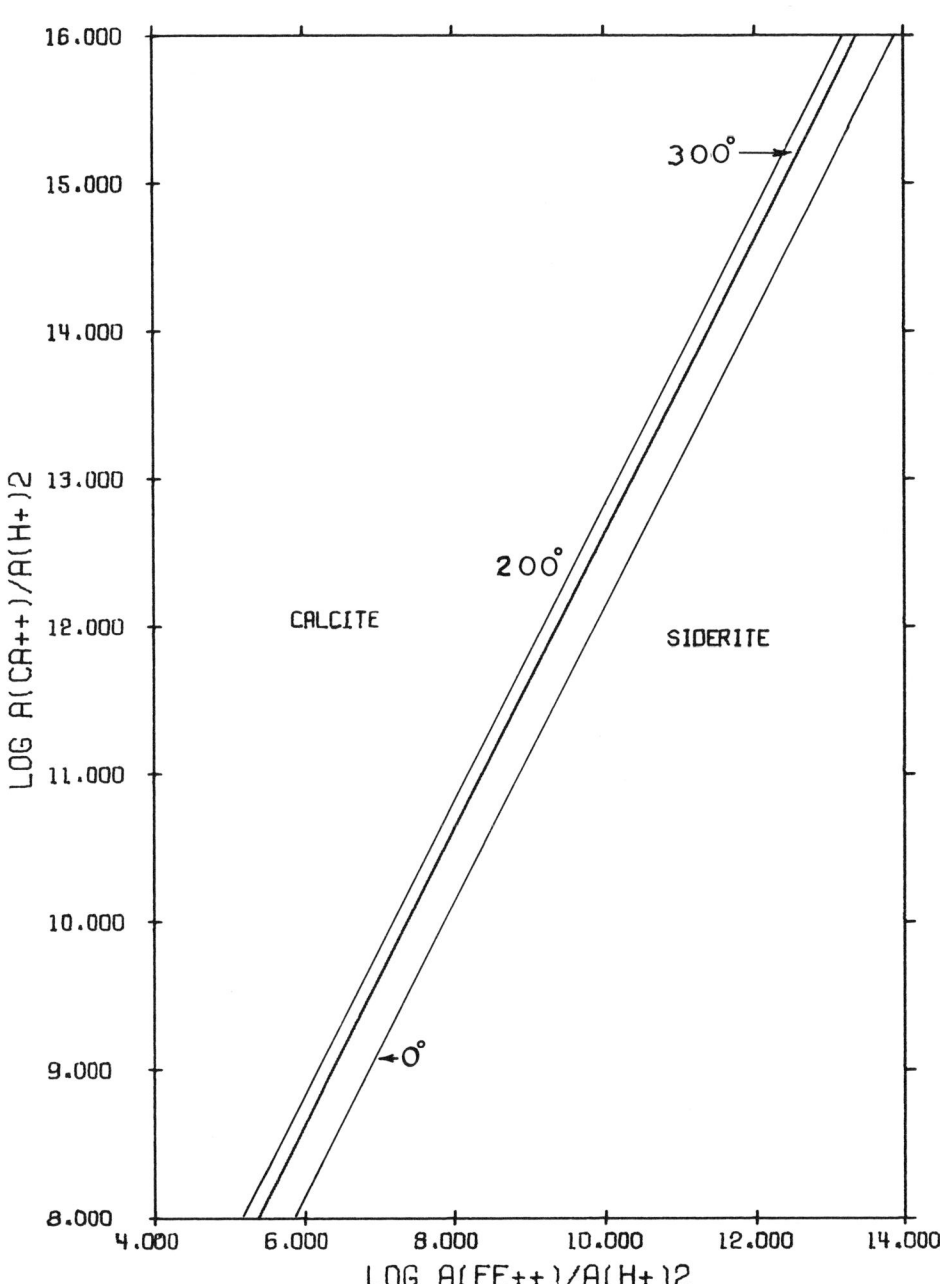

The System HCl—H$_2$O—CaO—CO$_2$—FeO at 0°, 200°, and 300°C.

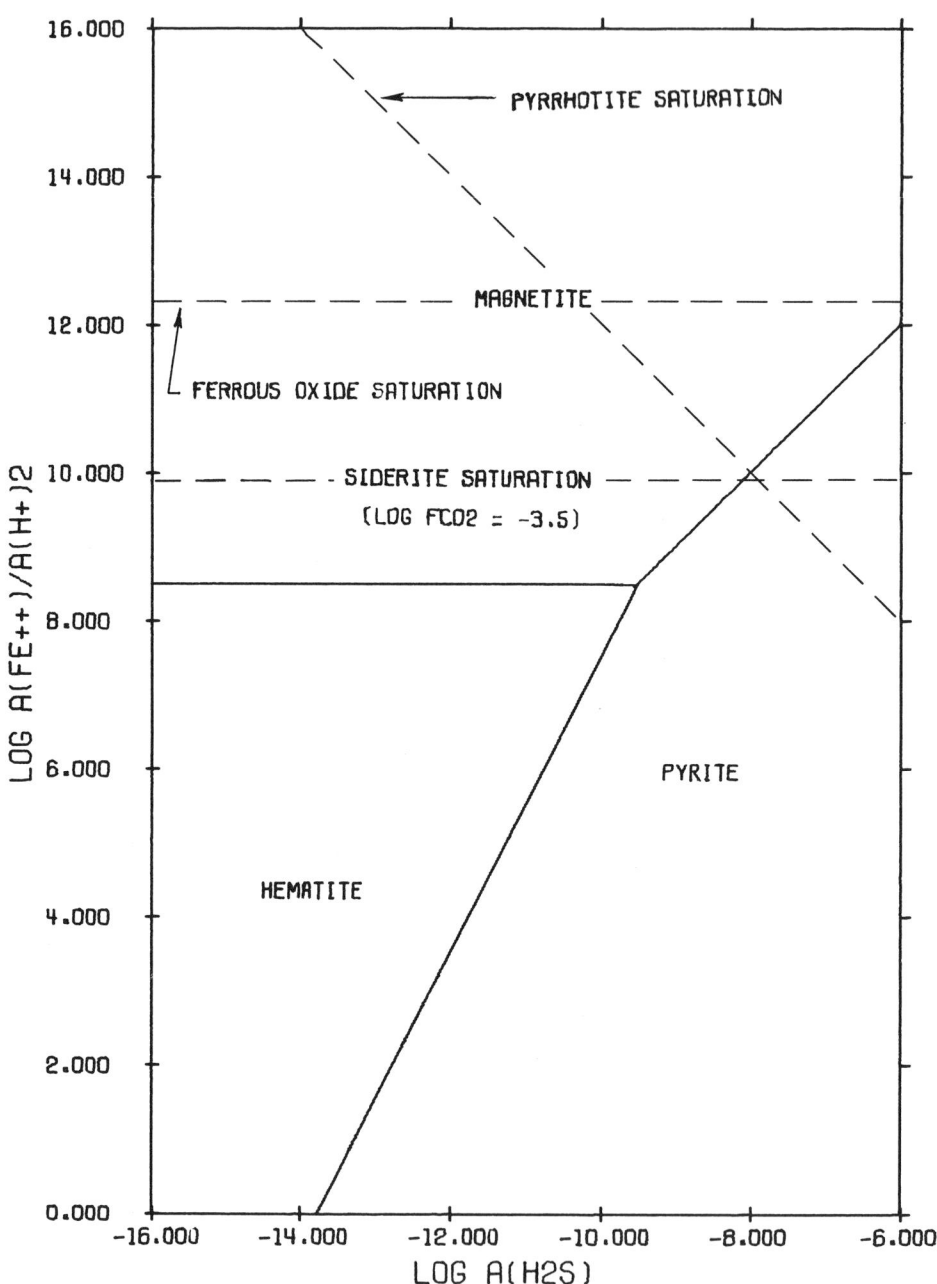

The System HCl—H₂O—FeS—H₂S—H₂SO₄ at 25°C.

Activity Diagrams

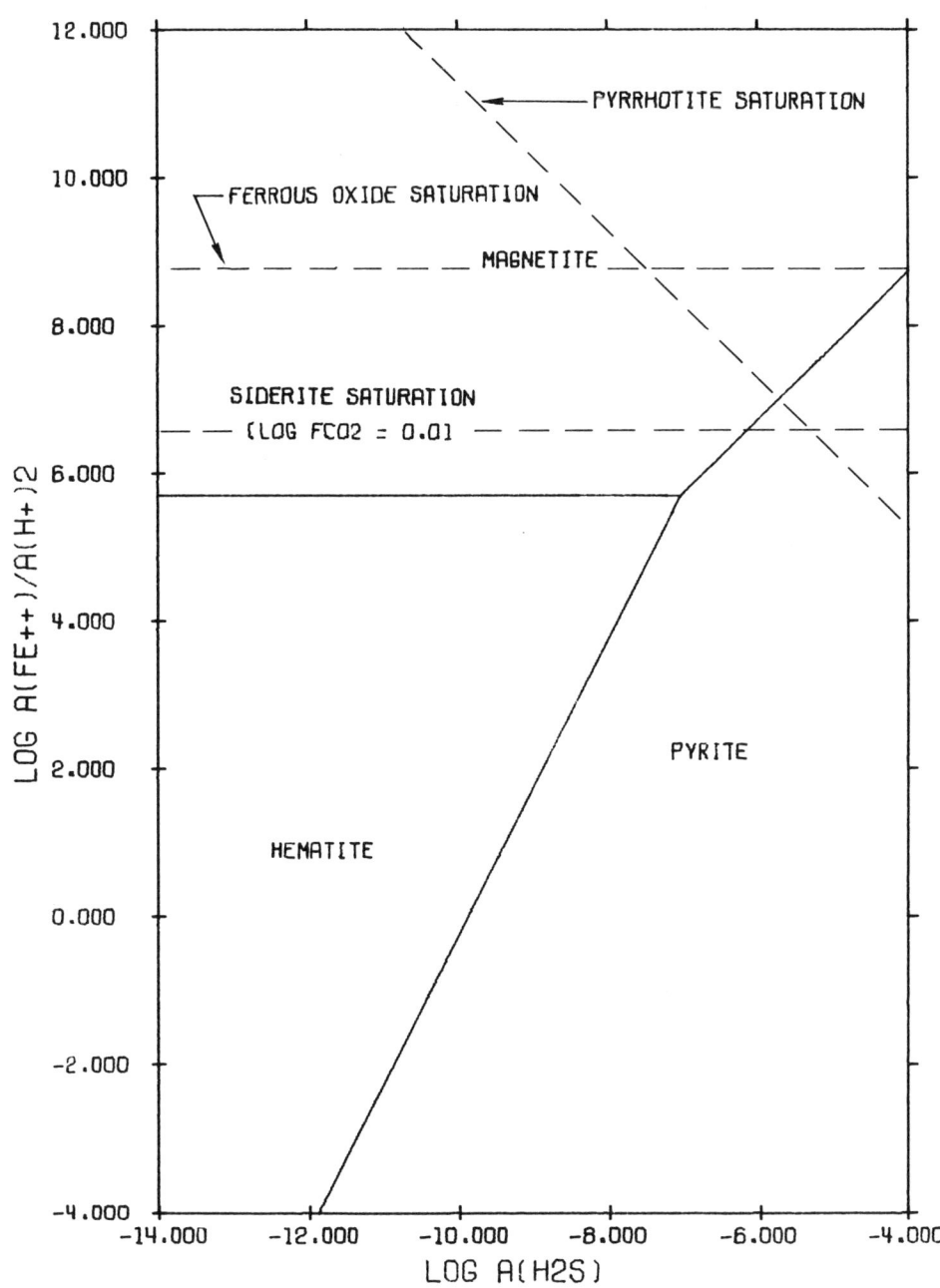

The System HCl—H$_2$O—FeS—H$_2$S—H$_2$SO$_4$ at 100°C.

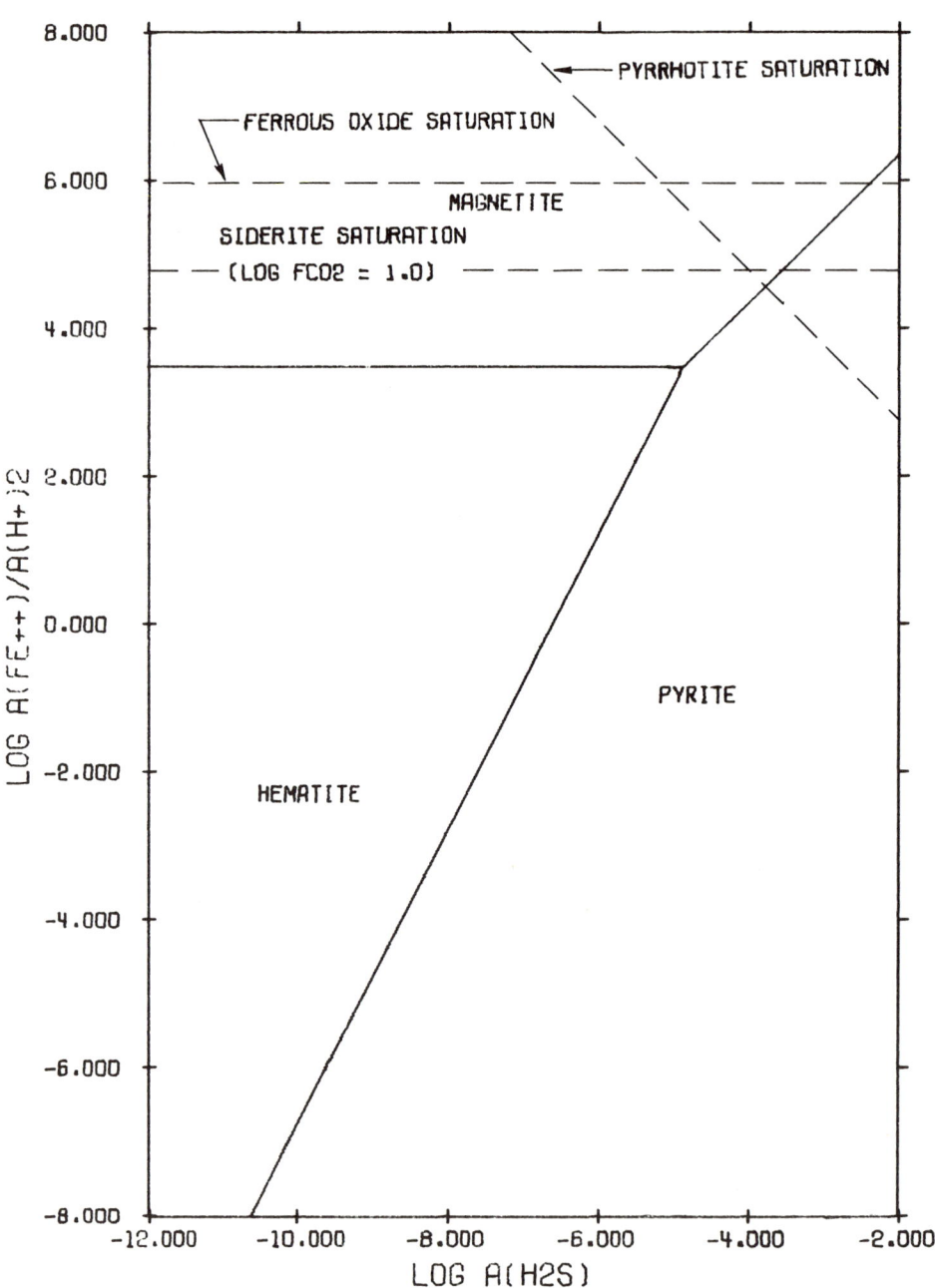

The System HCl—H$_2$O—FeS—H$_2$S—H$_2$SO$_4$ at 200°C.

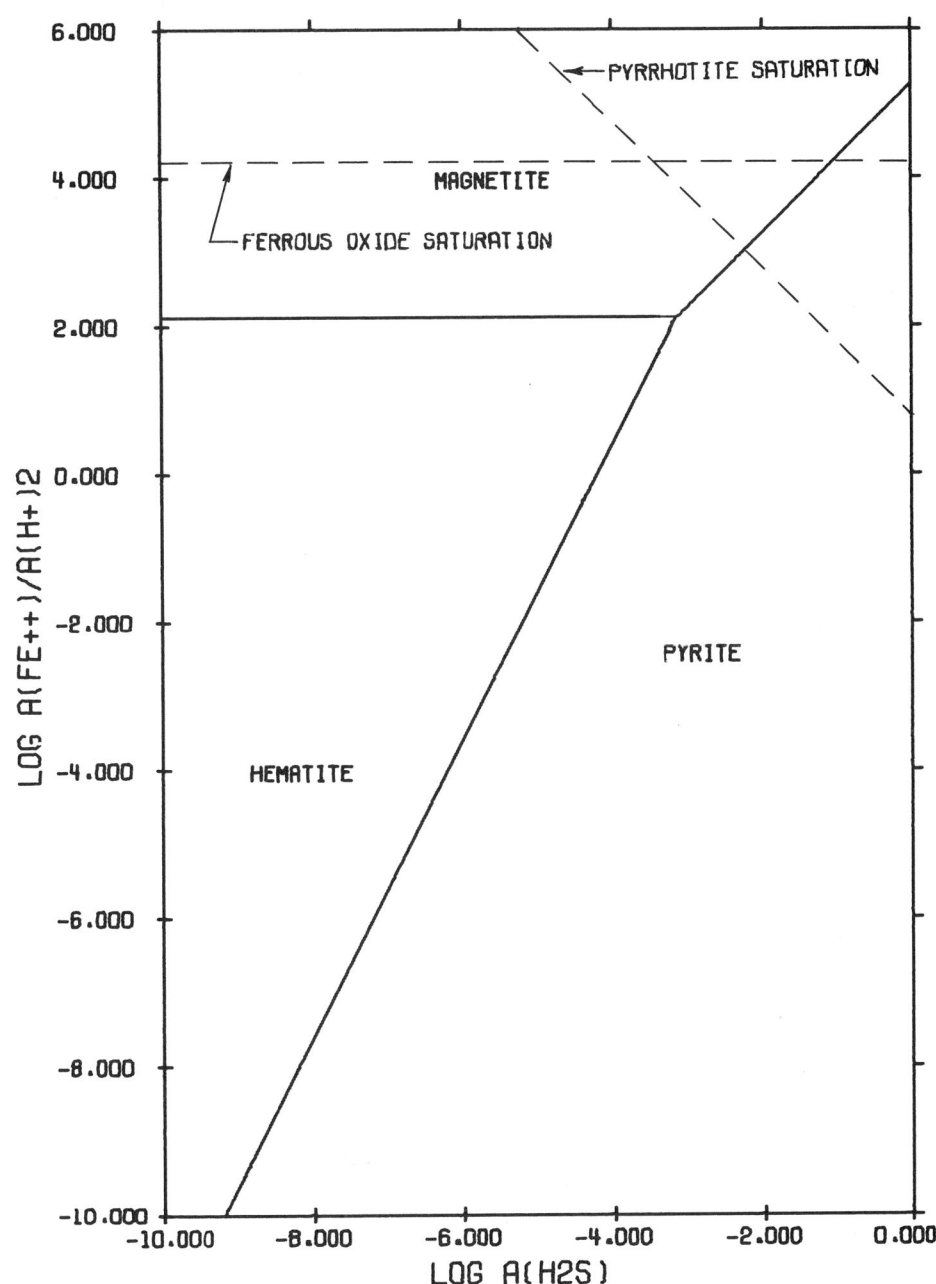

The System HCl—H₂O—FeS—H₂S—H₂SO₄ at 300°C.

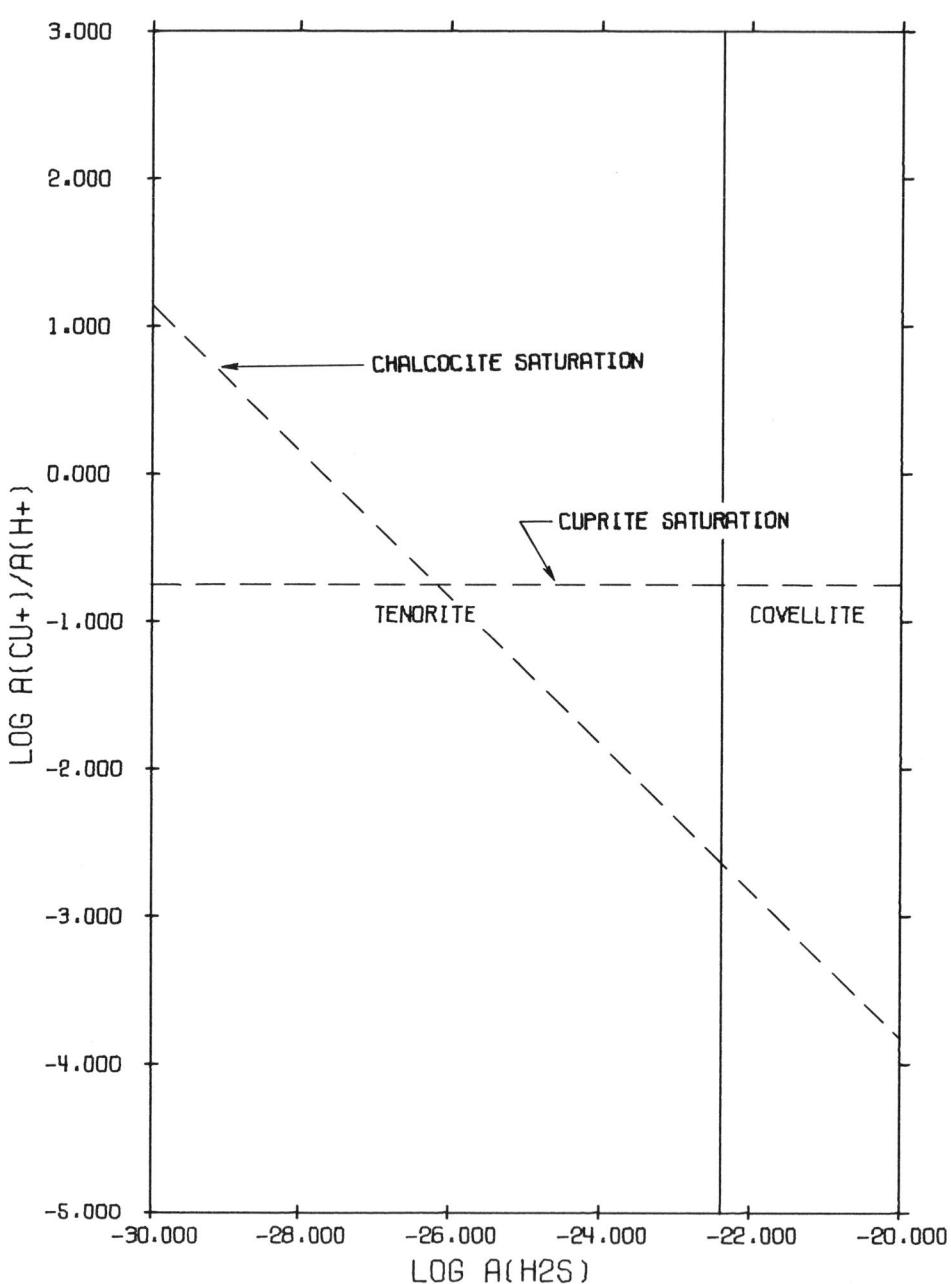

The System HCl—H$_2$O—Cu$_2$S—H$_2$S—H$_2$SO$_4$ at 25°C.

Activity Diagrams

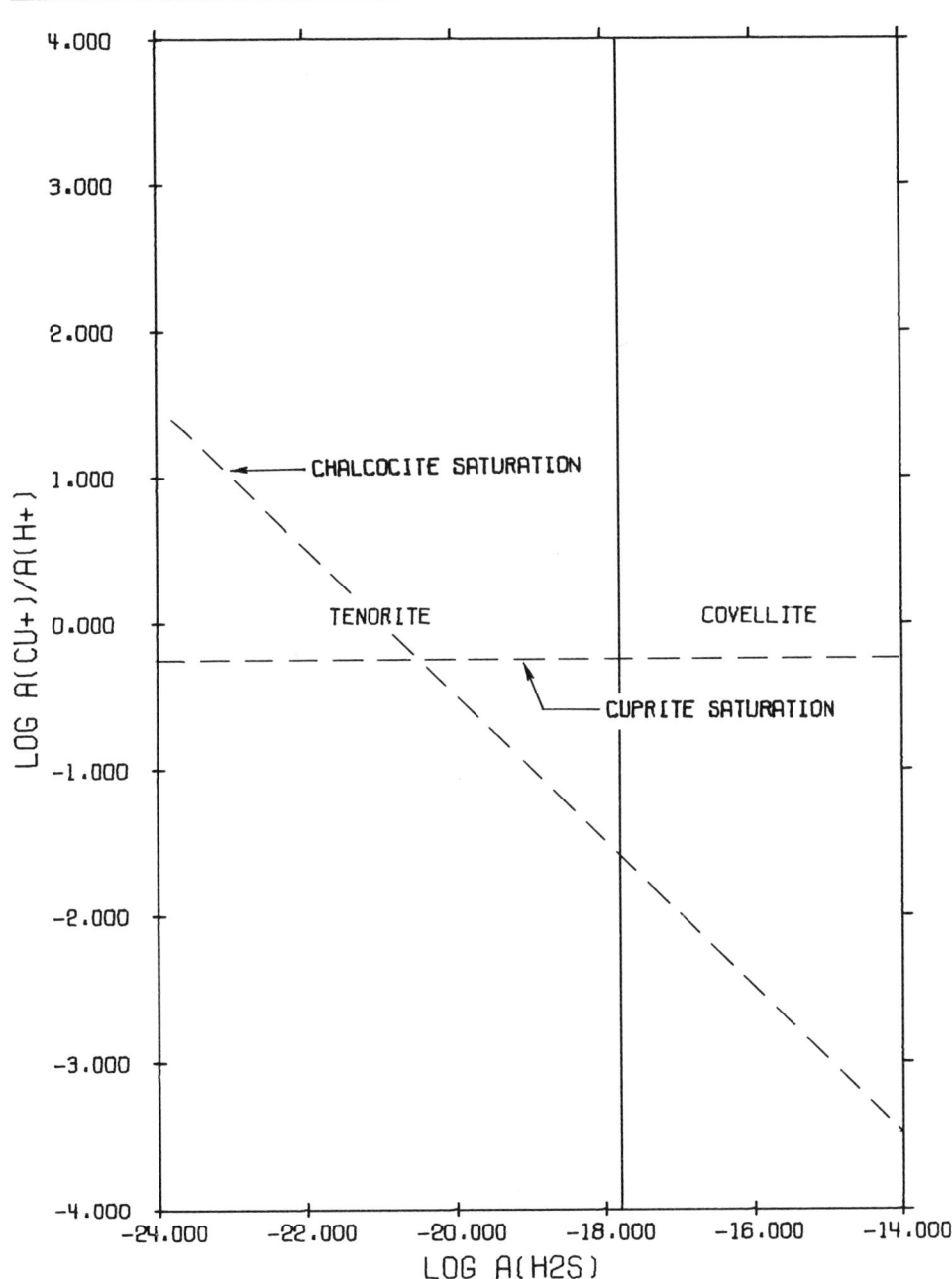

The System HCl—H$_2$O—Cu$_2$S—H$_2$S—H$_2$SO$_4$ at 100°C.

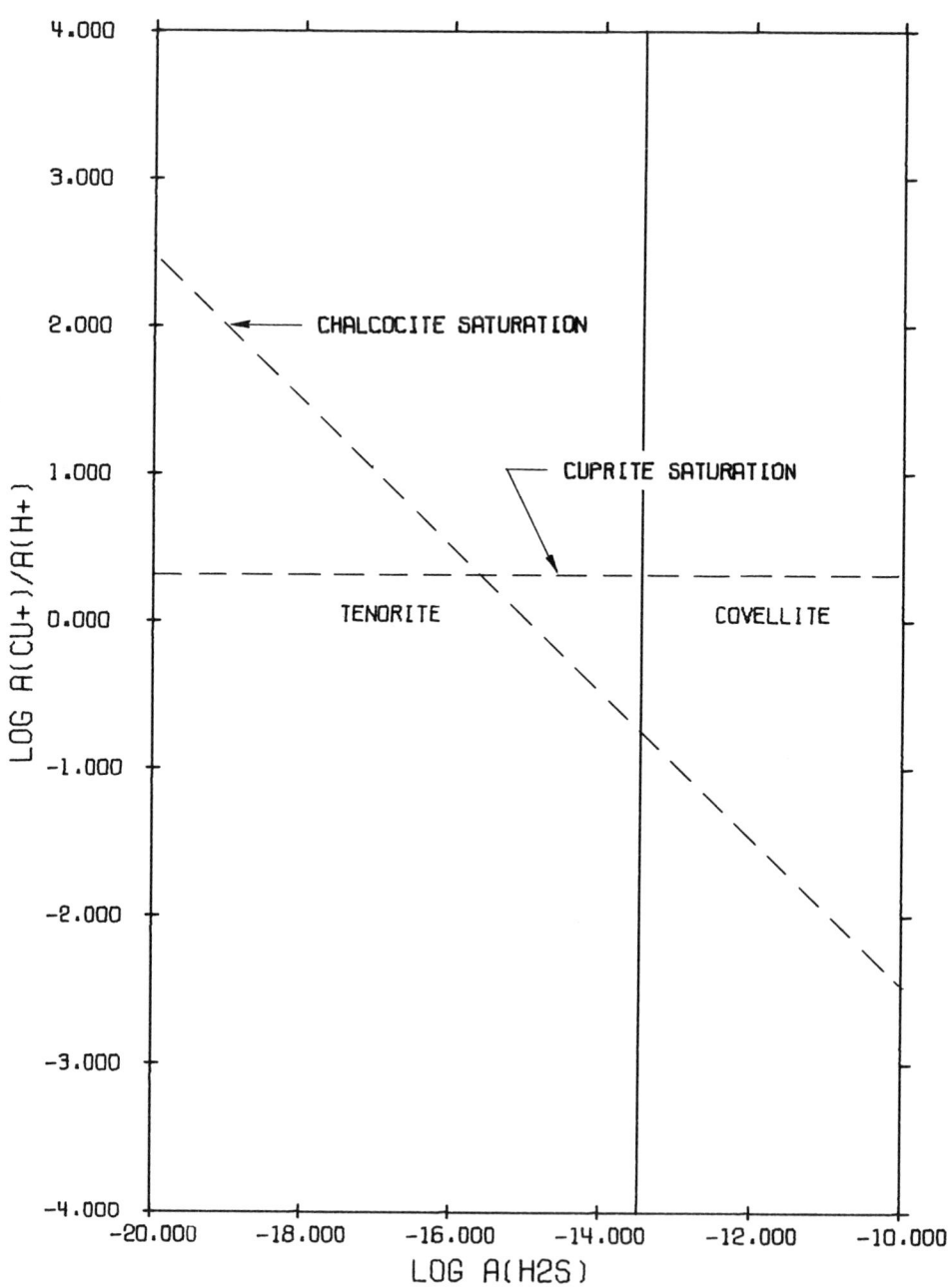

The System HCl—H$_2$O—Cu$_2$S—H$_2$S—H$_2$SO$_4$ at 200°C.

Activity Diagrams

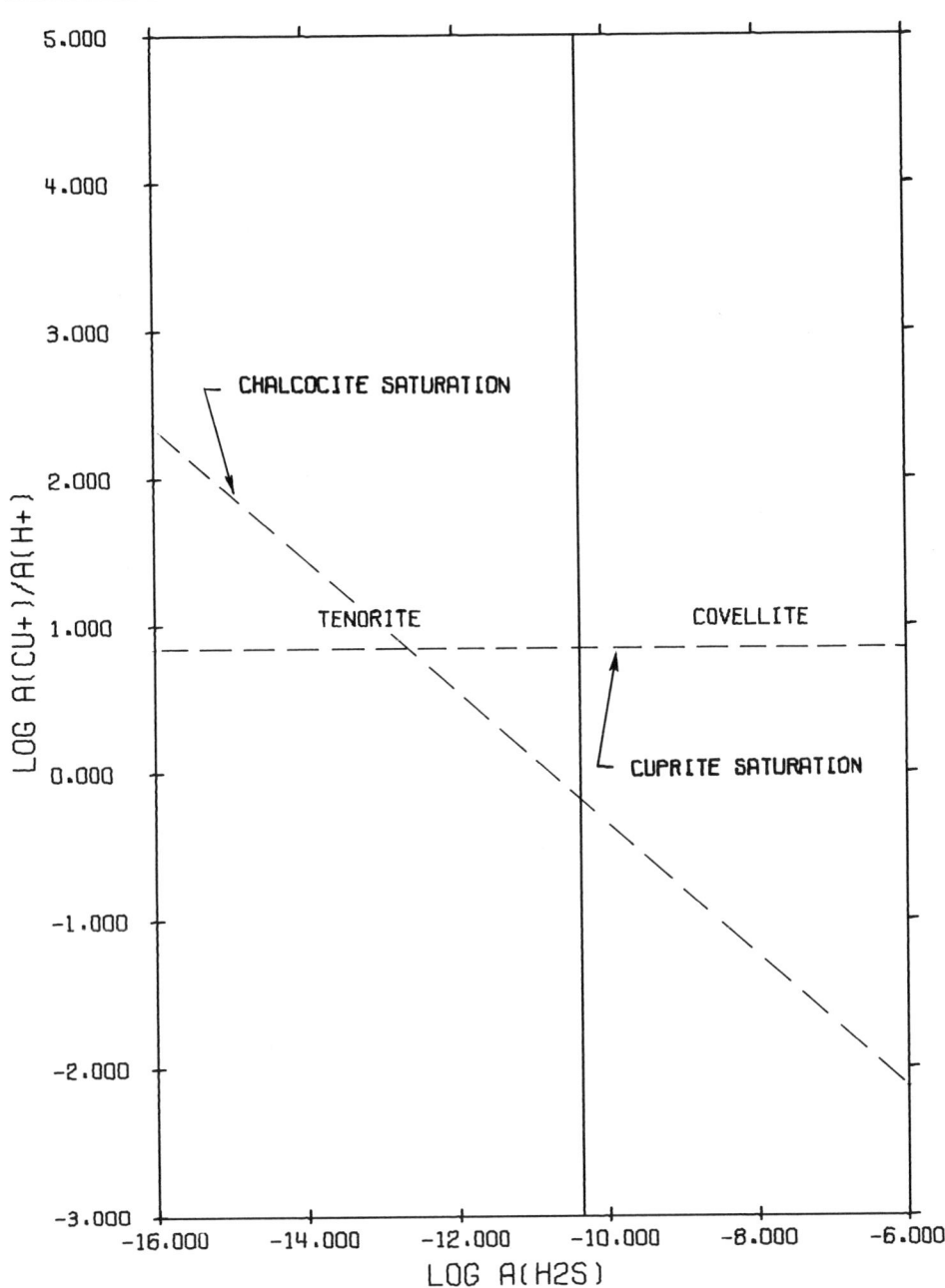

The System HCl—H$_2$O—Cu$_2$S—H$_2$S—H$_2$SO$_4$ at 300°C.

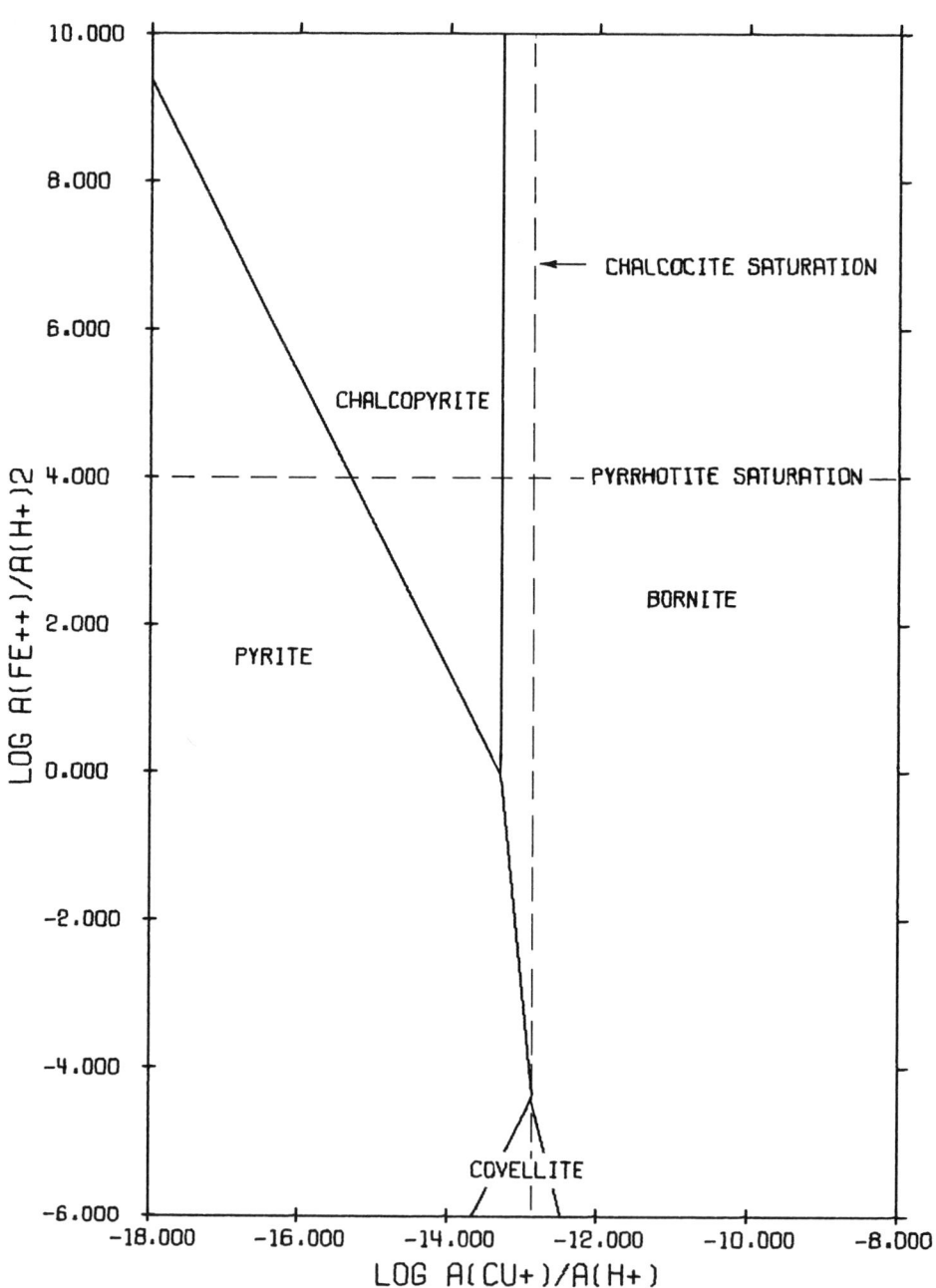

The System HCl—H$_2$O—Cu$_2$S—FeS—H$_2$S—H$_2$SO$_4$ at 25°C; log a_{H_2S} = −2.00.

Activity Diagrams

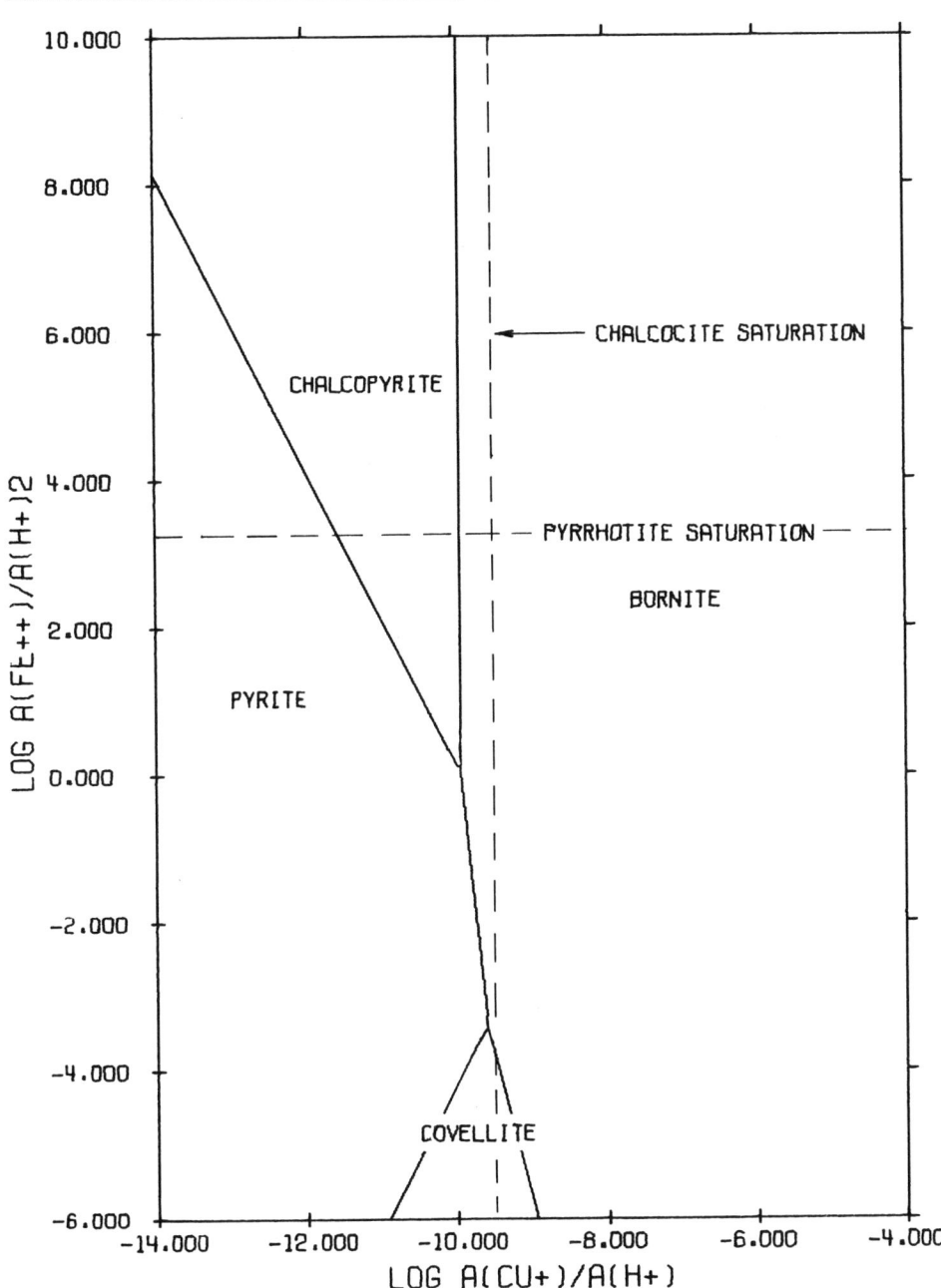

The System HCl—H$_2$O—Cu$_2$S—FeS—H$_2$S—H$_2$SO$_4$ at 100°C; log a_{H_2S} = −2.00.

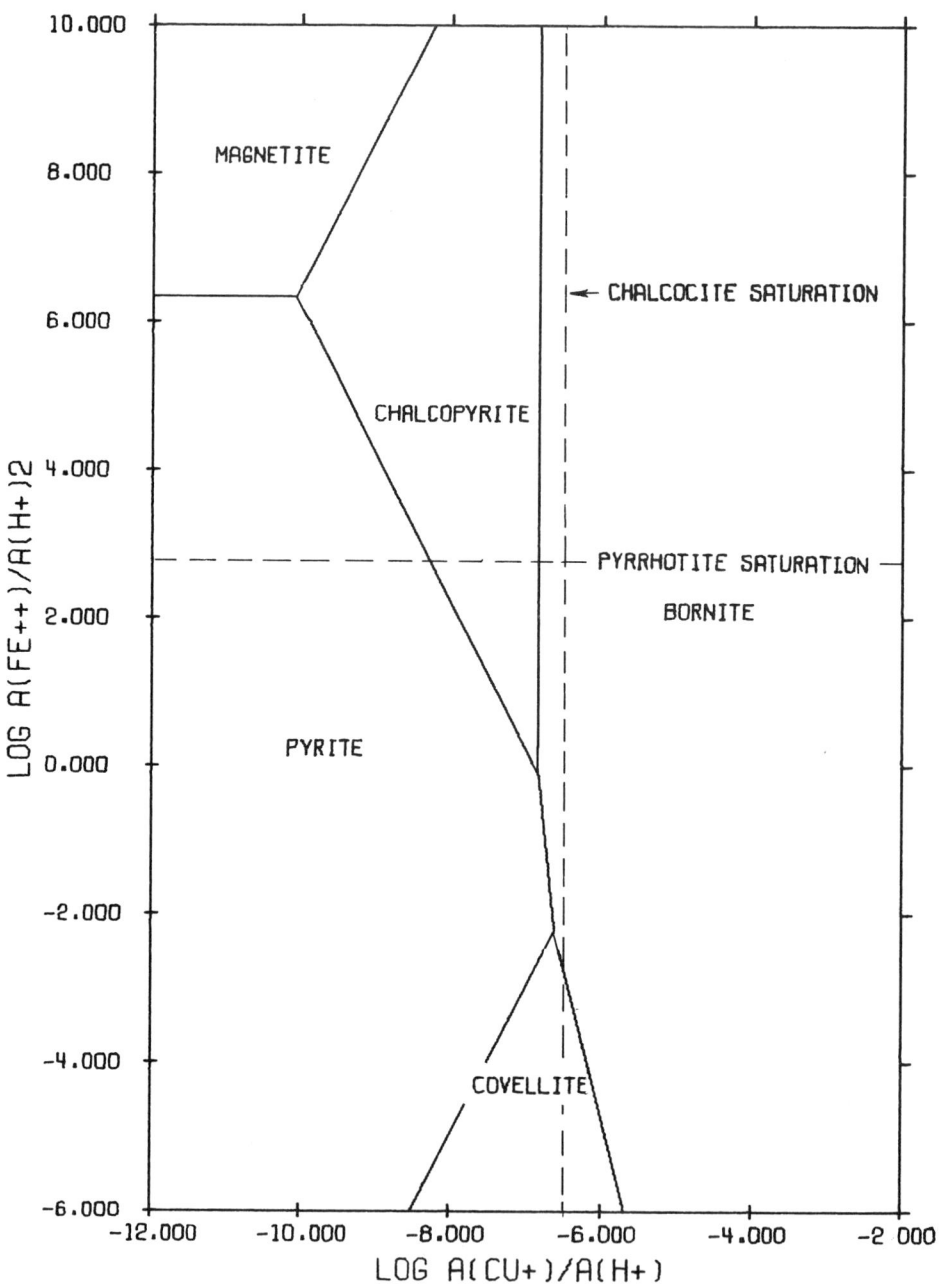

The System HCl—H$_2$O—Cu$_2$S—FeS—H$_2$S—H$_2$SO$_4$ at 200°C; log a_{H_2S} = −2.00.

Activity Diagrams

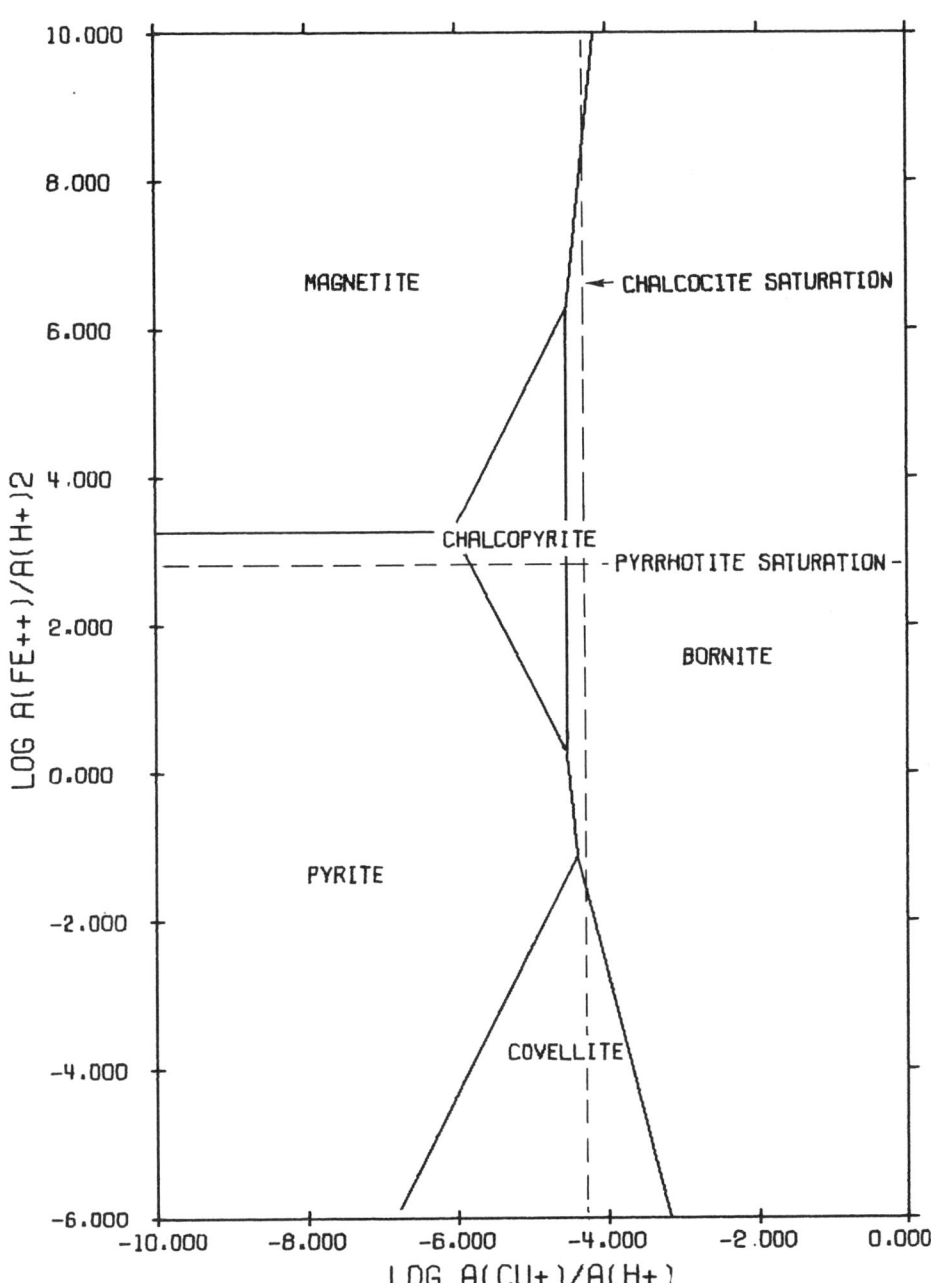

The System HCl—H$_2$O—Cu$_2$S—FeS—H$_2$S—H$_2$SO$_4$ at 300°C; log a_{H_2S} = −2.00.

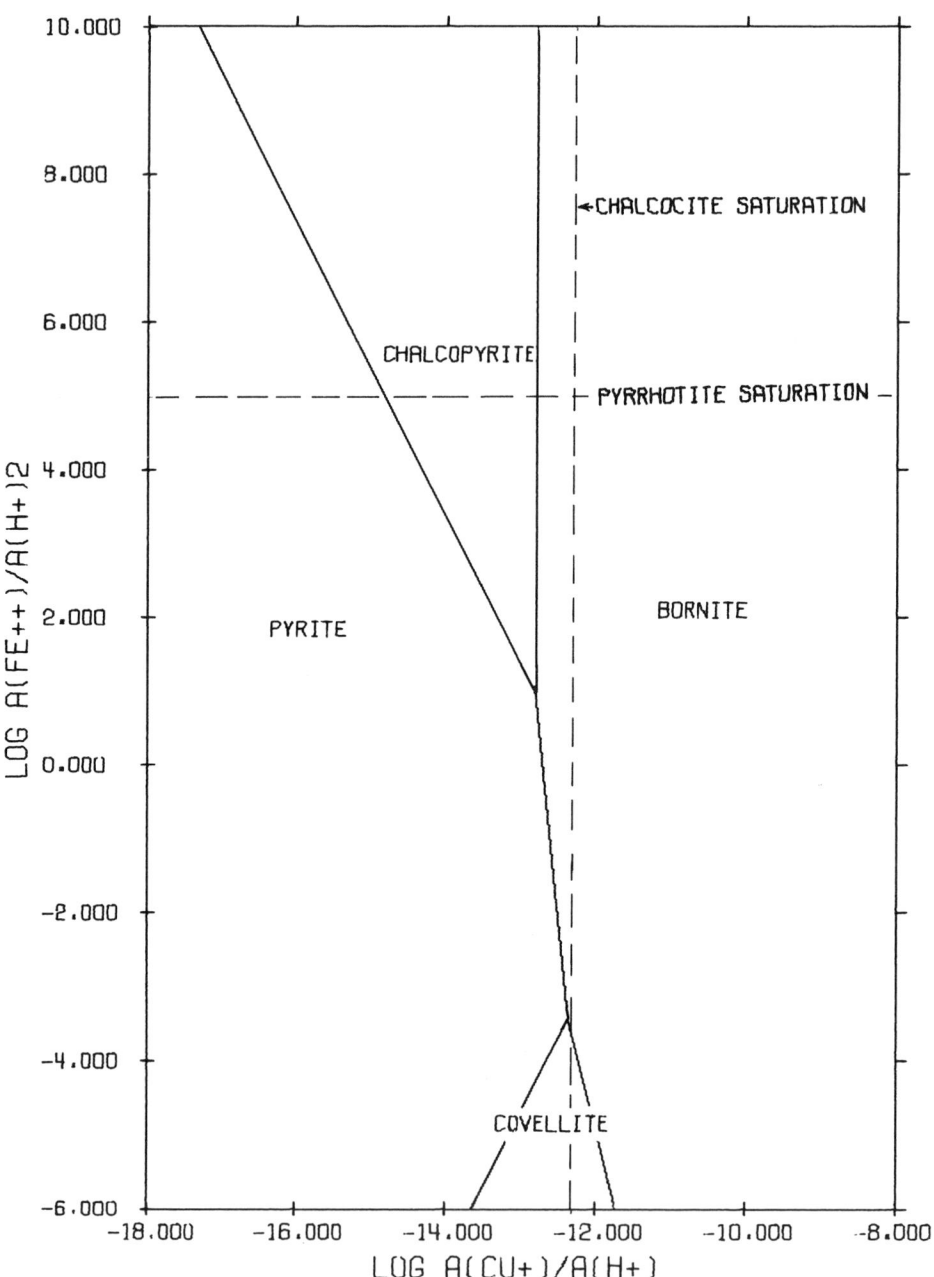

The System HCl—H$_2$O—Cu$_2$S—FeS—H$_2$S—H$_2$SO$_4$ at 25°C; log a_{H_2S} = −3.00.

Activity Diagrams

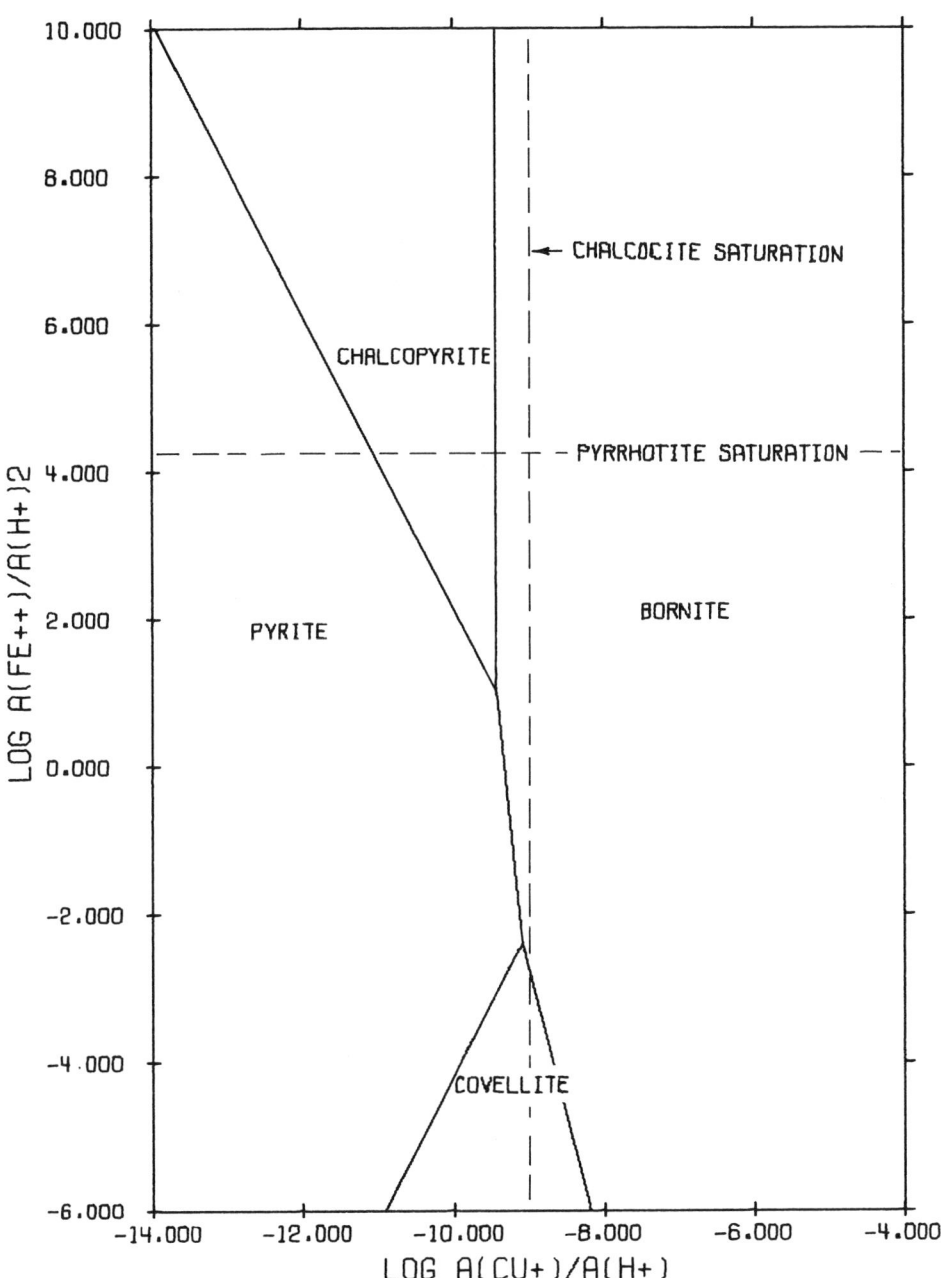

The System HCl—H$_2$O—Cu$_2$S—FeS—H$_2$S—H$_2$SO$_4$ at 100°C; log a_{H_2S} = −3.00.

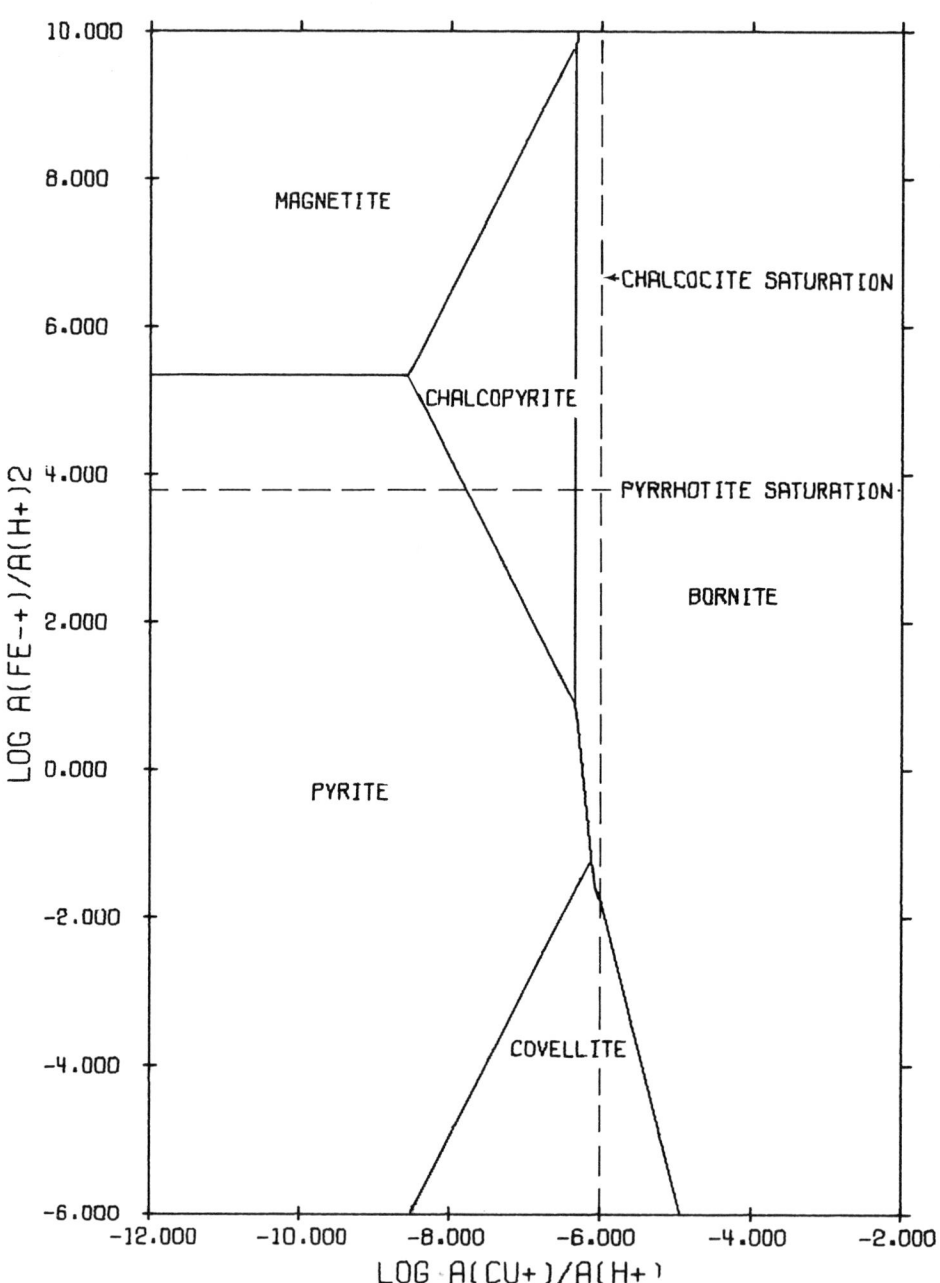

The System HCl—H$_2$O—Cu$_2$S—FeS—H$_2$S—H$_2$SO$_4$ at 200°C; log a_{H_2S} = −3.00.

Activity Diagrams

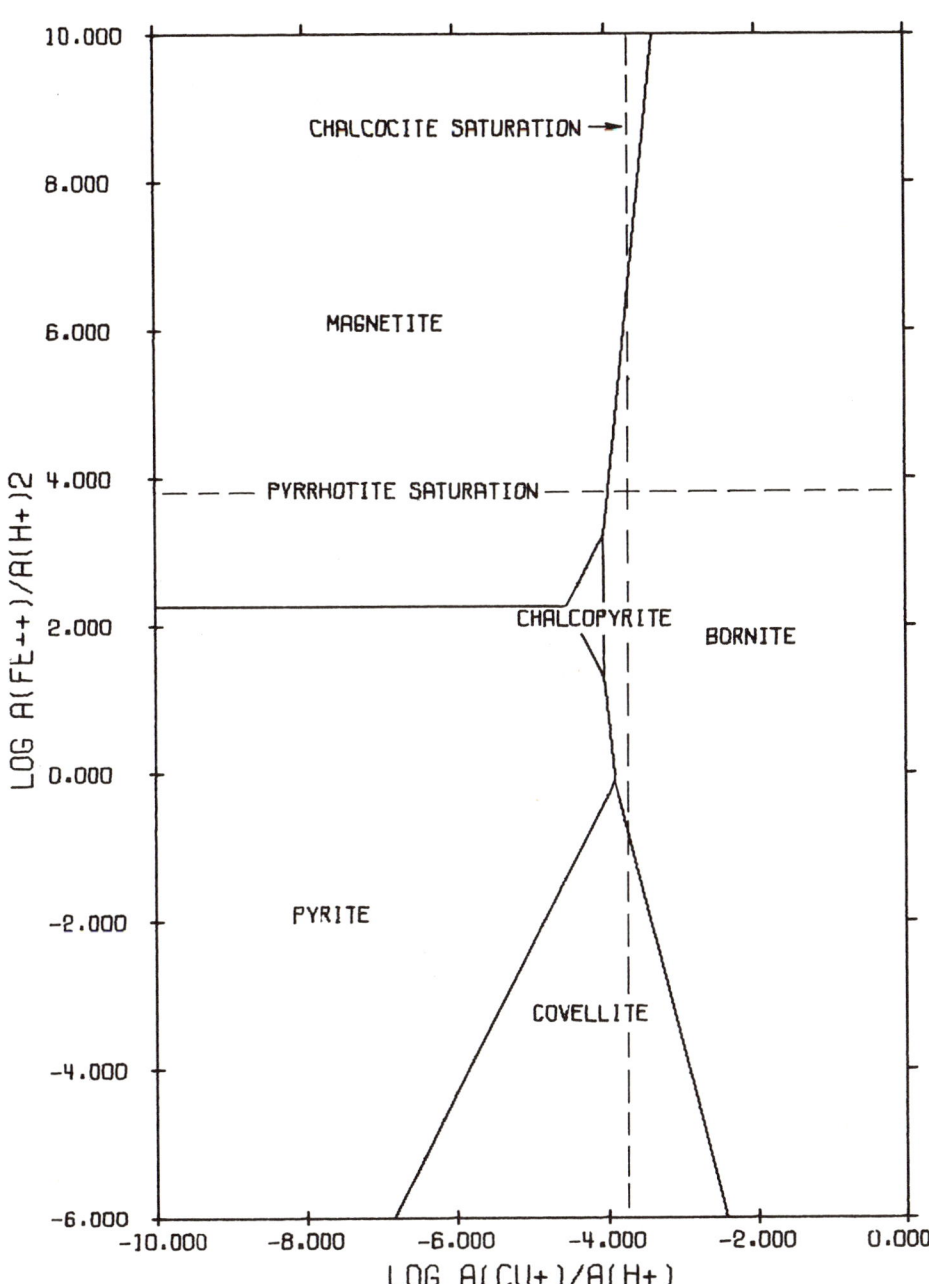

The System HCl—H$_2$O—Cu$_2$S—FeS—H$_2$S—H$_2$SO$_4$ at 300°C; log a_{H_2S} = −3.00.

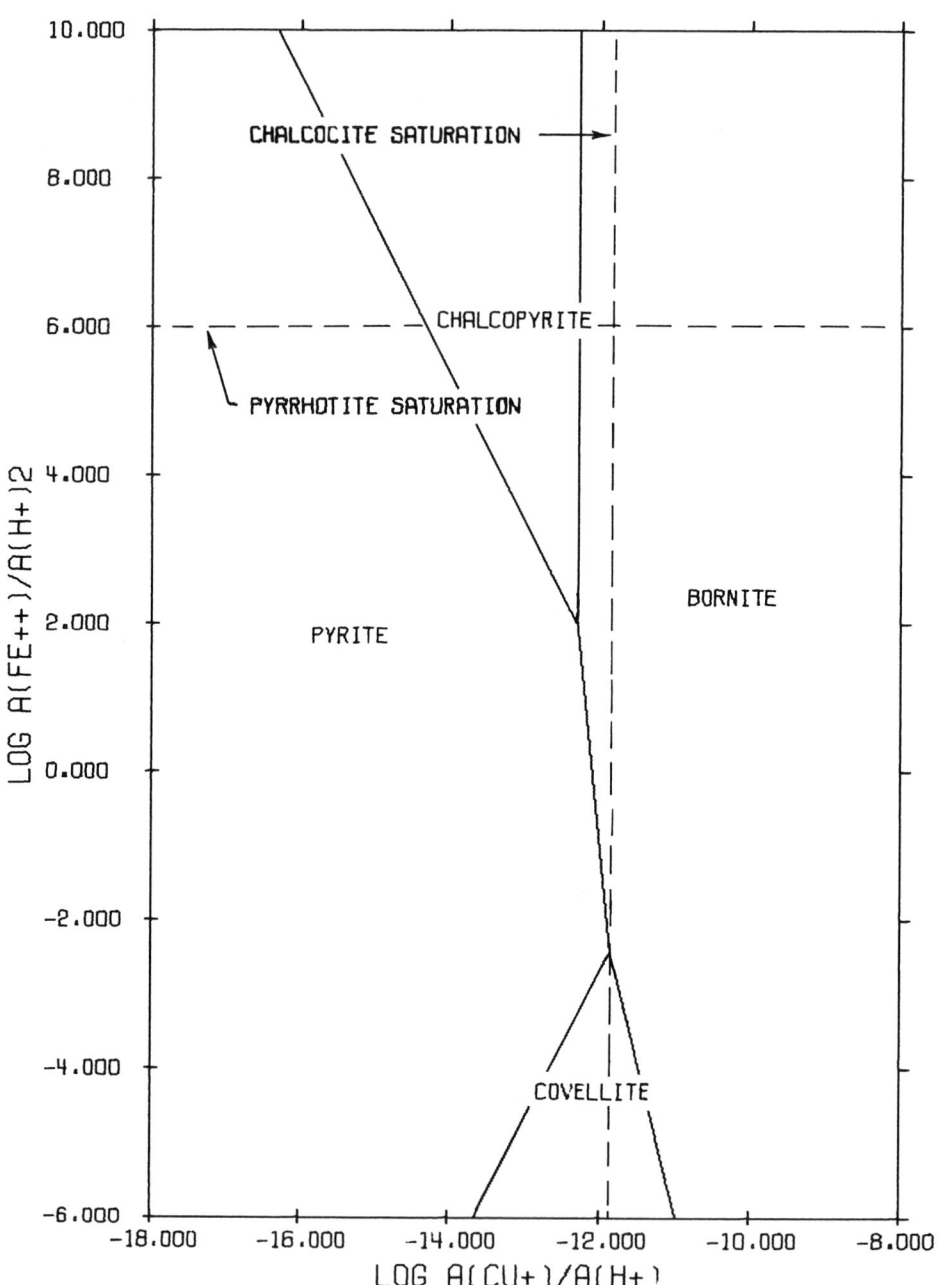

The System HCl—H$_2$O—Cu$_2$S—FeS—H$_2$S—H$_2$SO$_4$ at 25°C; log a_{H_2S} = -4.00.

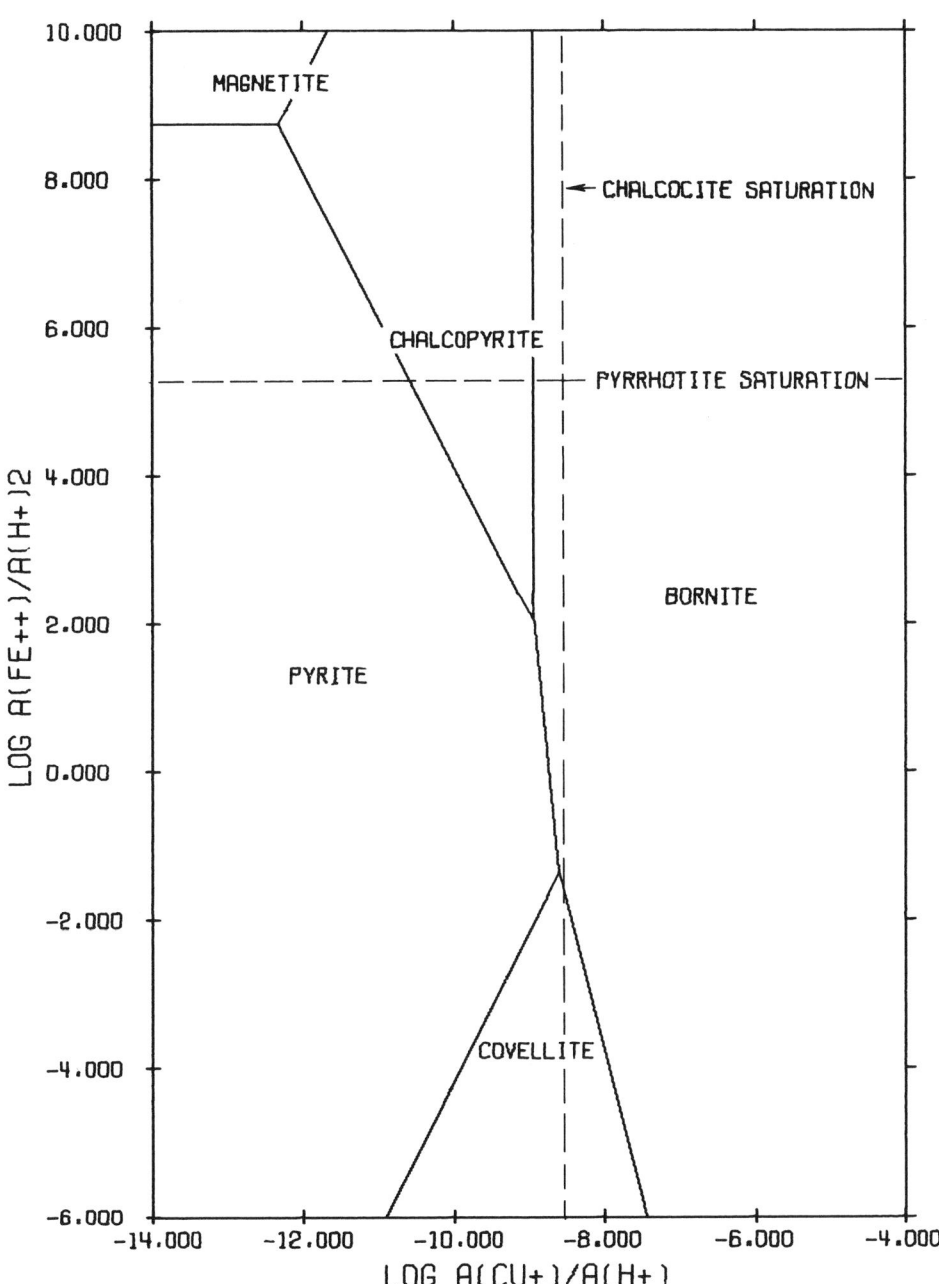

The System HCl—H$_2$O—Cu$_2$S—FeS—H$_2$S—H$_2$SO$_4$ at 100°C; log a_{H_2S} = −4.00.

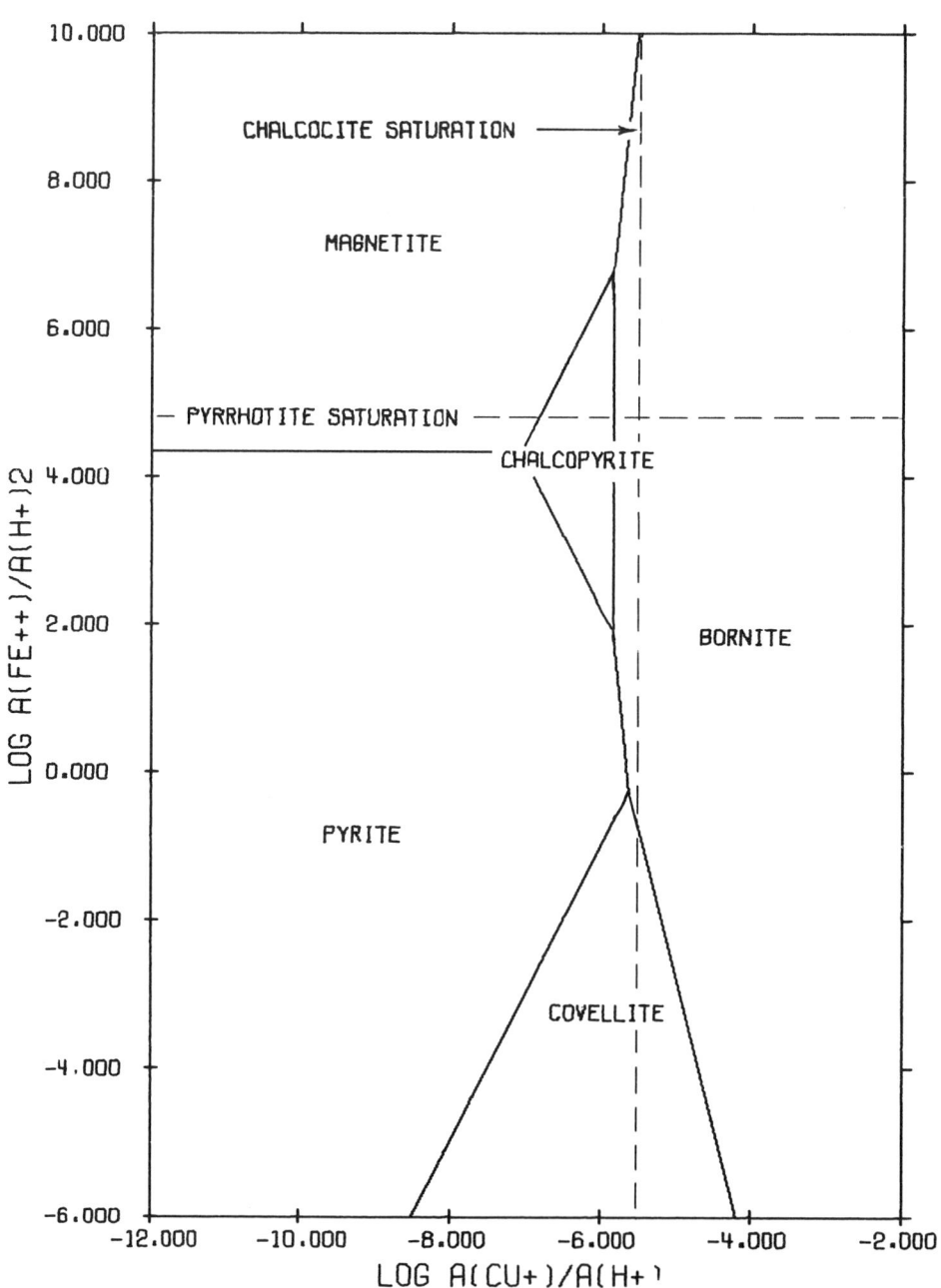

The System HCl—H$_2$O—Cu$_2$S—FeS—H$_2$S—H$_2$SO$_4$ at 200°C; log a_{H_2S} = −4.00.

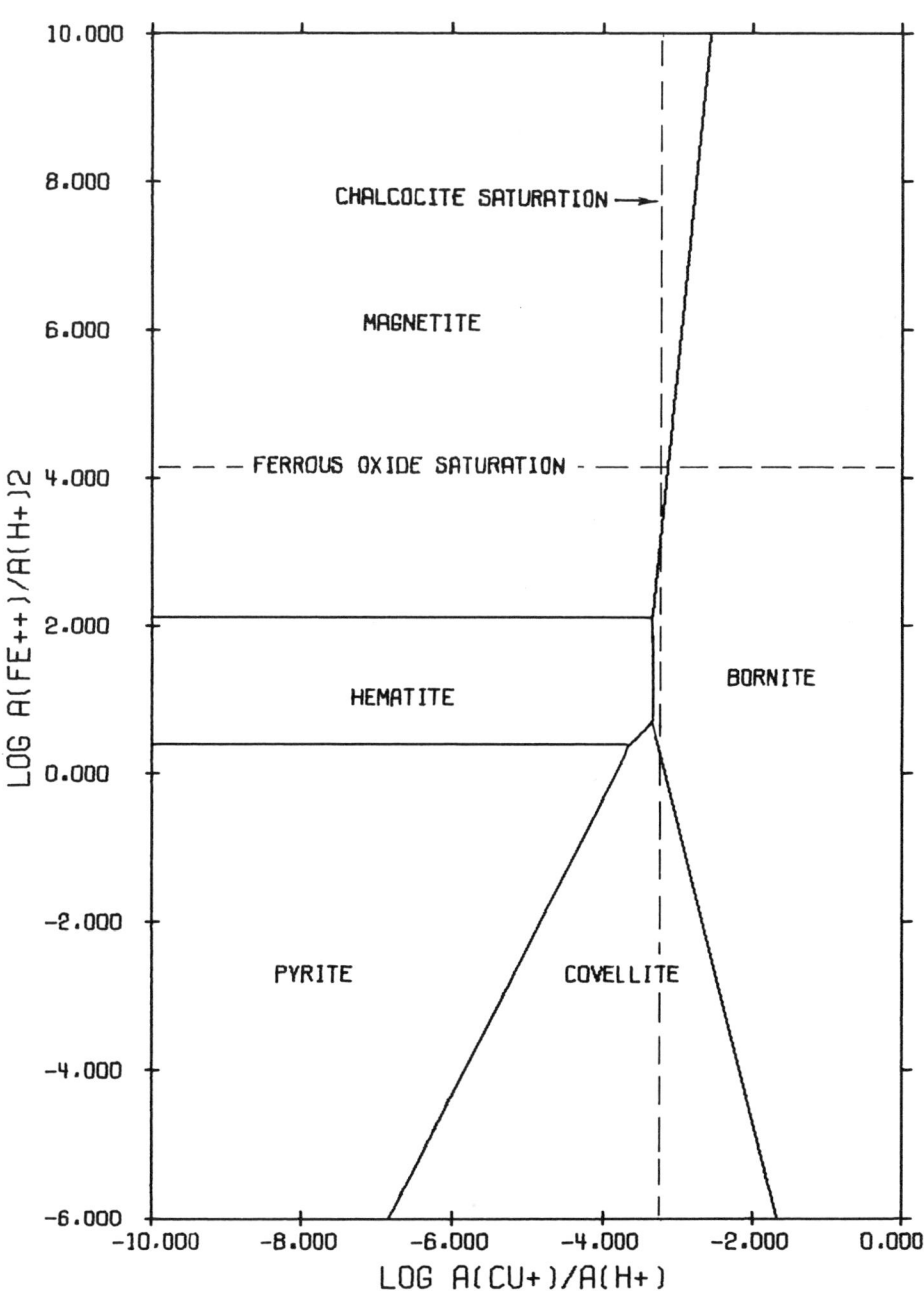

The System HCl—H_2O—Cu_2S—FeS—H_2S—H_2SO_4 at 300°C; log a_{H_2S} = −4.00.

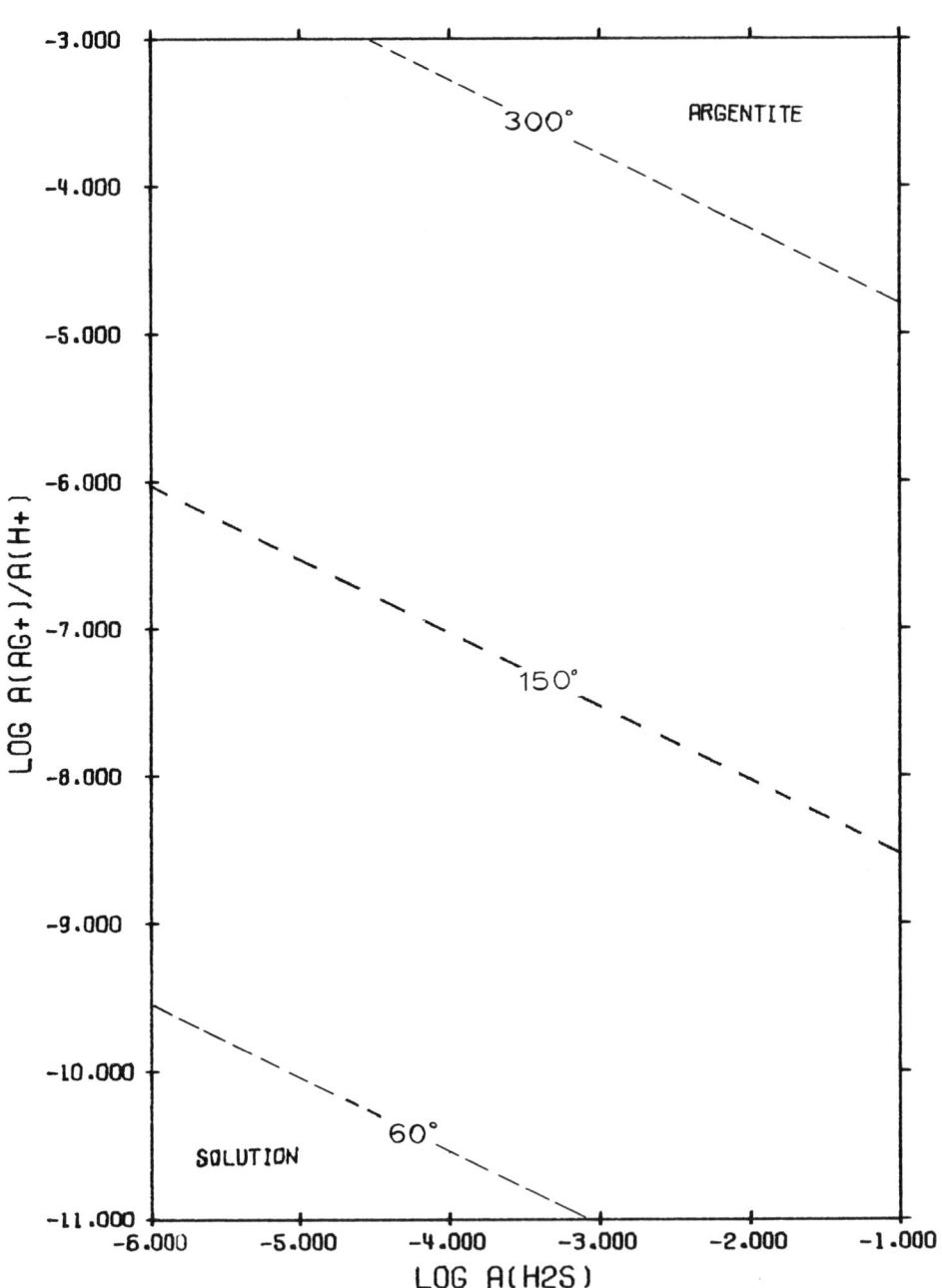

The System HCl—H$_2$O—Ag$_2$S—H$_2$S at 60°, 150°, and 300°C.

Activity Diagrams

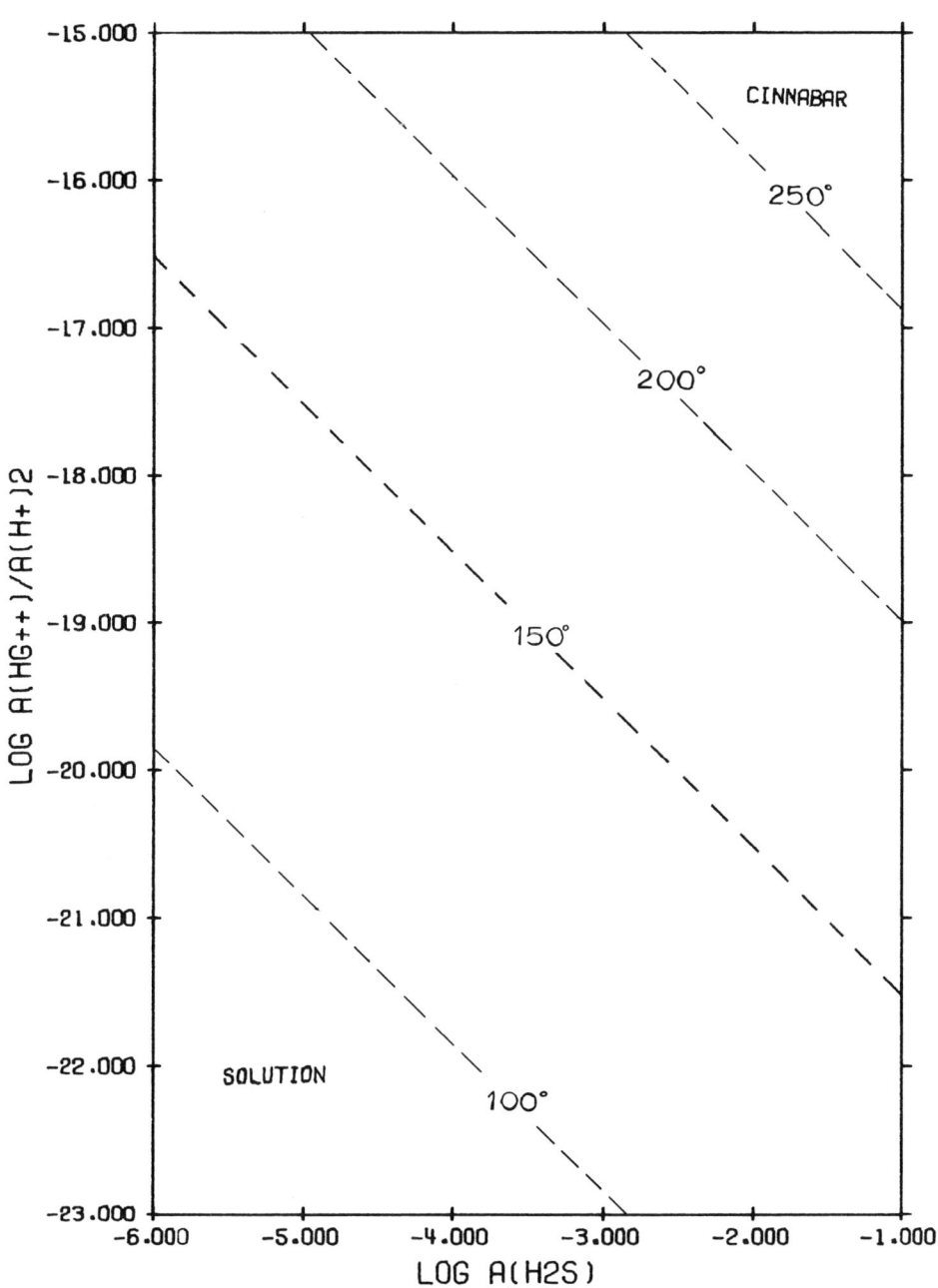

The System HCl—H₂O—HgS—H₂S at 100°, 150°, 200°, and 250°C.

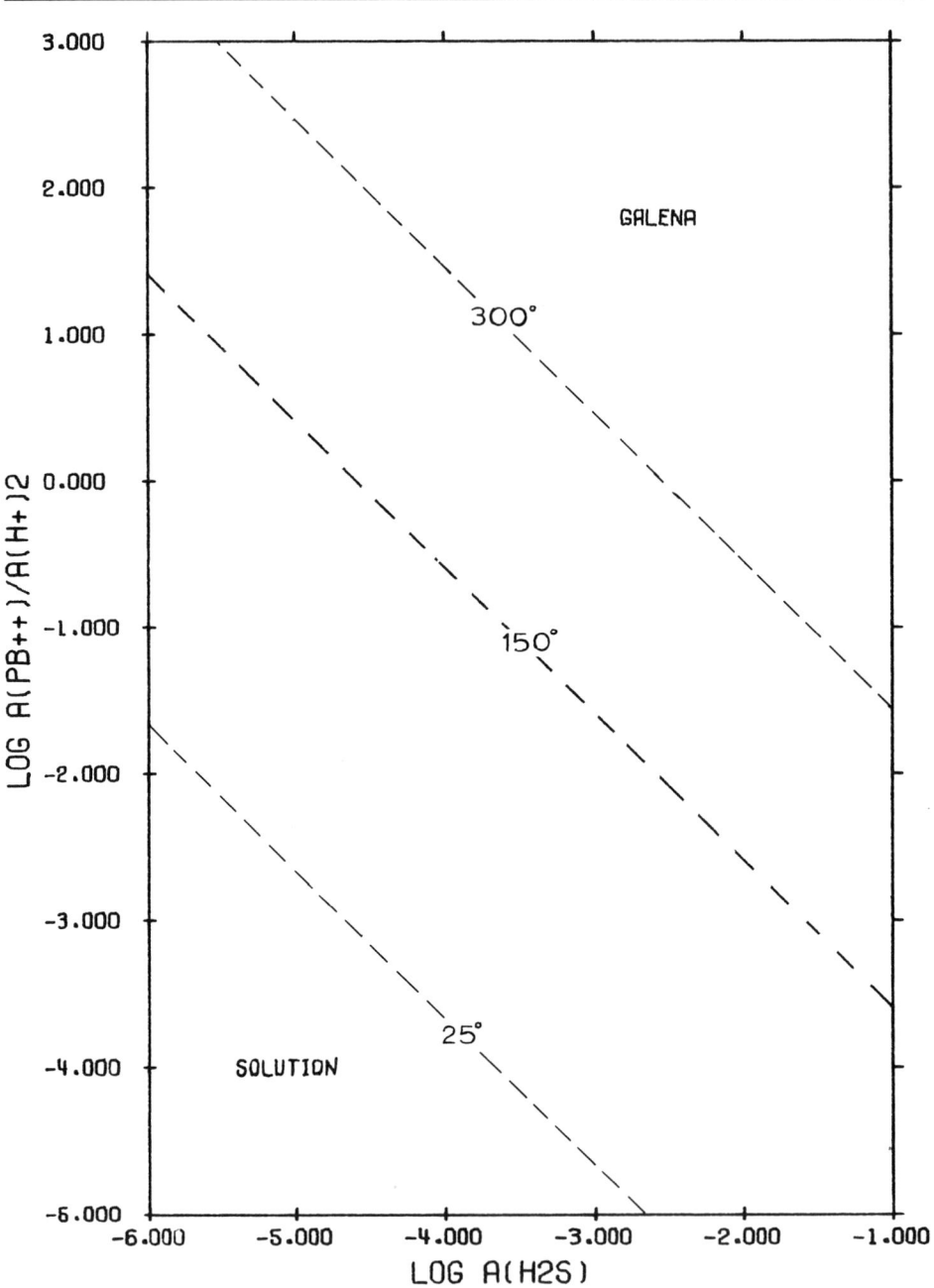

The System HCl—H$_2$O—H$_2$S—PbS at 25°, 150°, and 300°C.

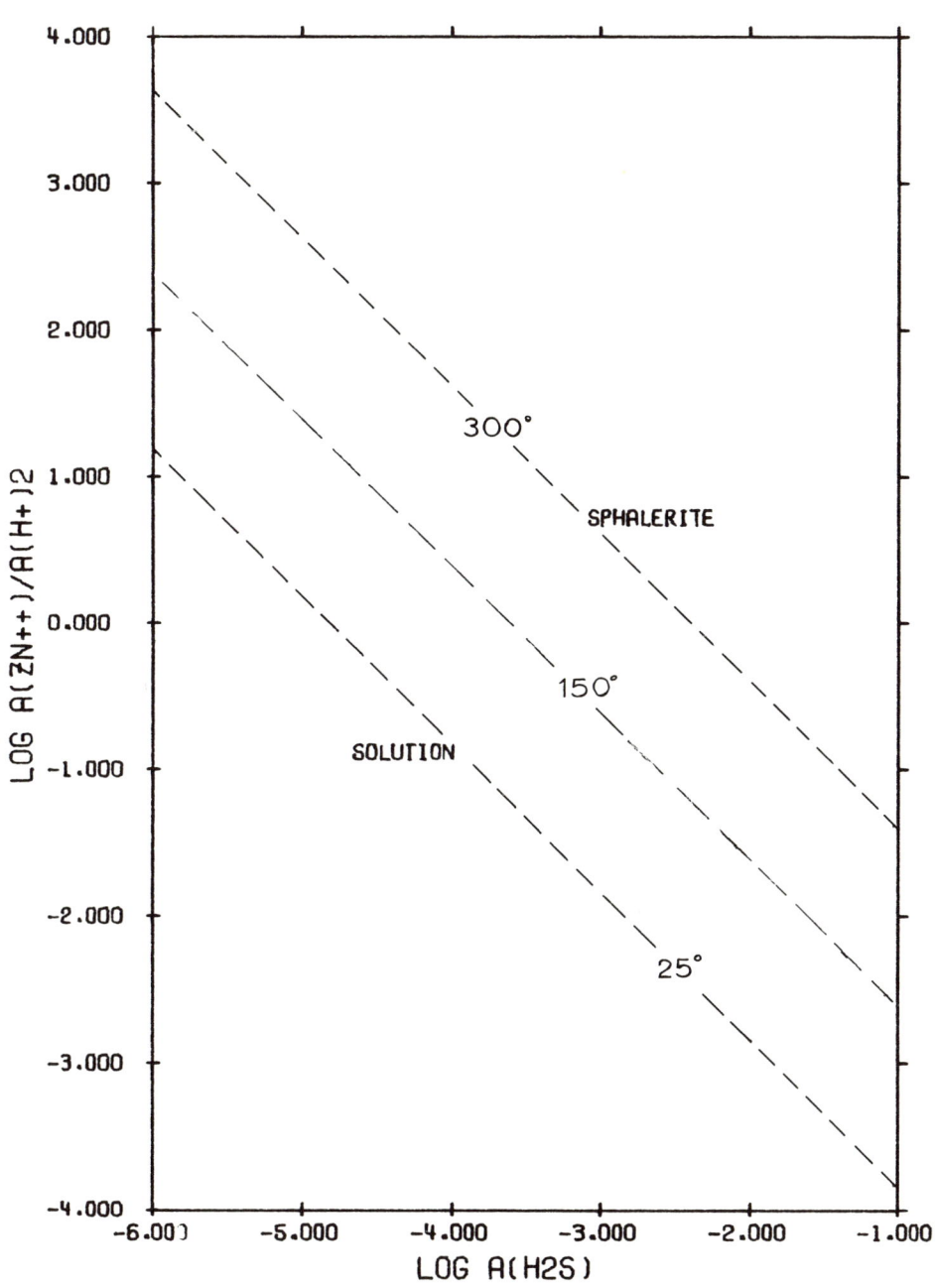

The System HCl—H$_2$O—H$_2$S—ZnS at 25°, 150°, and 300°C.

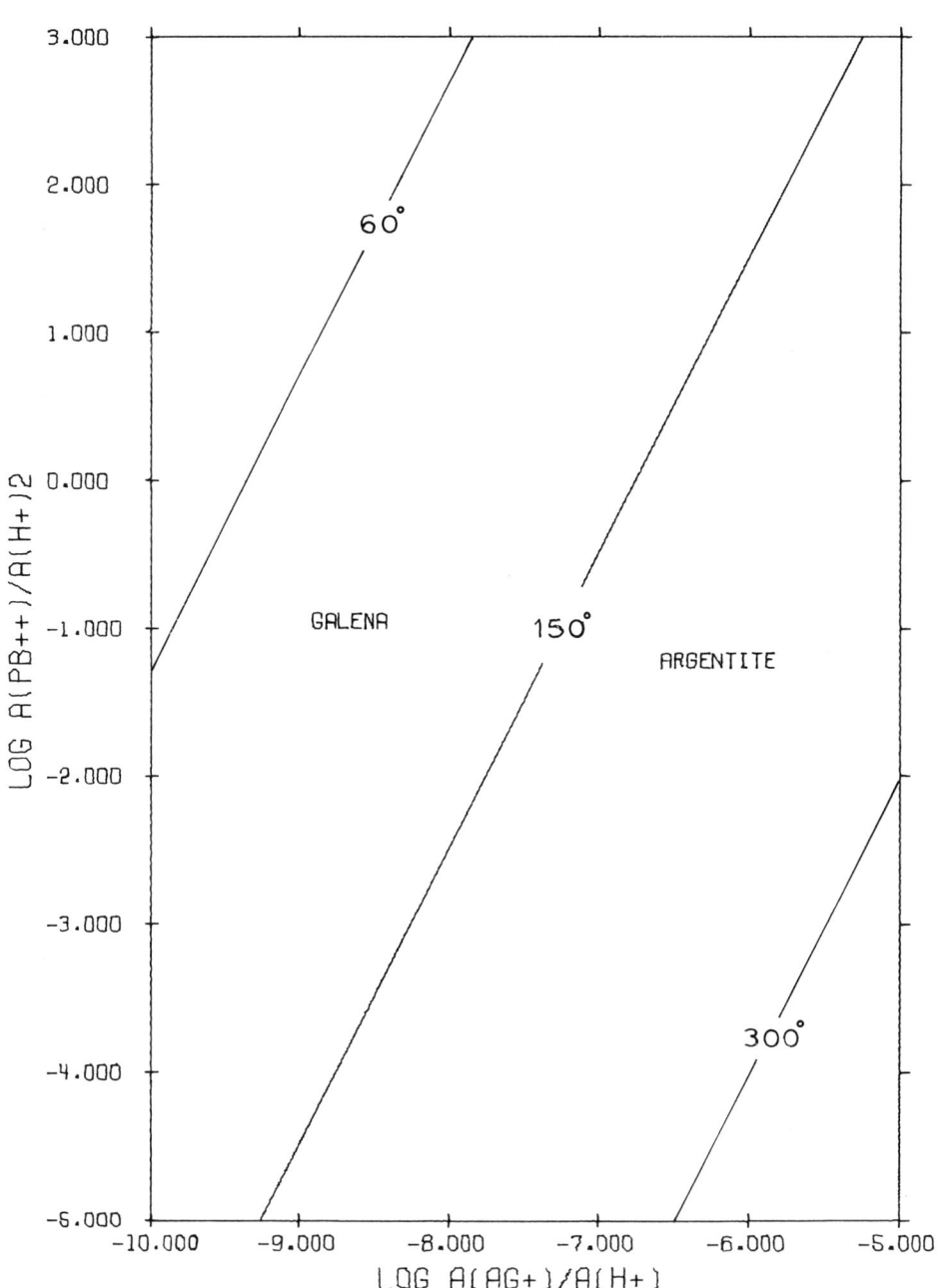

The System HCl—H₂O—Ag₂S—PbS at 60°, 150°, and 300°C.

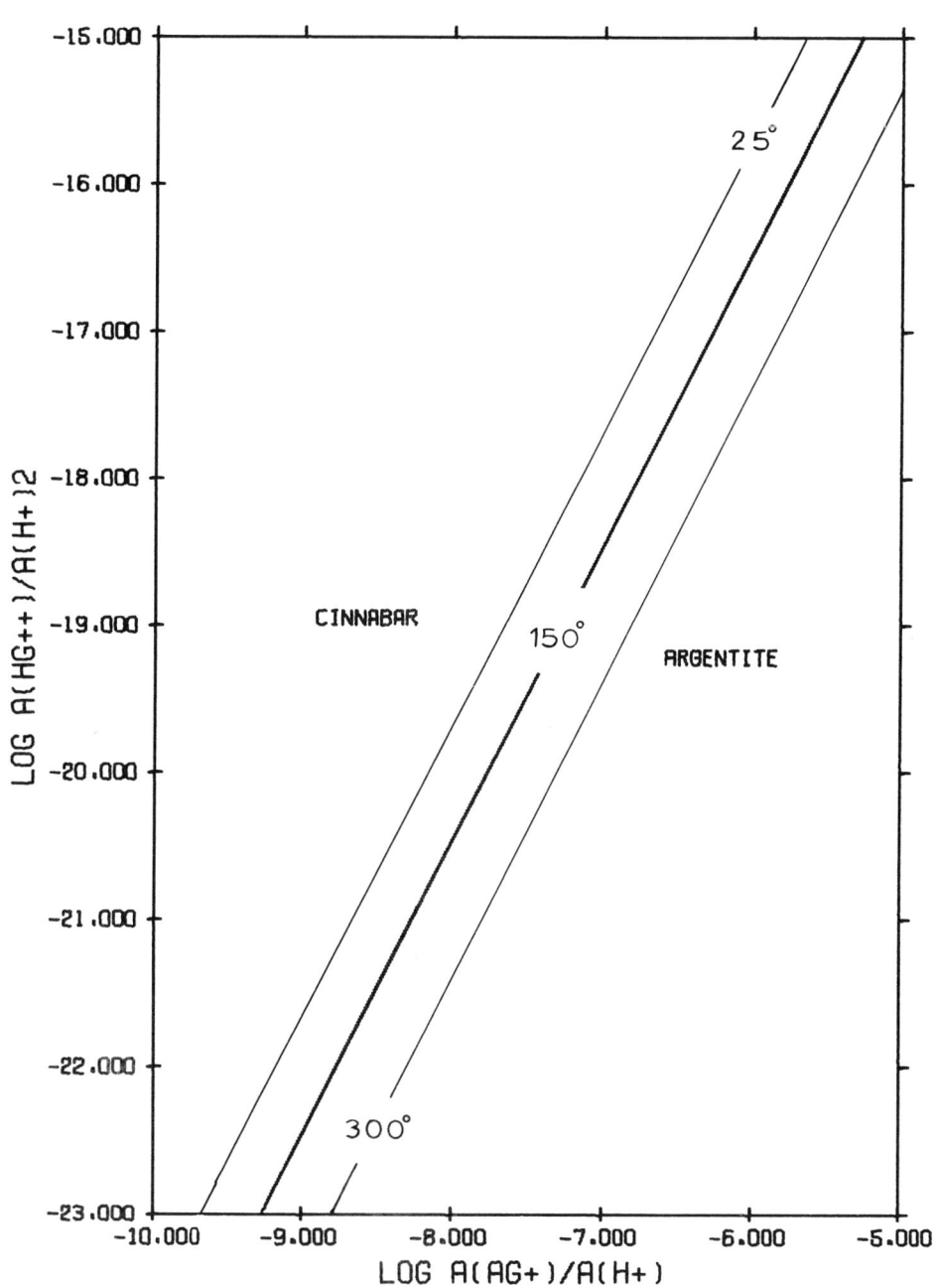

The System HCl—H₂O—Ag₂S—HgS at 25°, 150°, and 300°C.

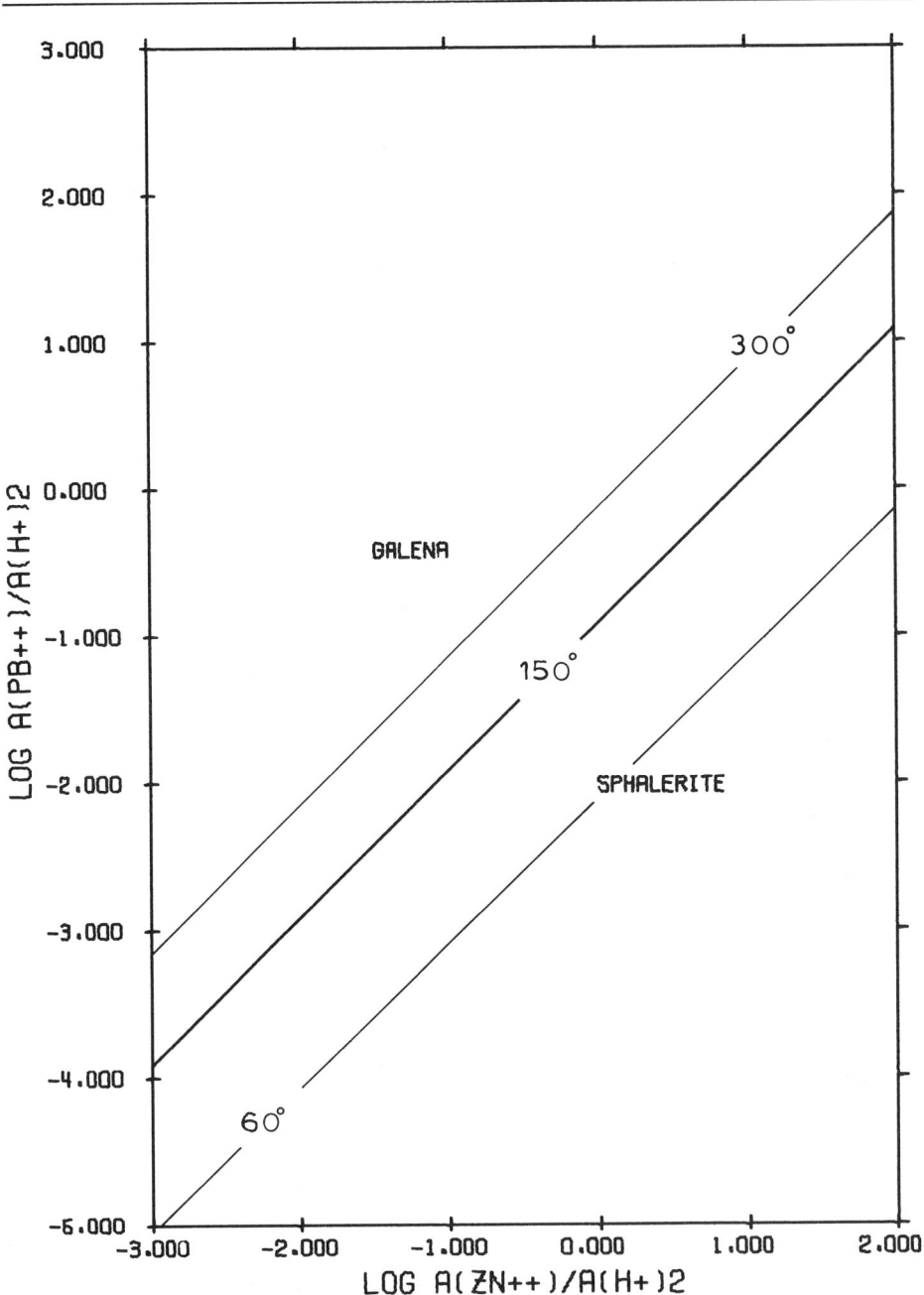

The System HCl—H$_2$O—PbS—ZnS at 60°, 150°, and 300°C.

Activity Diagrams

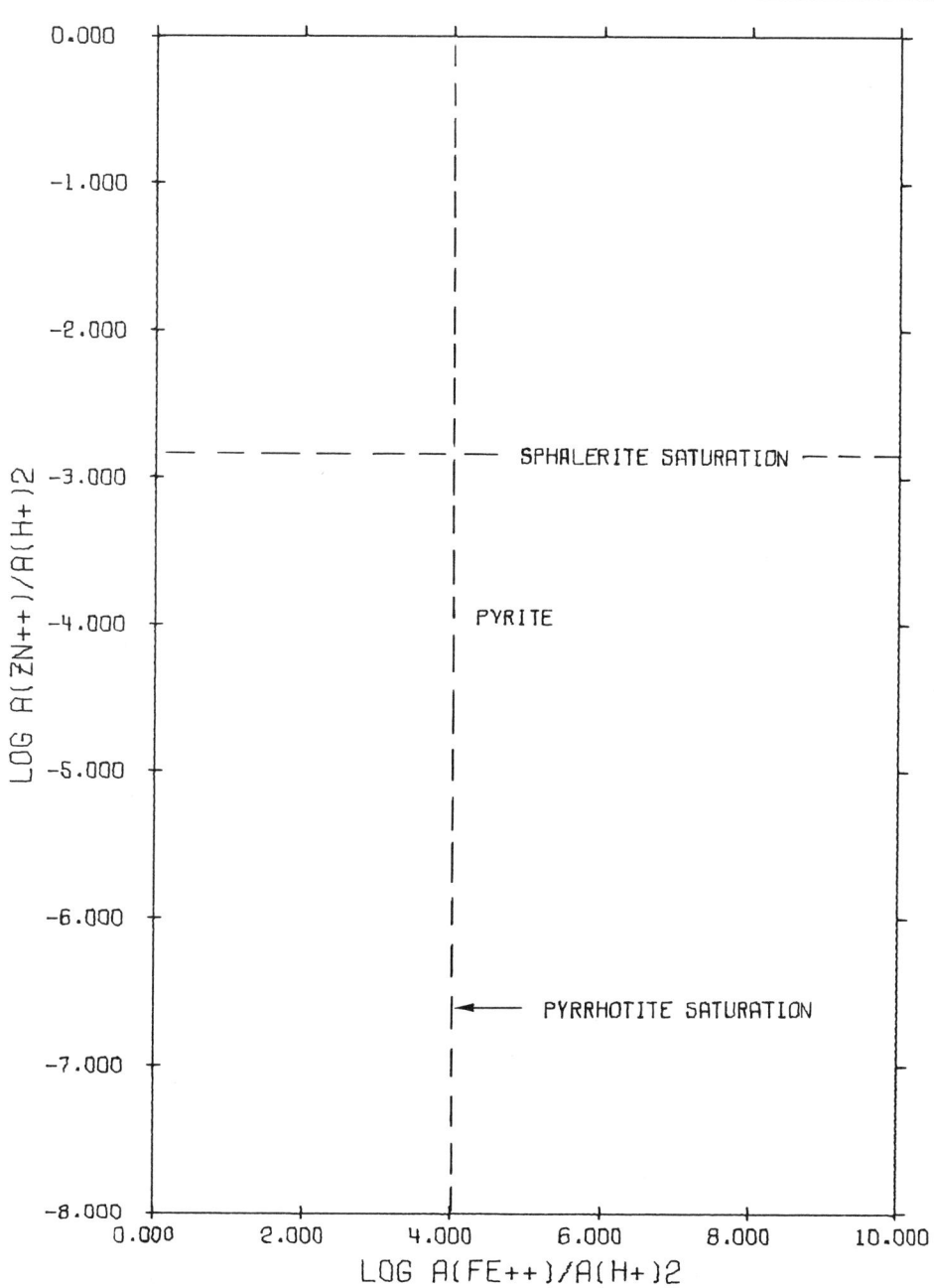

The System HCl—H$_2$O—FeS—H$_2$S—H$_2$SO$_4$—ZnS at 25°C; log a_{H_2S} = −2.00.

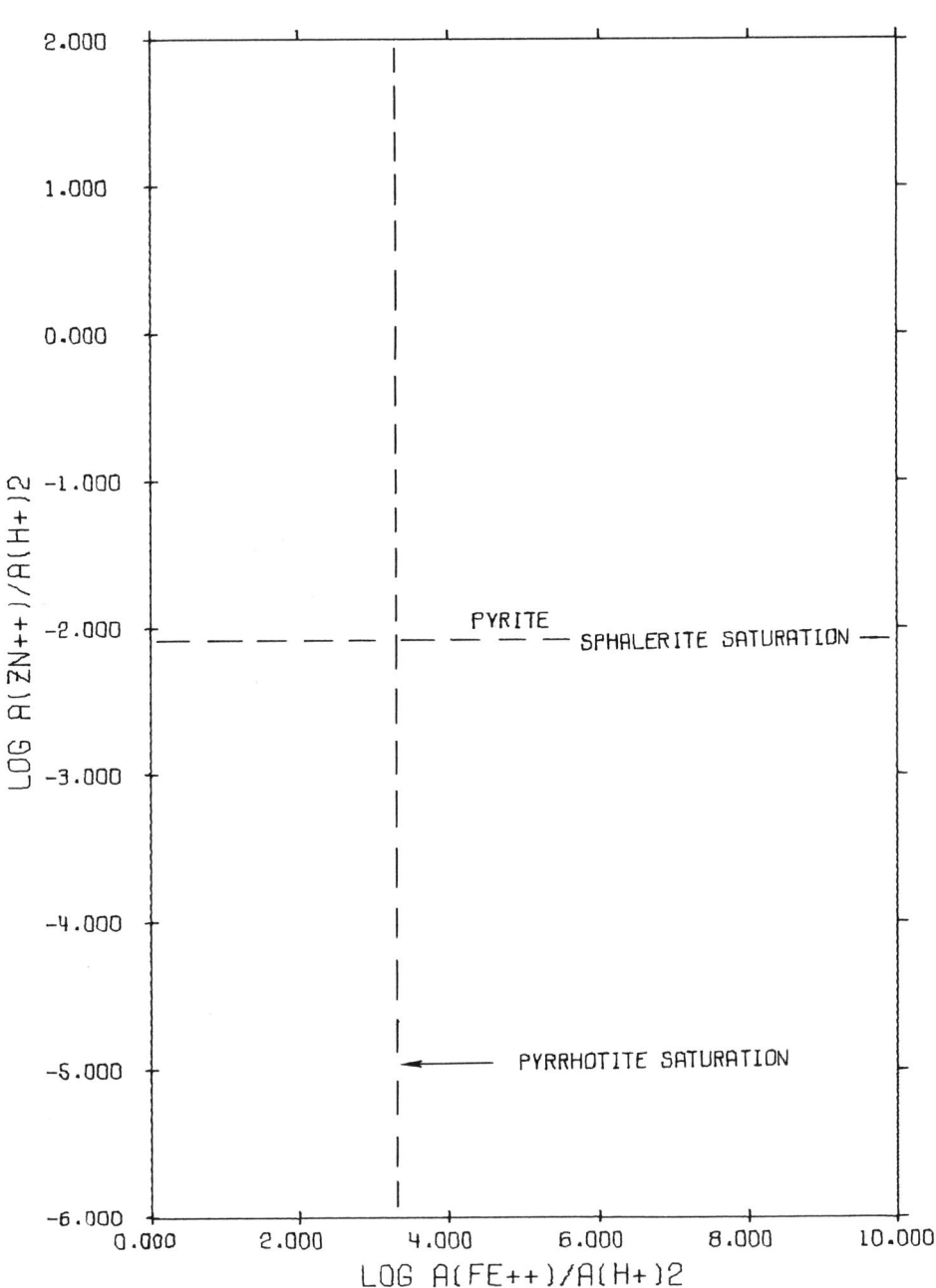

The System HCl—H$_2$O—FeS—H$_2$S—H$_2$SO$_4$—ZnS at 100°C; log a_{H_2S} = −2.00.

Activity Diagrams

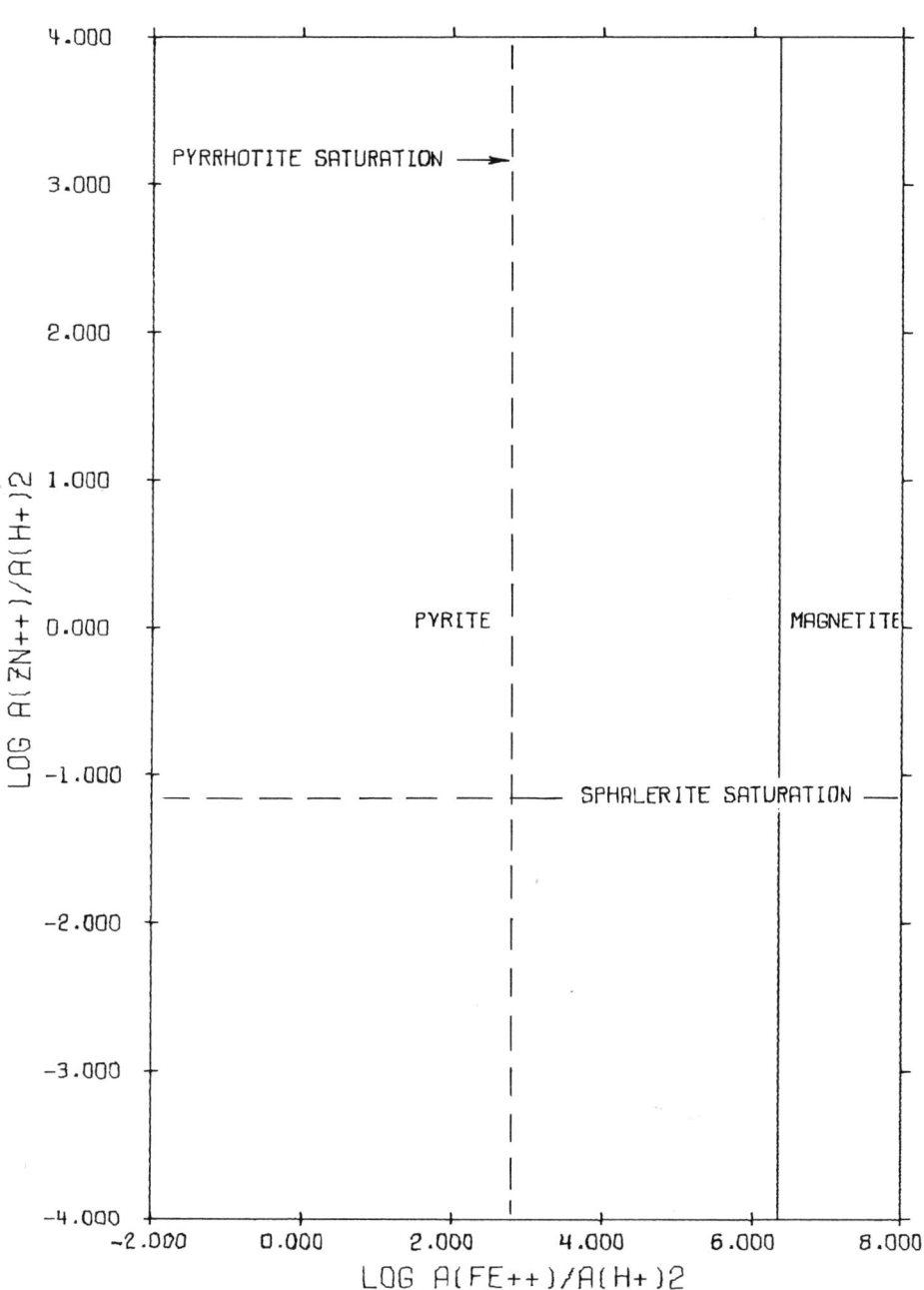

The System HCl—H$_2$O—FeS—H$_2$S—H$_2$SO$_4$—ZnS at 200°C; $\log a_{H_2S} = -2.00$.

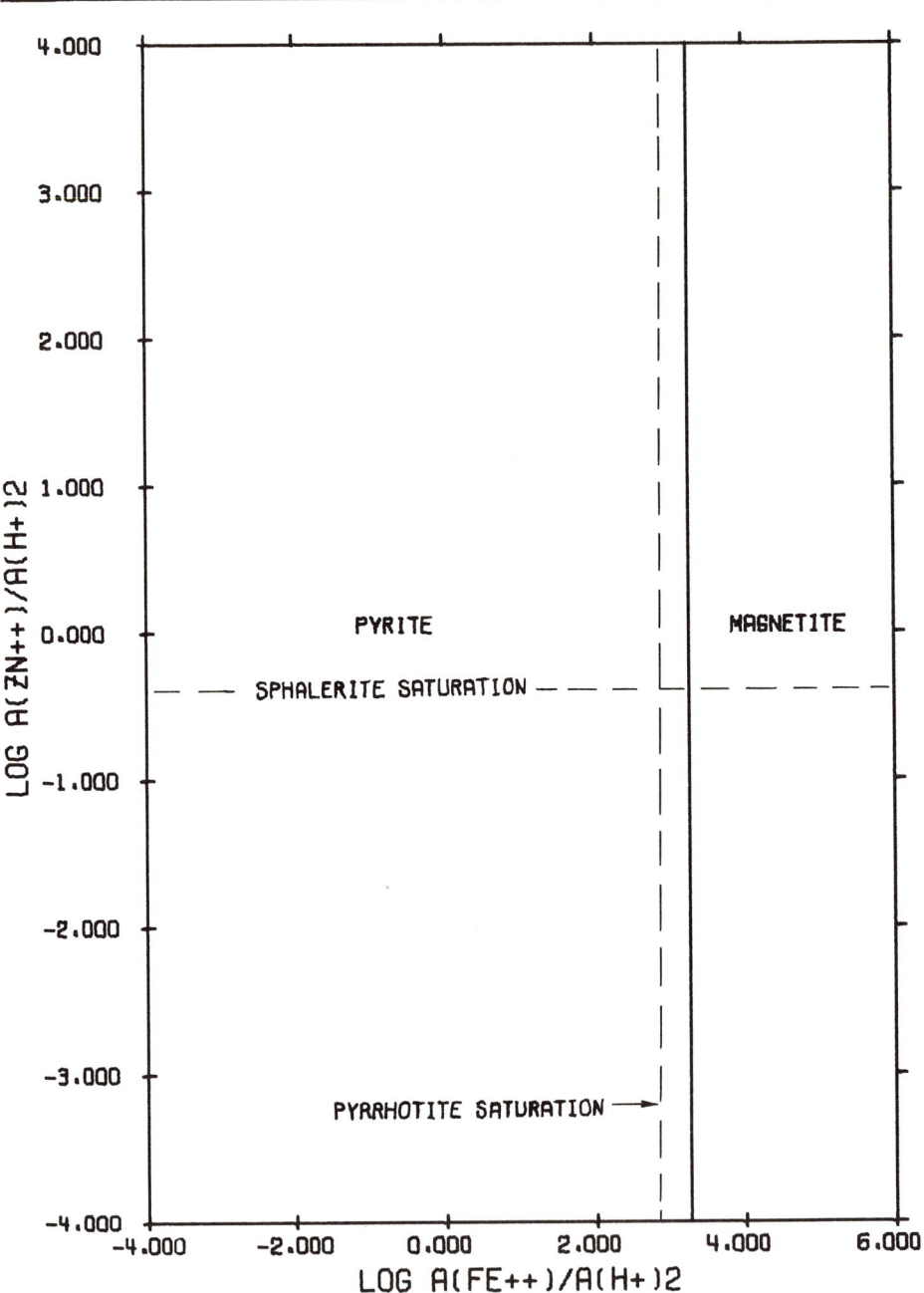

The System HCl—H$_2$O—FeS—H$_2$S—H$_2$SO$_4$—ZnS at 300°C; log a_{H_2S} = −2.00.

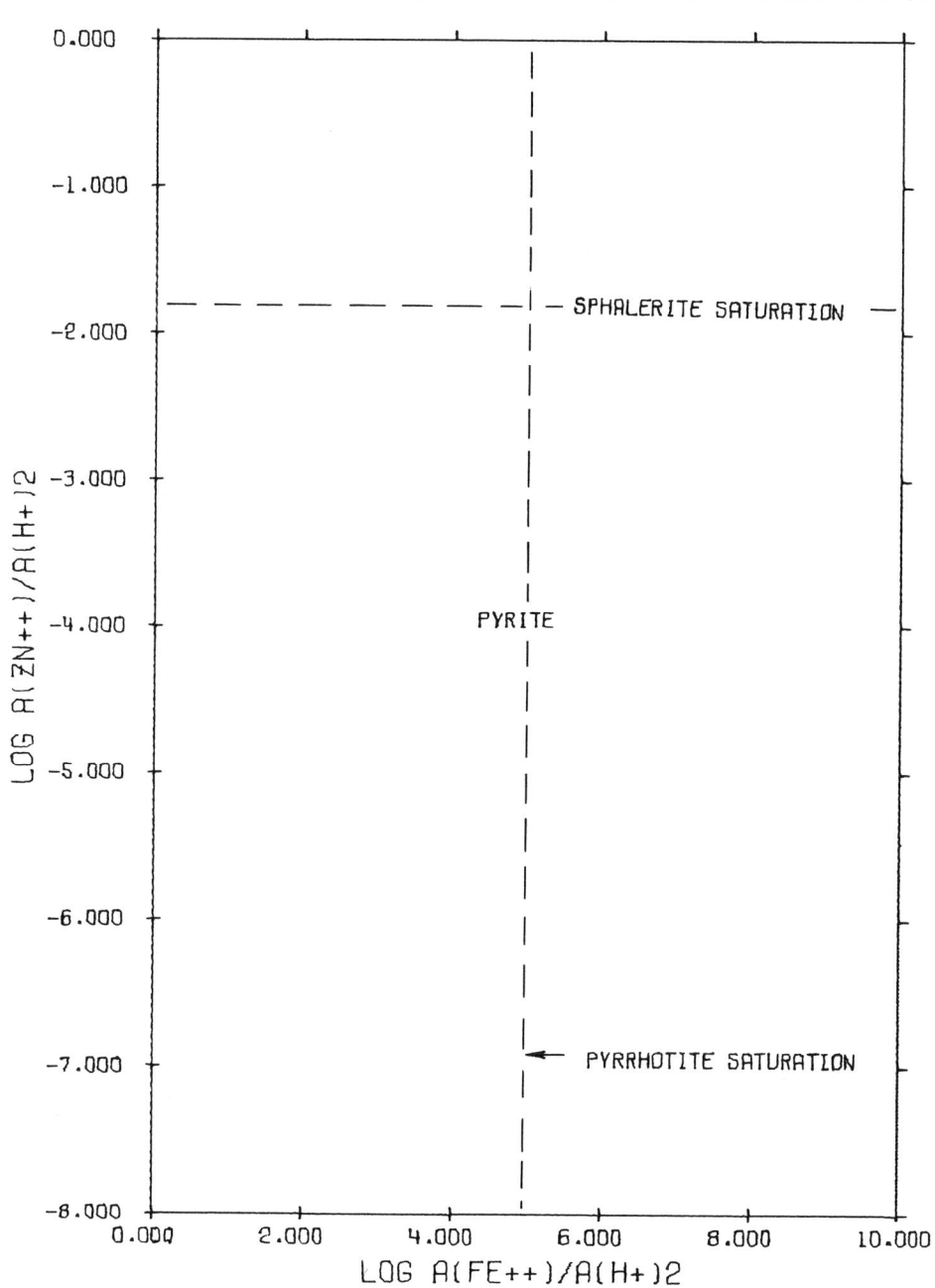

The System HCl—H_2O—FeS—H_2S—H_2SO_4—ZnS at 25°C; log a_{H_2S} = −3.00.

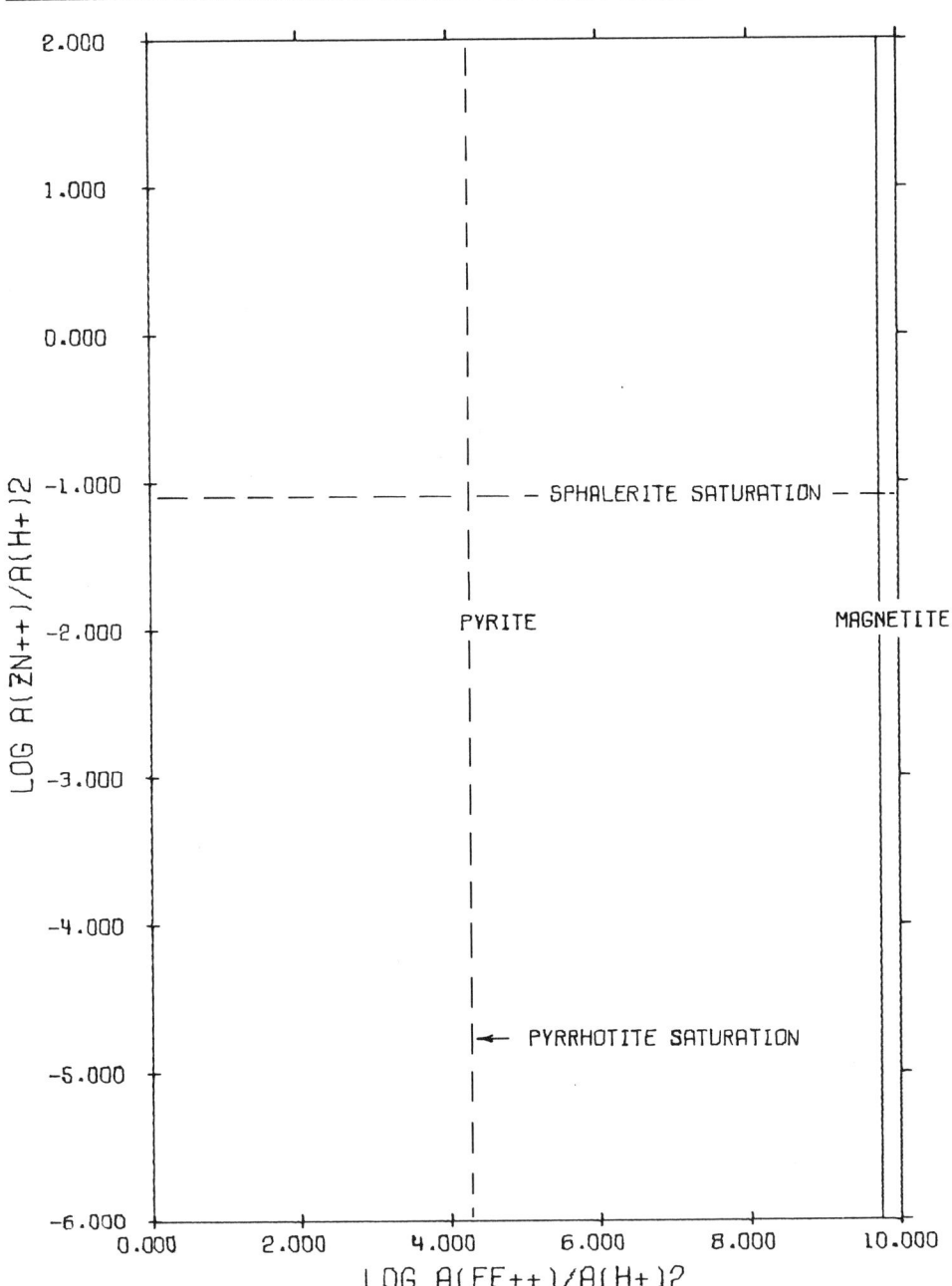

The System HCl—H$_2$O—FeS—H$_2$S—H$_2$SO$_4$—ZnS at 100°C; log a_{H_2S} = −3.00.

Activity Diagrams

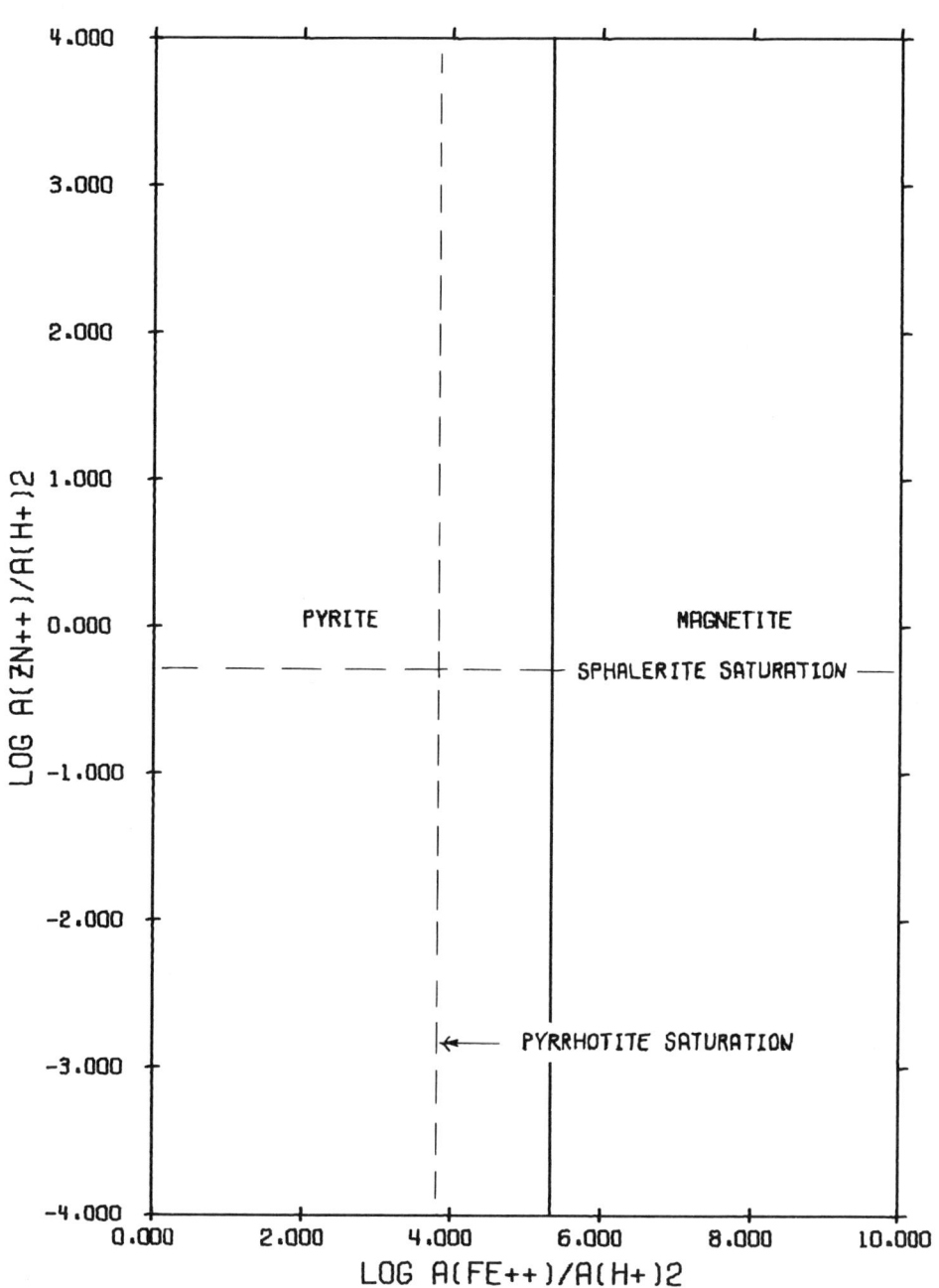

The System HCl—H$_2$O—FeS—H$_2$S—H$_2$SO$_4$—ZnS at 200°C; log a_{H_2S} = −3.00.

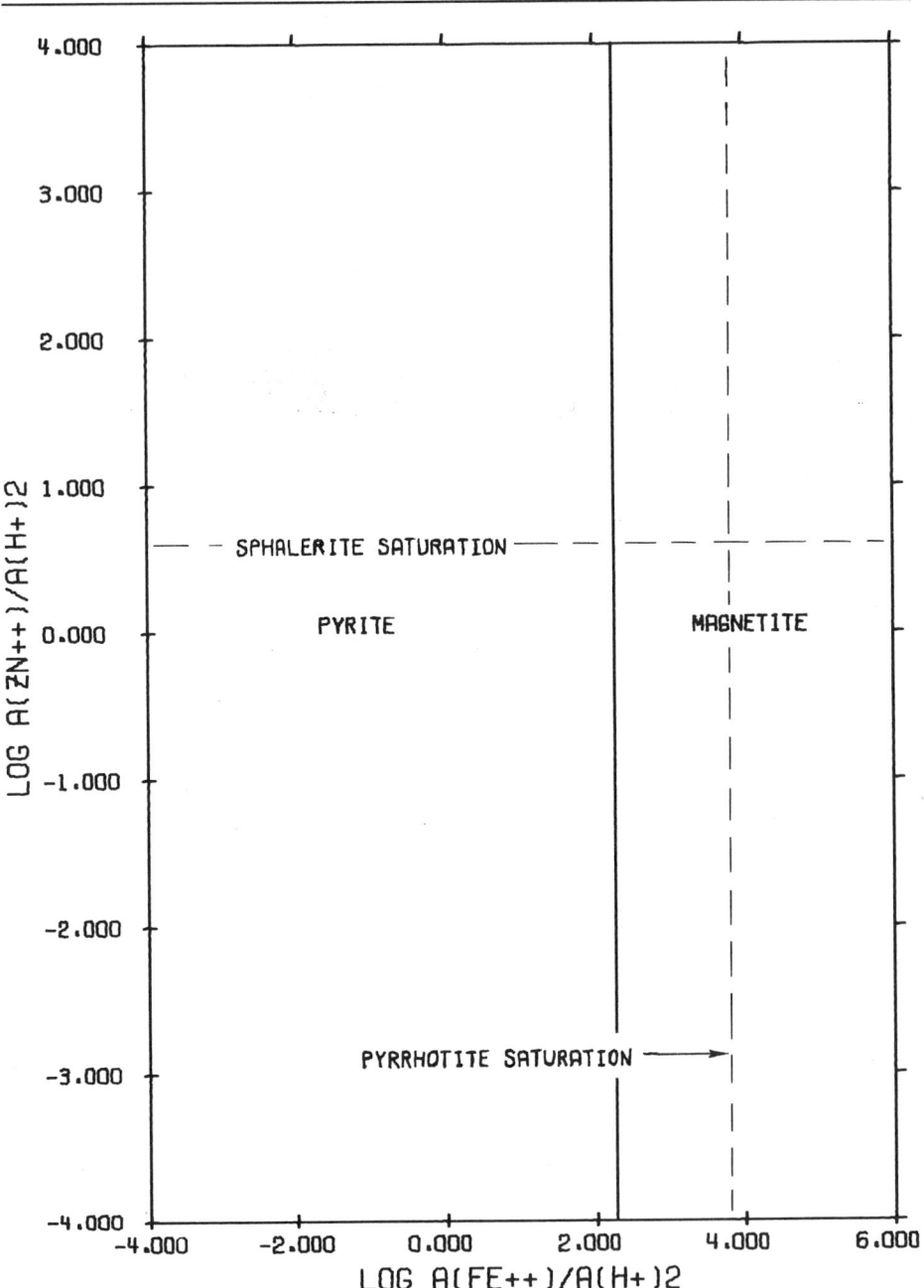

The System HCl—H$_2$O—FeS—H$_2$S—H$_2$SO$_4$—ZnS at 300°C; log $a_{H_2S} = -3.00$.

Activity Diagrams

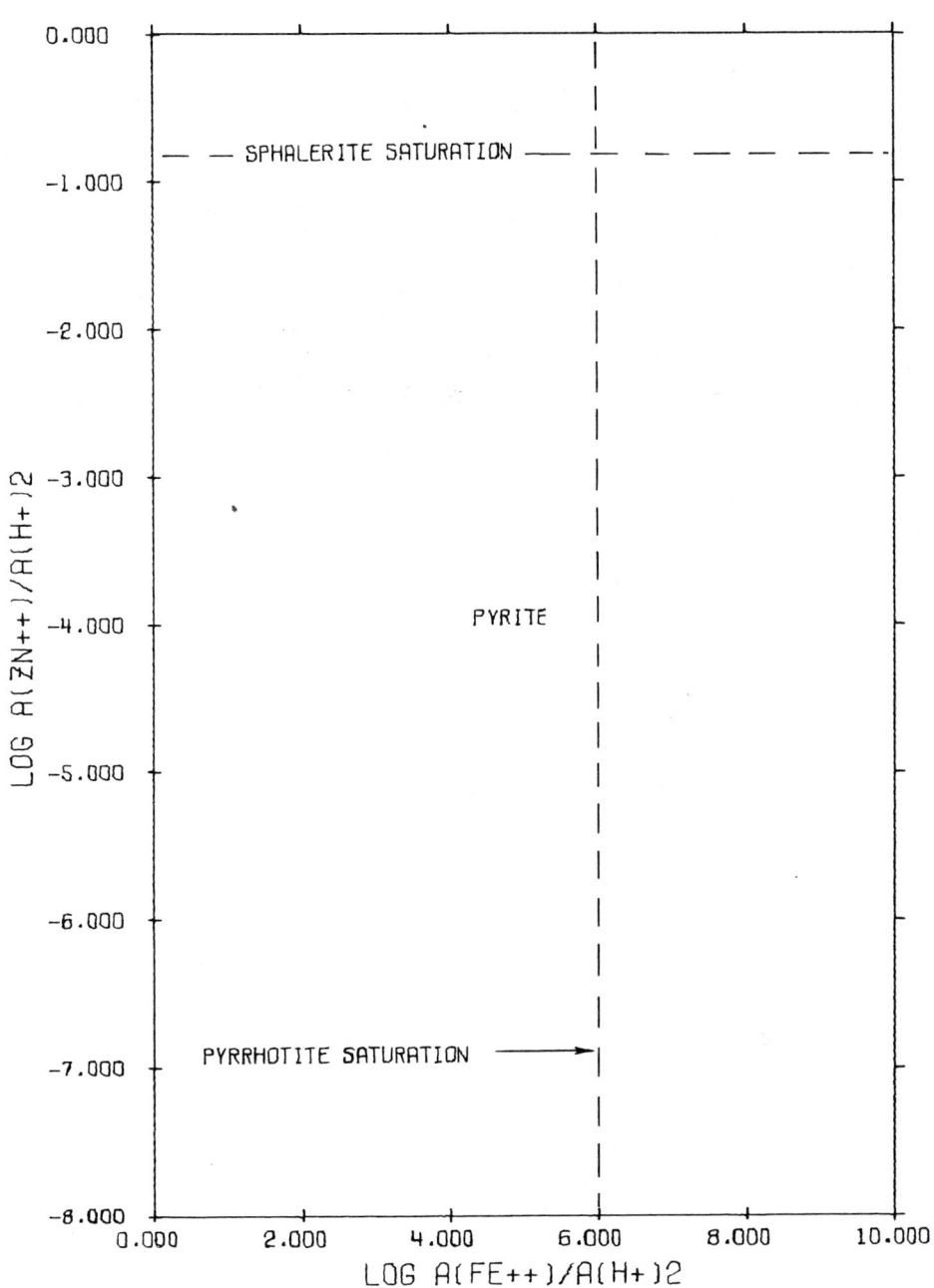

The System HCl—H$_2$O—FeS—H$_2$S—H$_2$SO$_4$—ZnS at 25°C; log $a_{H_2S} = -4.00$.

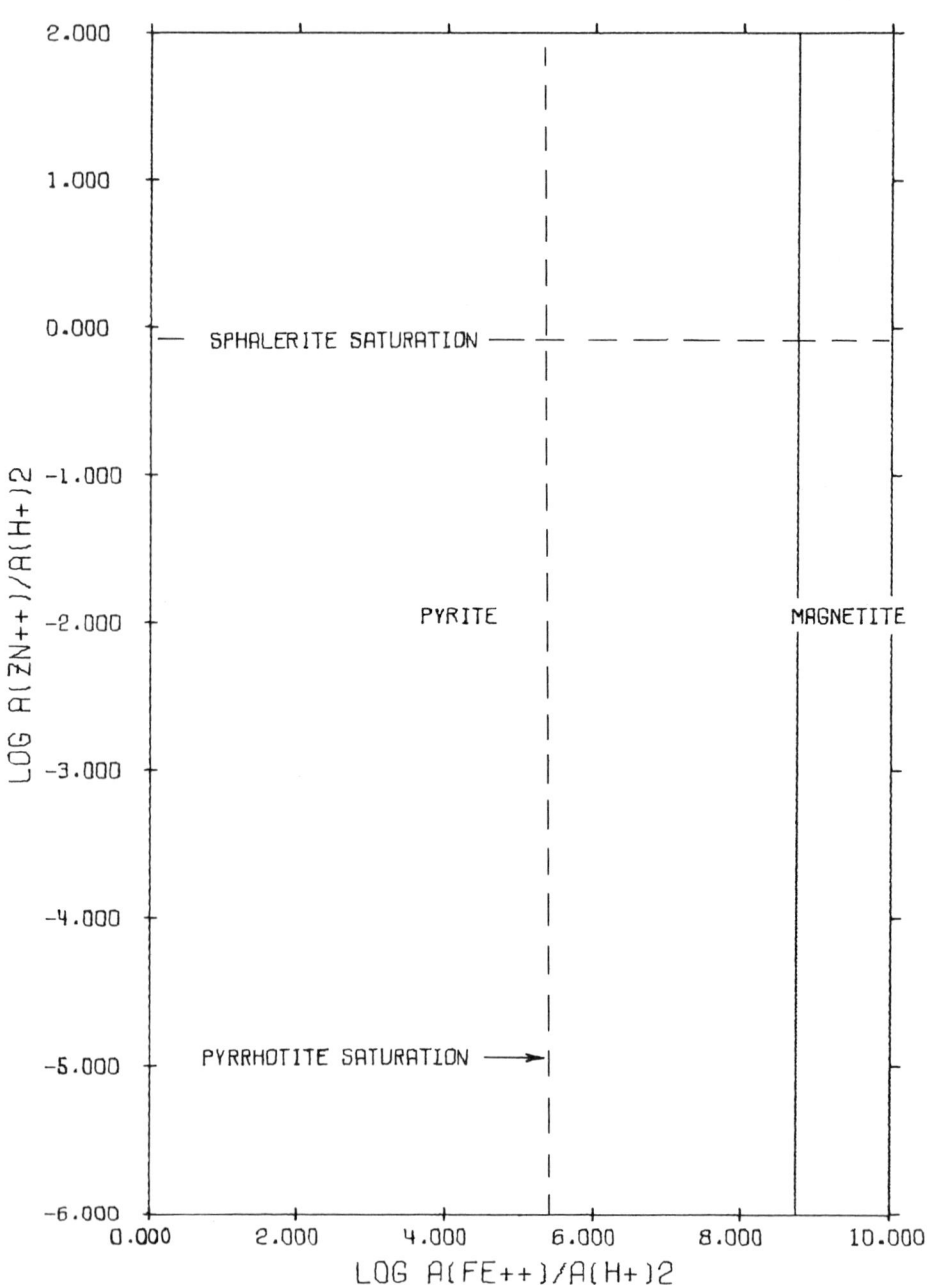

The System HCl—H_2O—FeS—H_2S—H_2SO_4—ZnS at 100°C; $\log a_{H_2S} = -4.00$.

Activity Diagrams

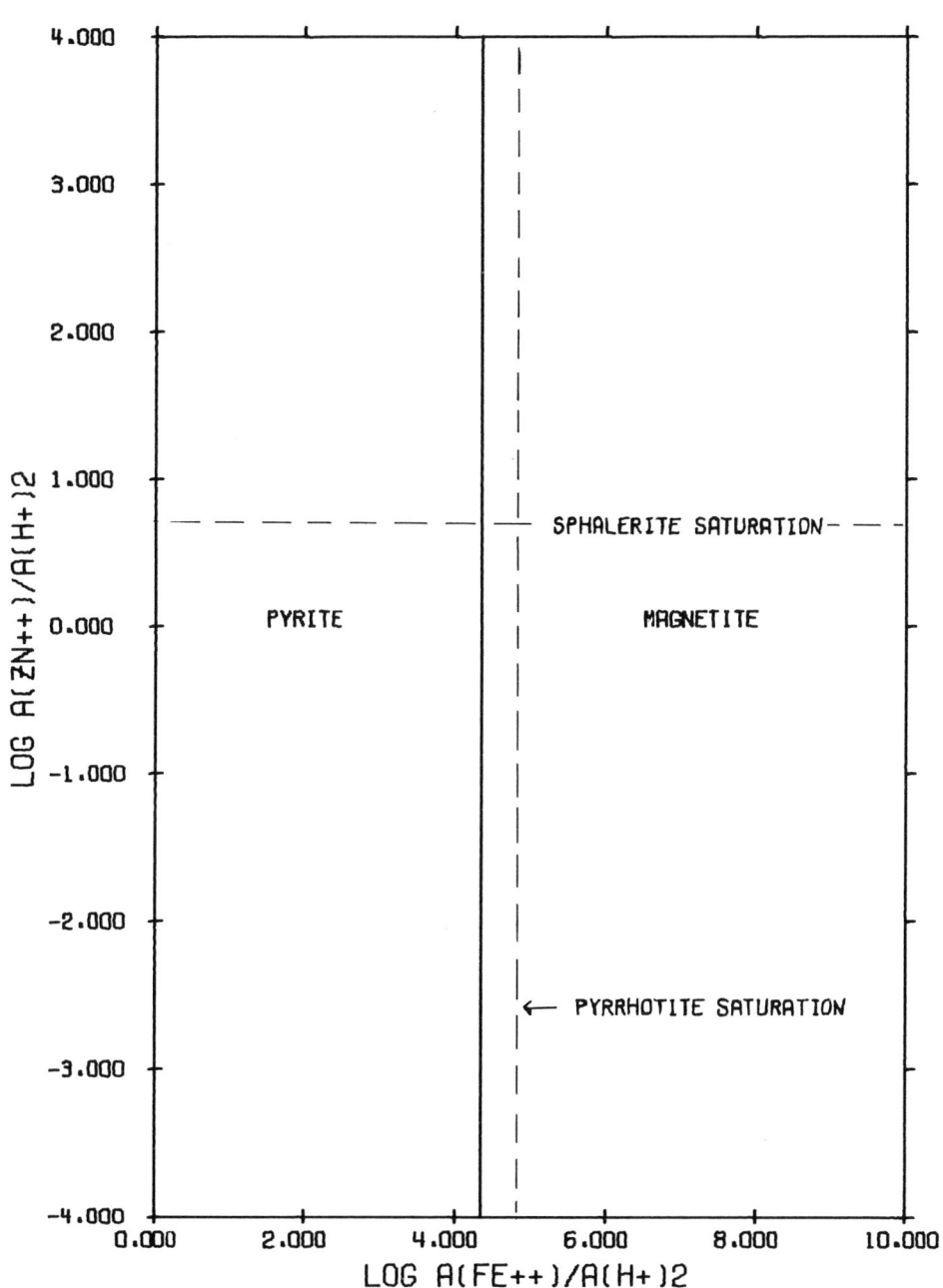

The System HCl—H$_2$O—FeS—H$_2$S—H$_2$SO$_4$—ZnS at 200°C; log $a_{\text{H}_2\text{S}}$ = −4.00.

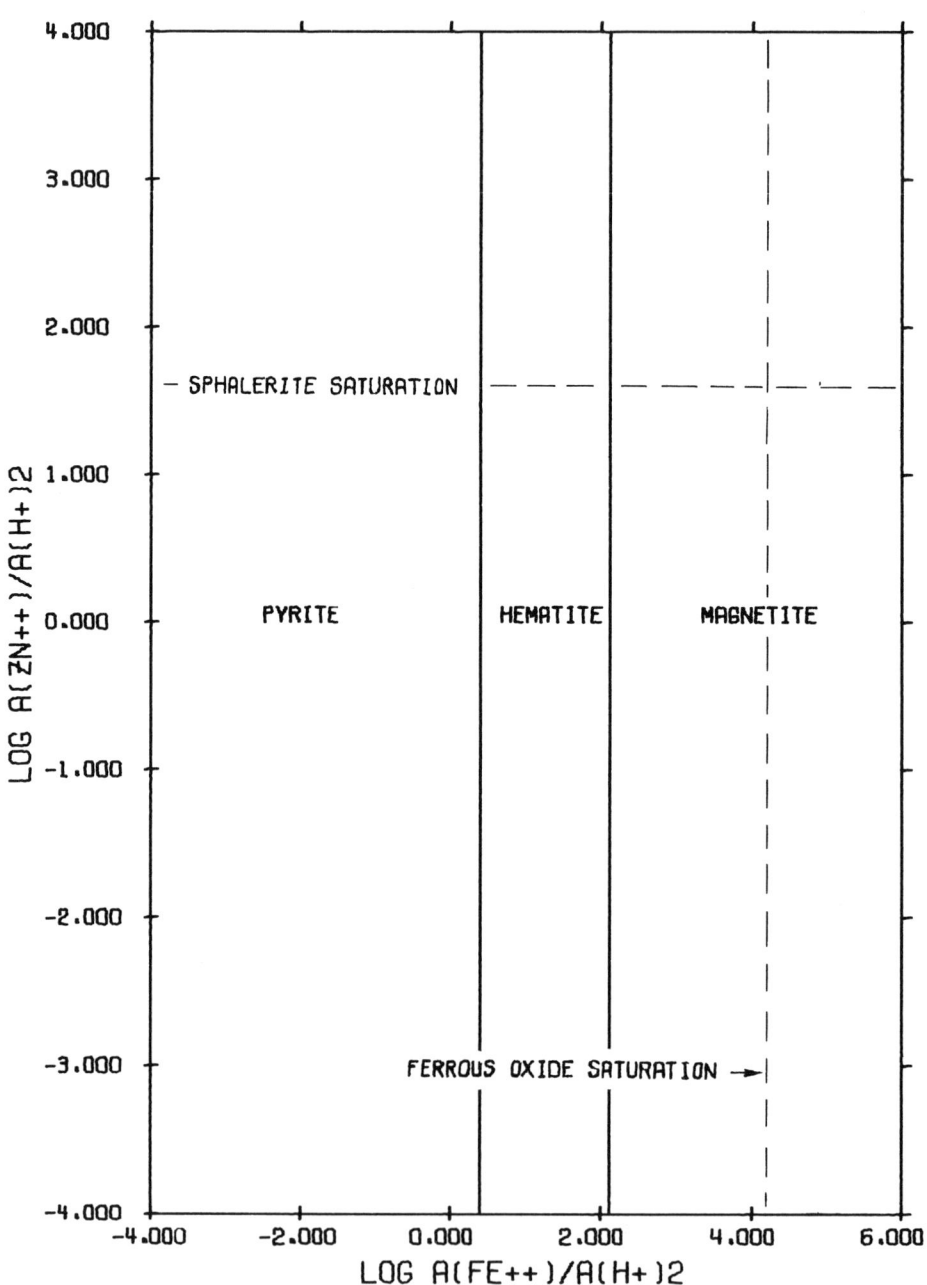

The System HCl—H$_2$O—FeS—H$_2$S—H$_2$SO$_4$—ZnS at 300°C; log $a_{H_2S} = -4.00$.

FUGACITY DIAGRAMS

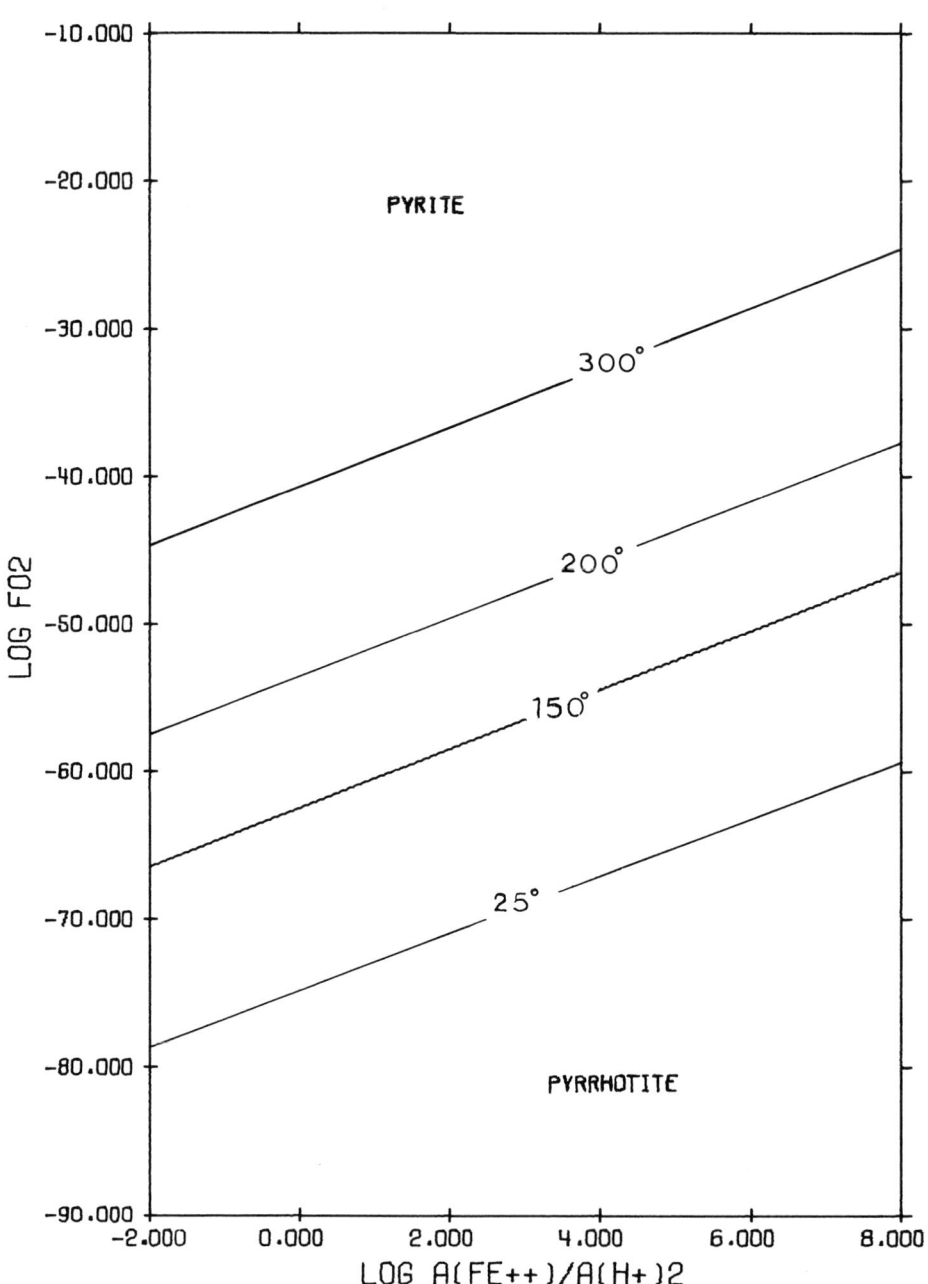

The System HCl—H₂O—FeS—H₂S—H₂SO₄ at 25°, 150°, 200°, and 300°C.

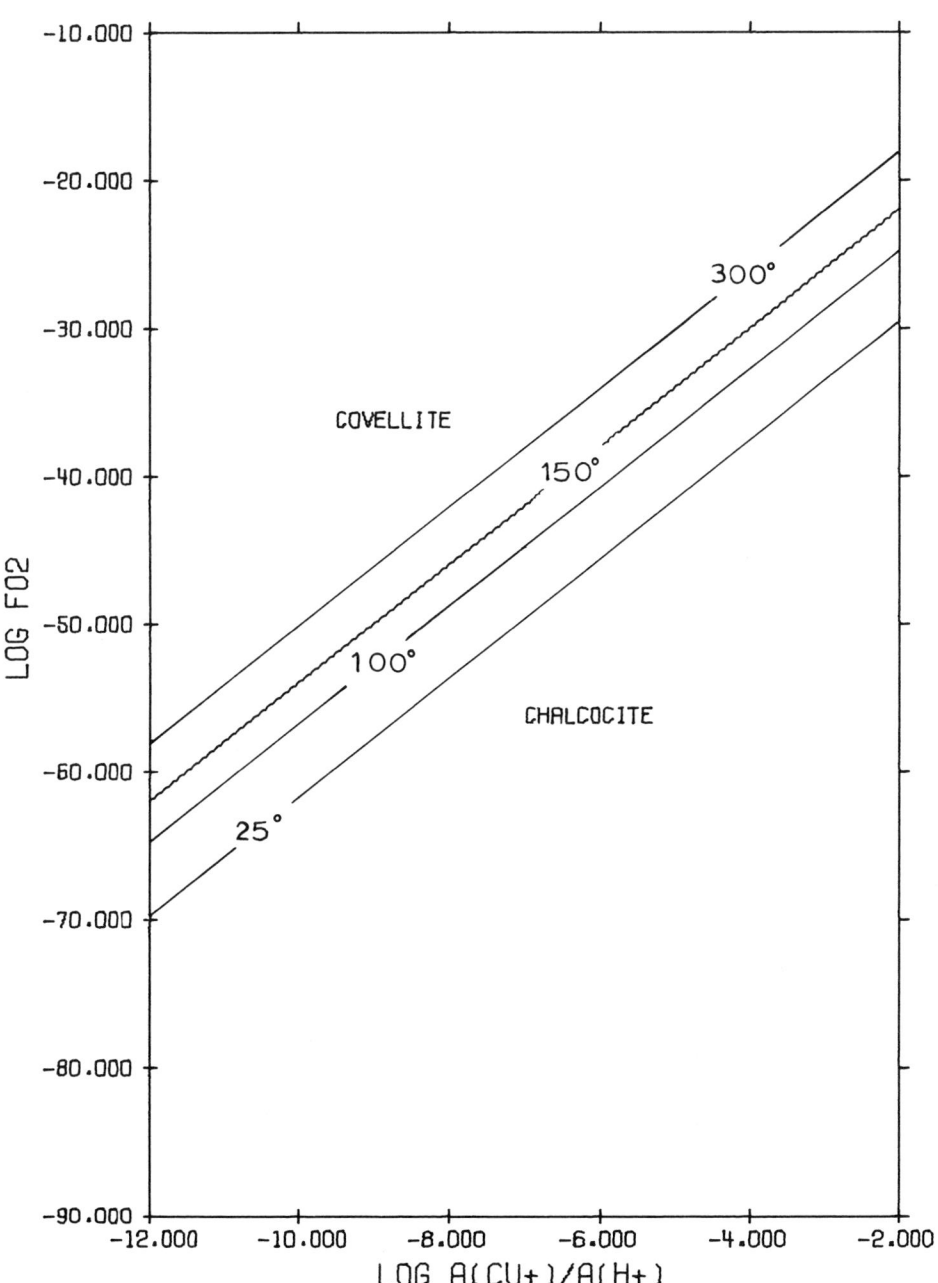

The System HCl—H_2O—Cu_2S—H_2S—H_2SO_4 at 25°, 100°, 150°, and 300°C.

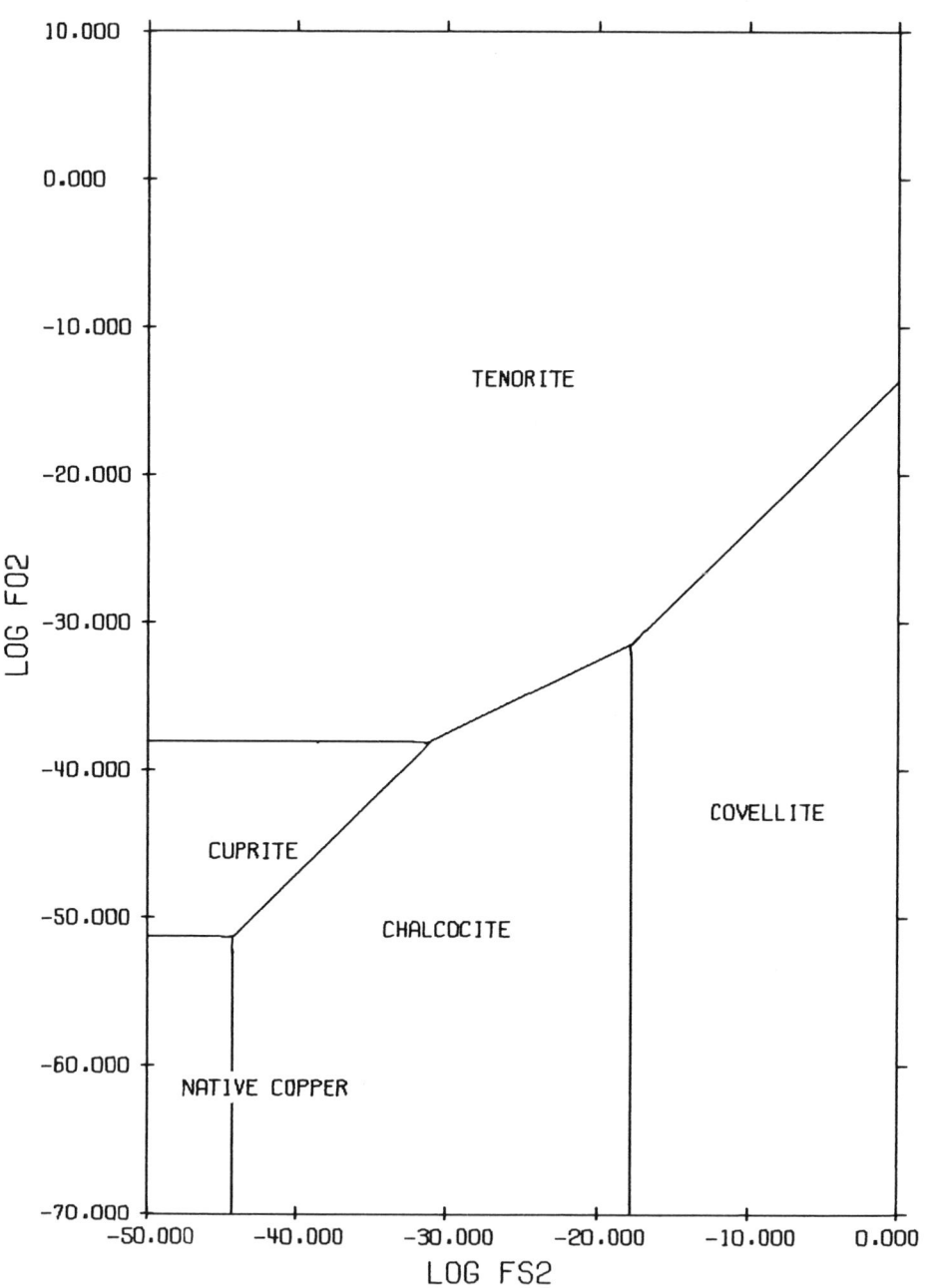

The System HCl—H$_2$O—Cu$_2$S—H$_2$S—H$_2$SO$_4$ at 25°C.

Fugacity Diagrams

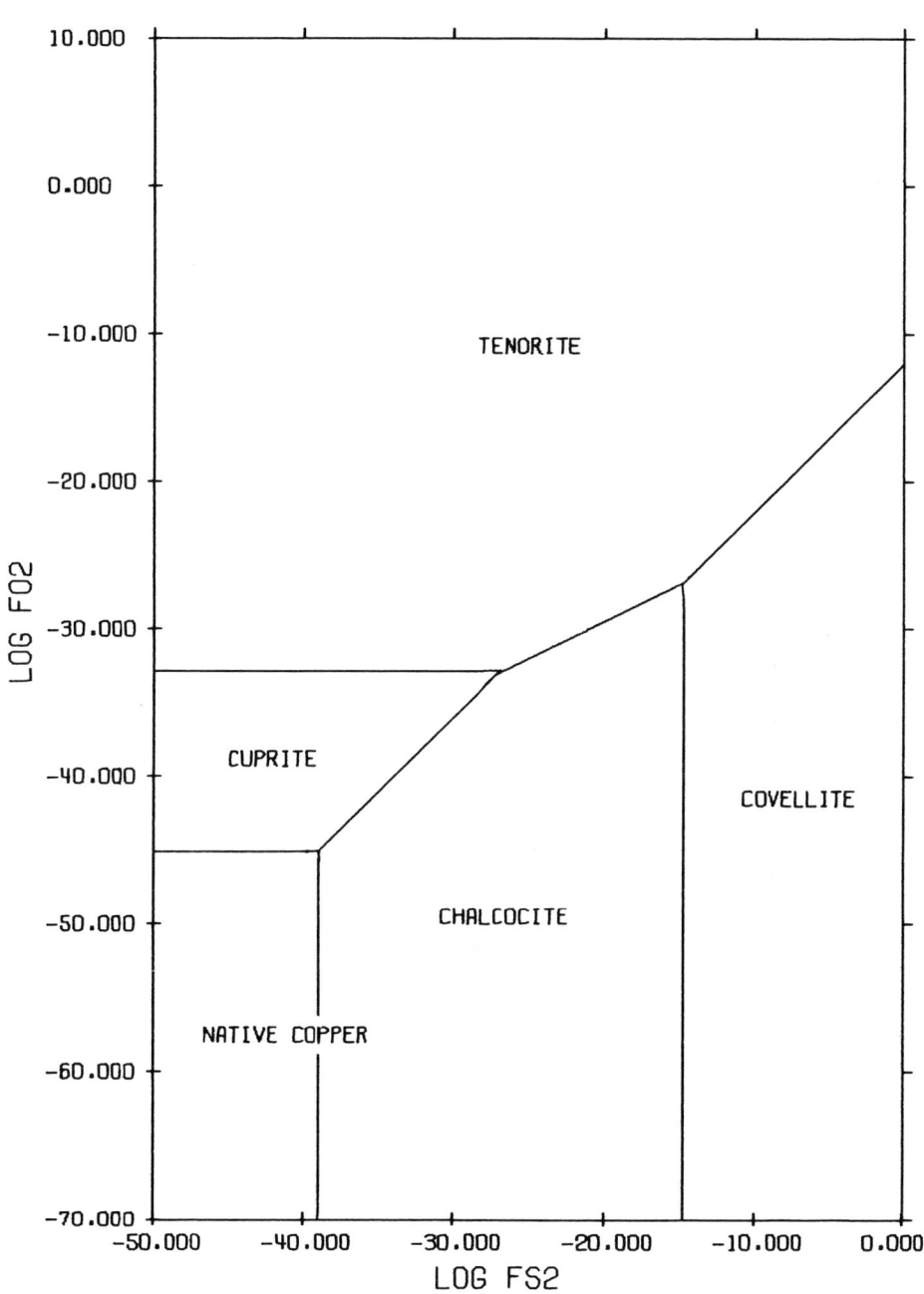

The System HCl—H_2O—Cu_2S—H_2S—H_2SO_4 at 60°C.

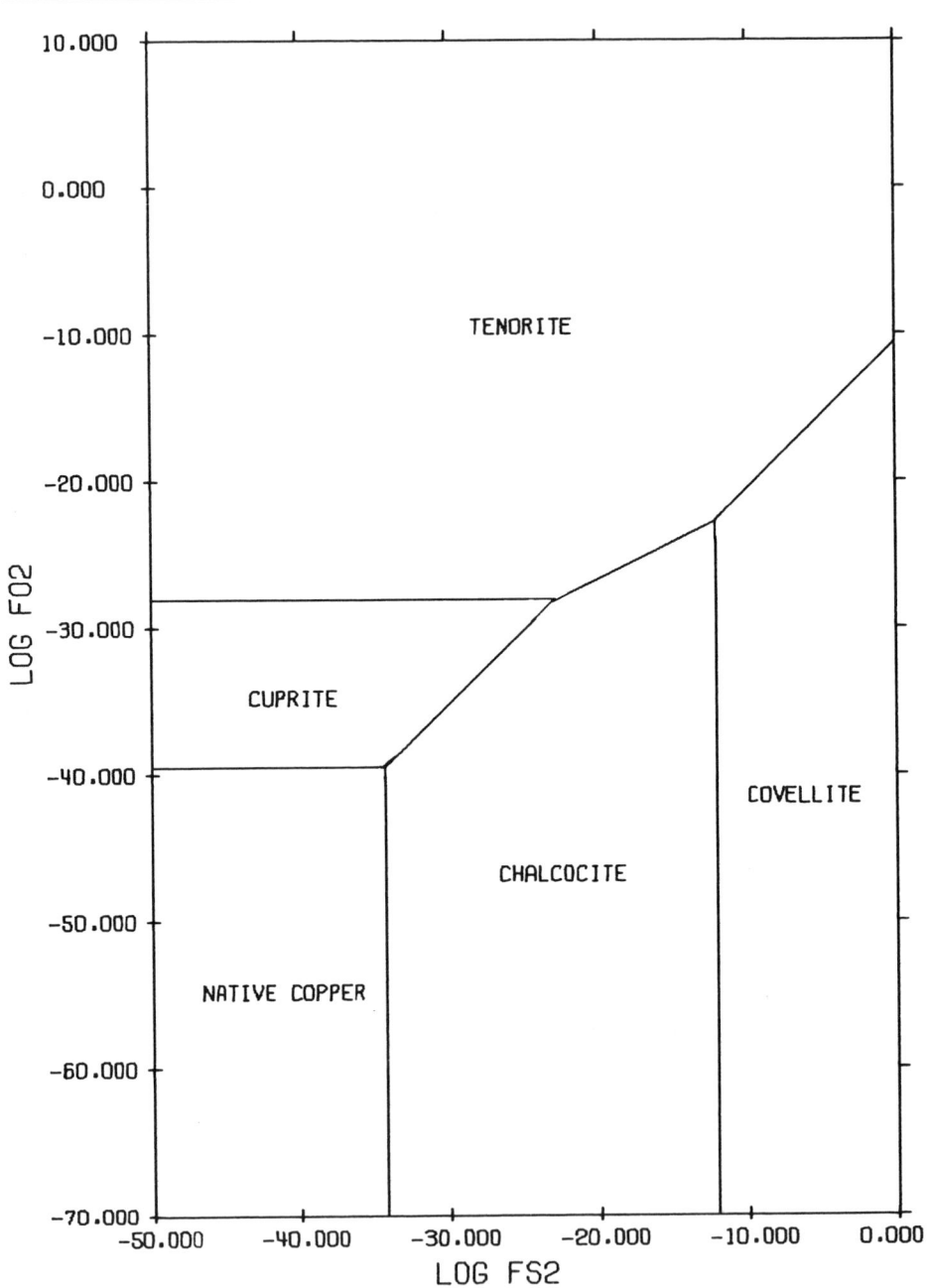

The System HCl—H$_2$O—Cu$_2$S—H$_2$S—H$_2$SO$_4$ at 100°C.

Fugacity Diagrams

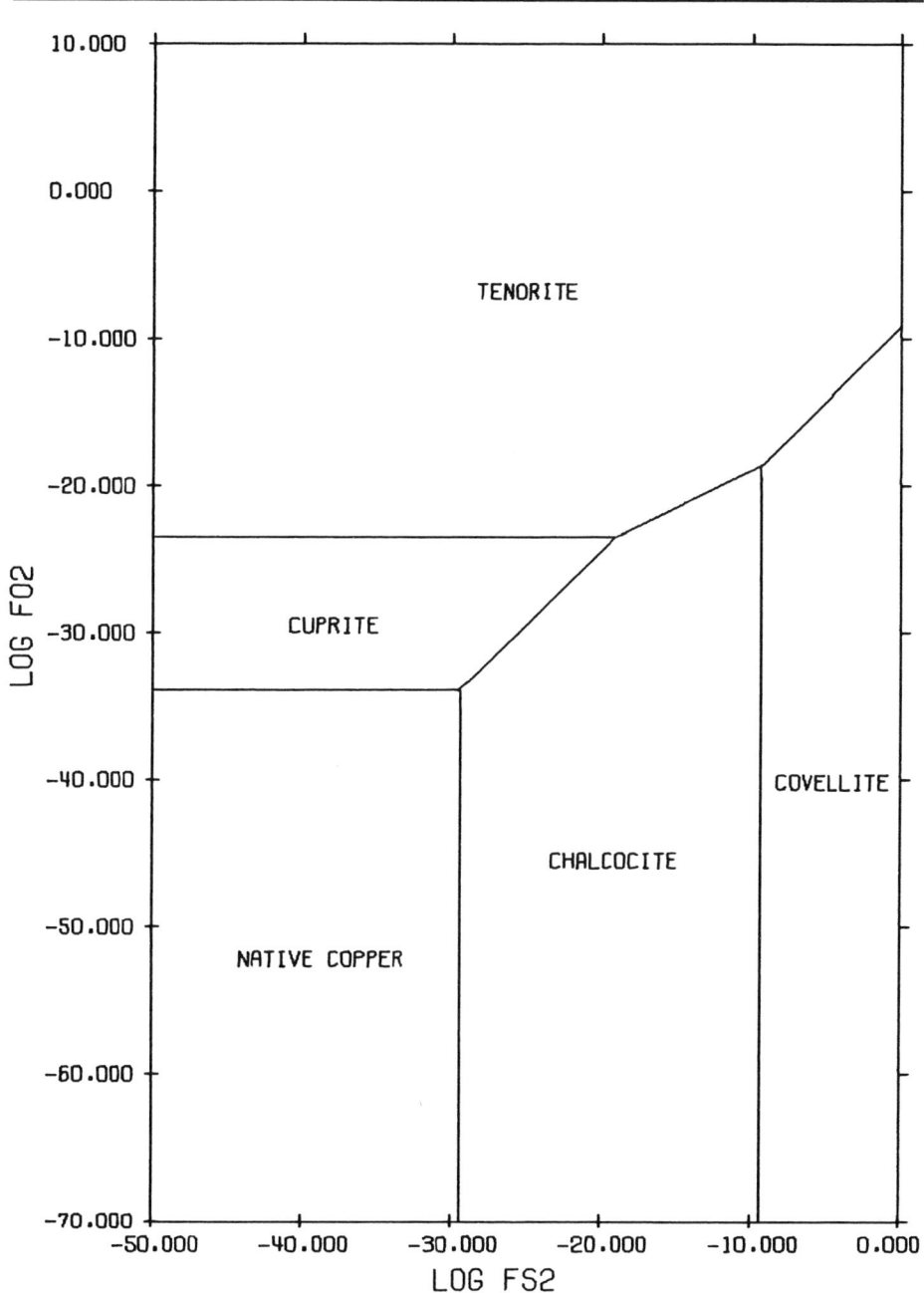

The System HCl—H$_2$O—Cu$_2$S—H$_2$S—H$_2$SO$_4$ at 150°C.

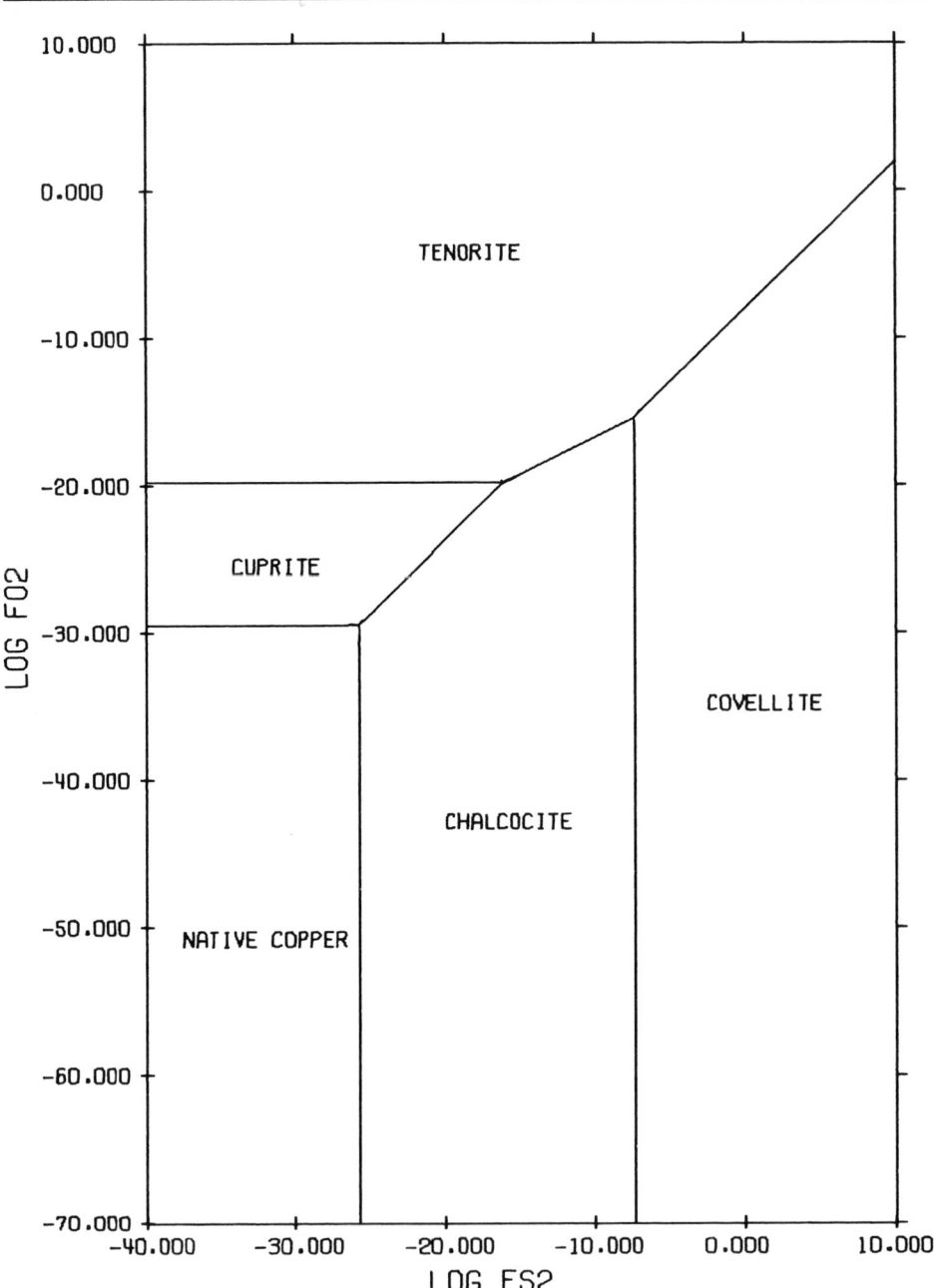

The System HCl—H$_2$O—Cu$_2$S—H$_2$S—H$_2$SO$_4$ at 200°C.

Fugacity Diagrams

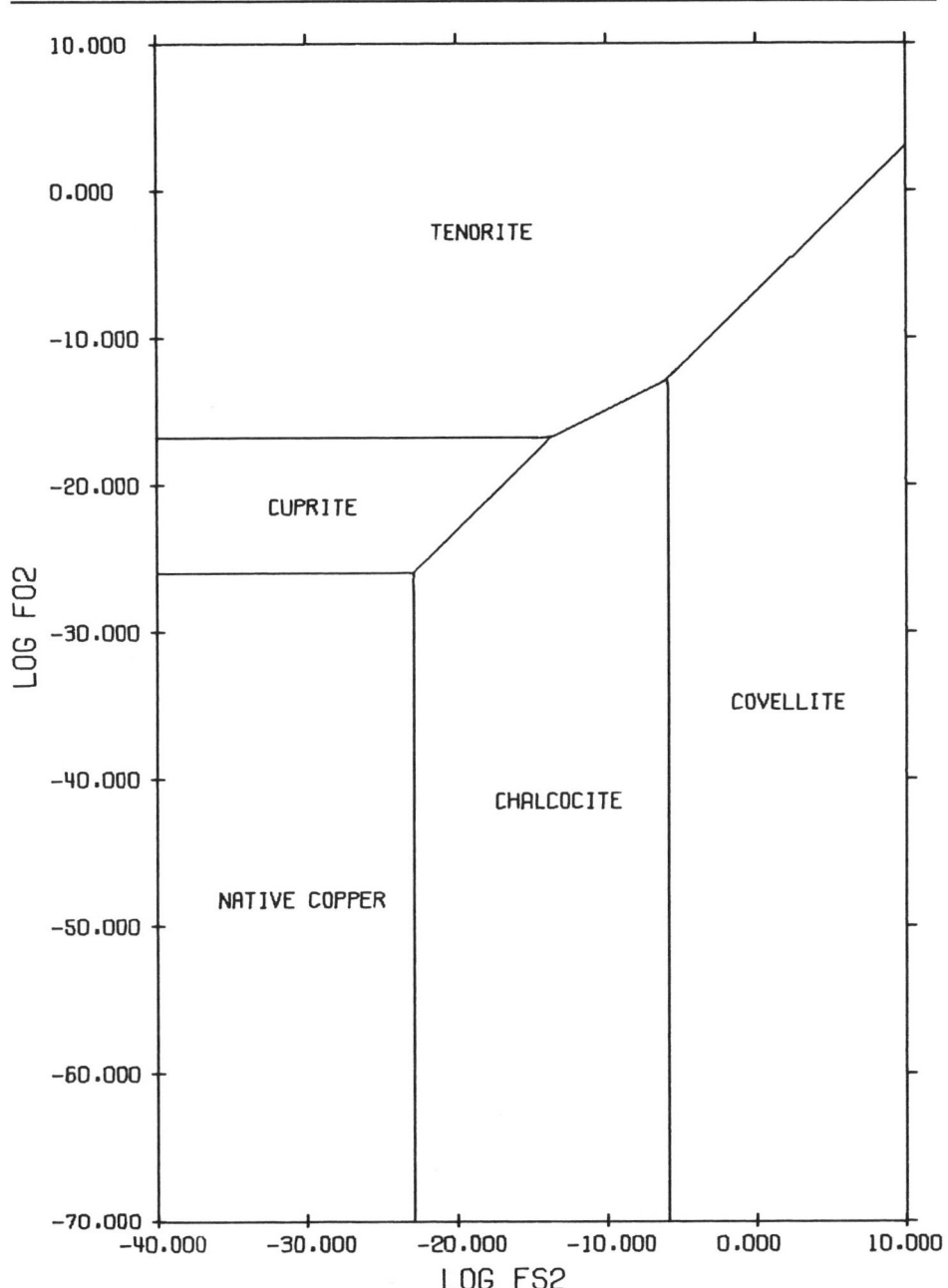

The System HCl—H$_2$O—Cu$_2$S—H$_2$S—H$_2$SO$_4$ at 250°C.

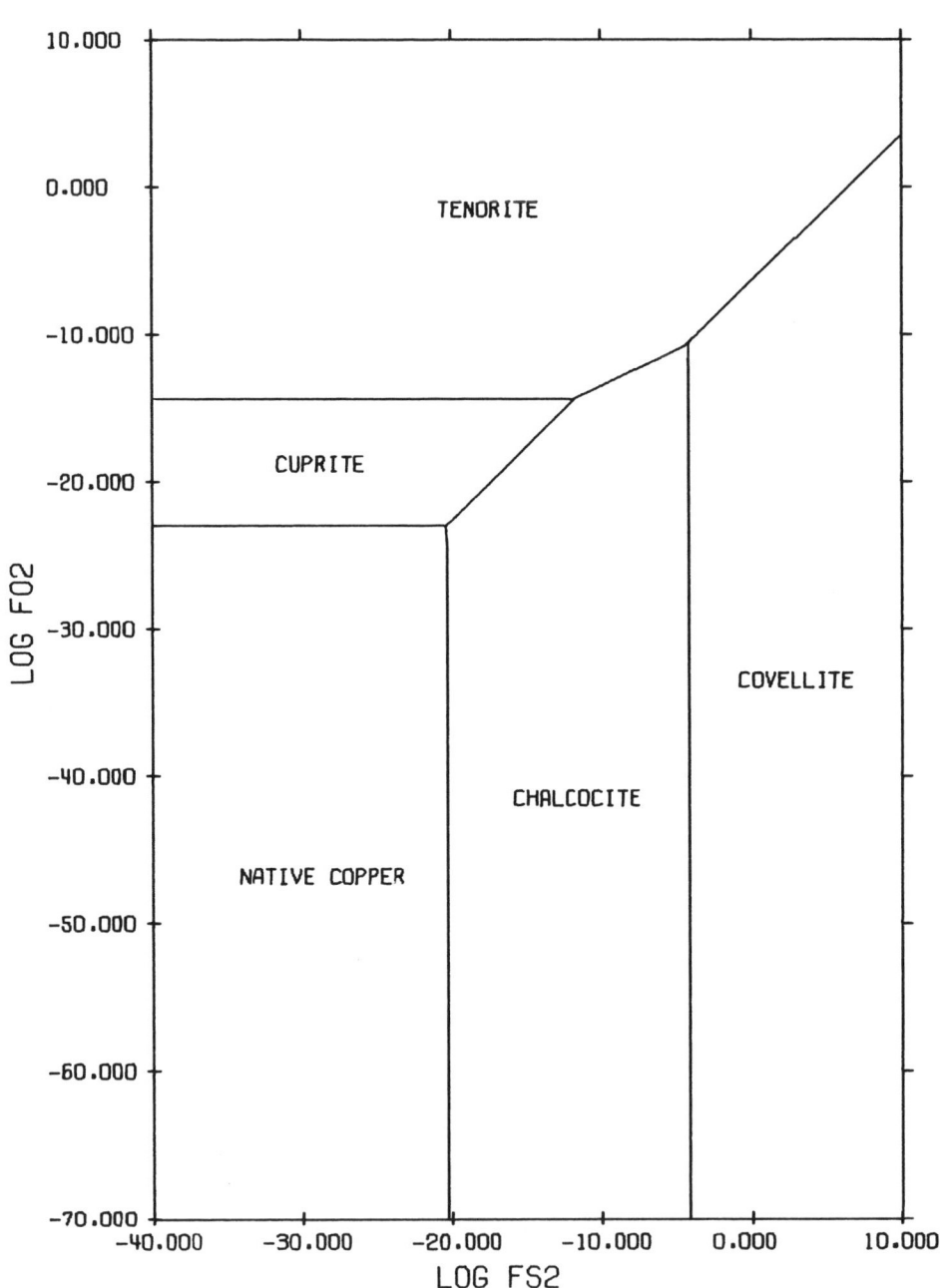

The System HCl—H_2O—Cu_2S—H_2S—H_2SO_4 at 300°C.

Fugacity Diagrams

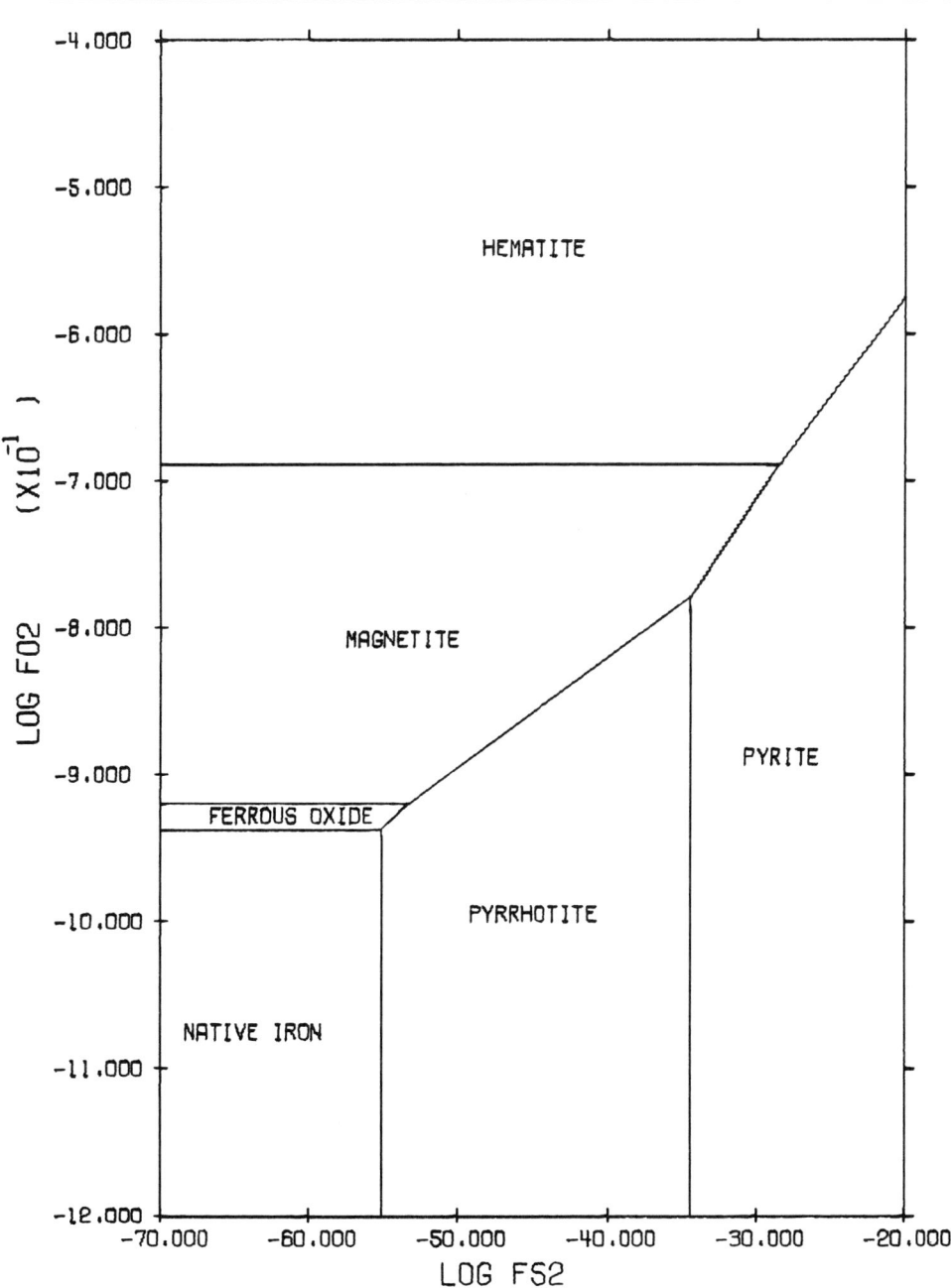

The System HCl—H_2O—FeS—H_2S—H_2SO_4 at 25°C.

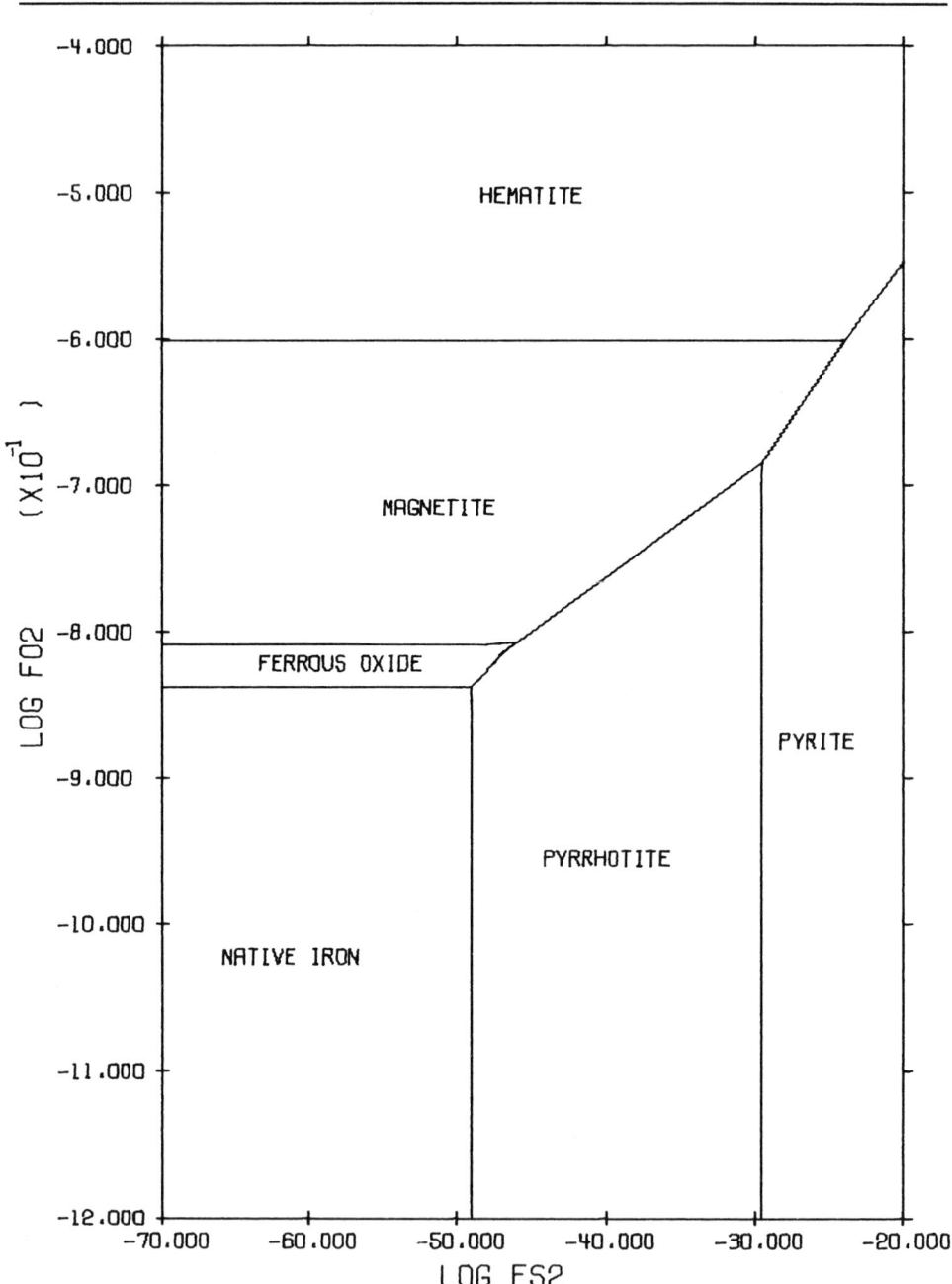

The System HCl—H$_2$O—FeS—H$_2$S—H$_2$SO$_4$ at 60°C.

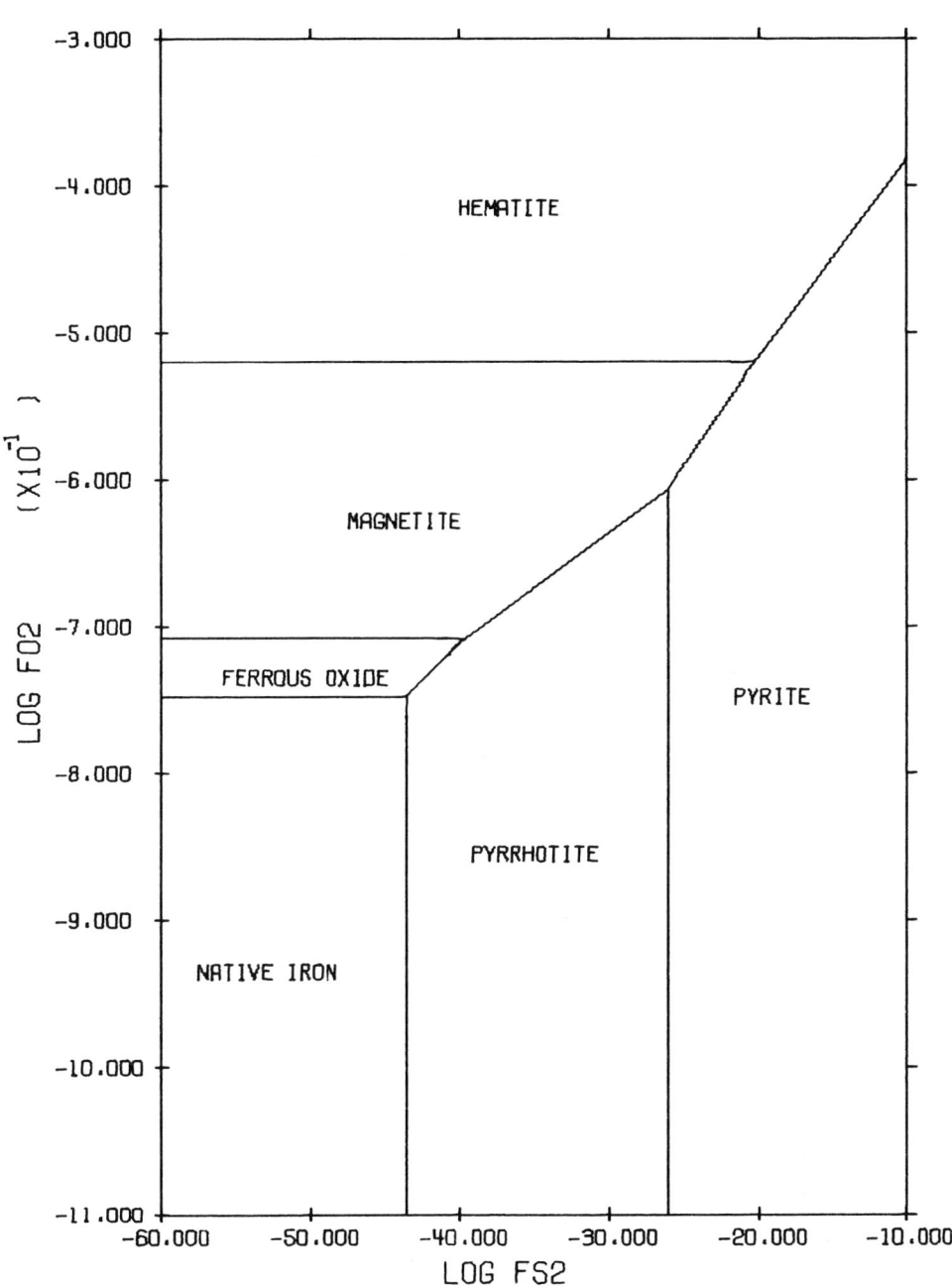

The System HCl—H$_2$O—FeS—H$_2$S—H$_2$SO$_4$ at 100°C.

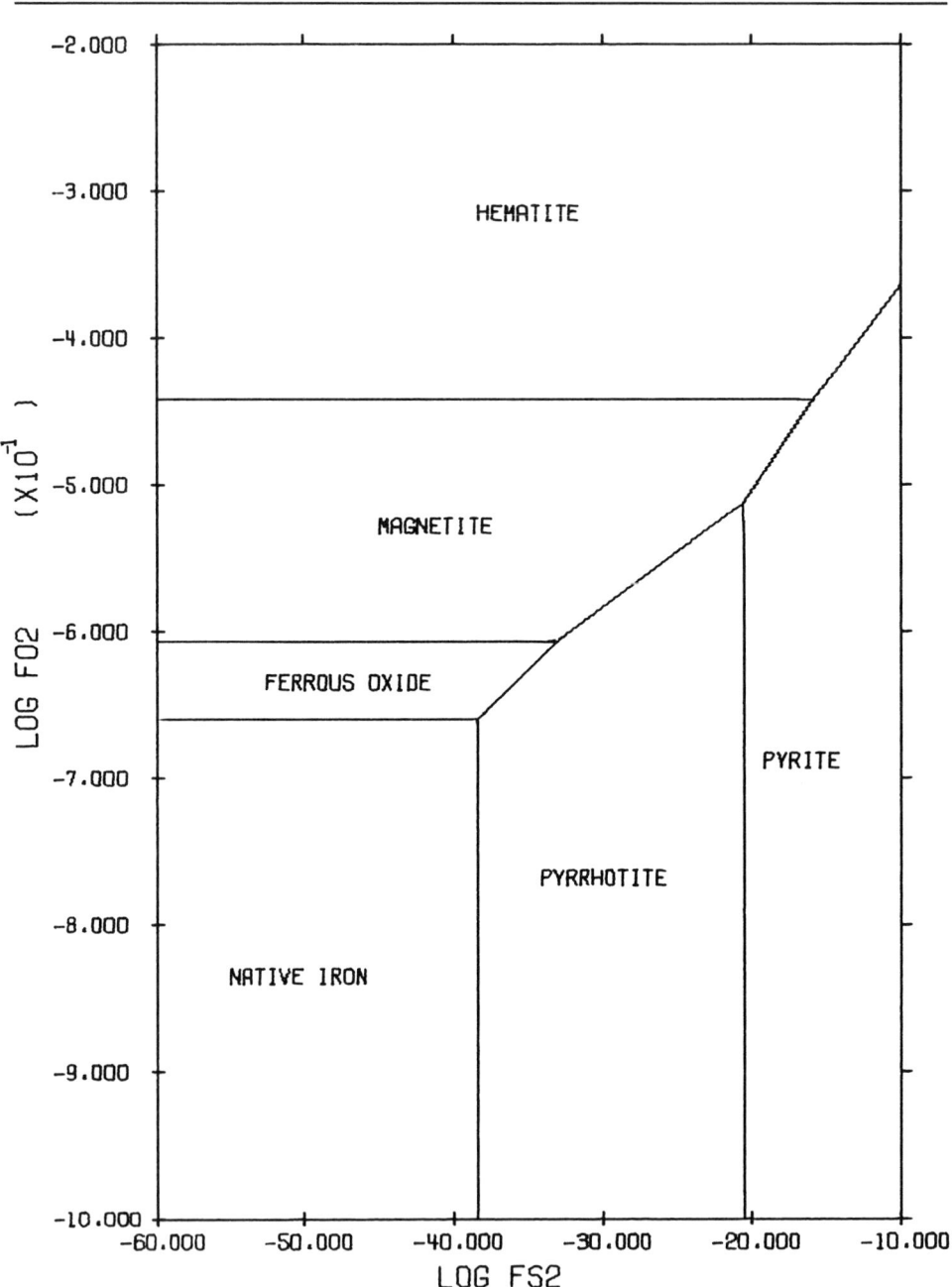

The System HCl—H₂O—FeS—H₂S—H₂SO₄ at 150°C.

Fugacity Diagrams

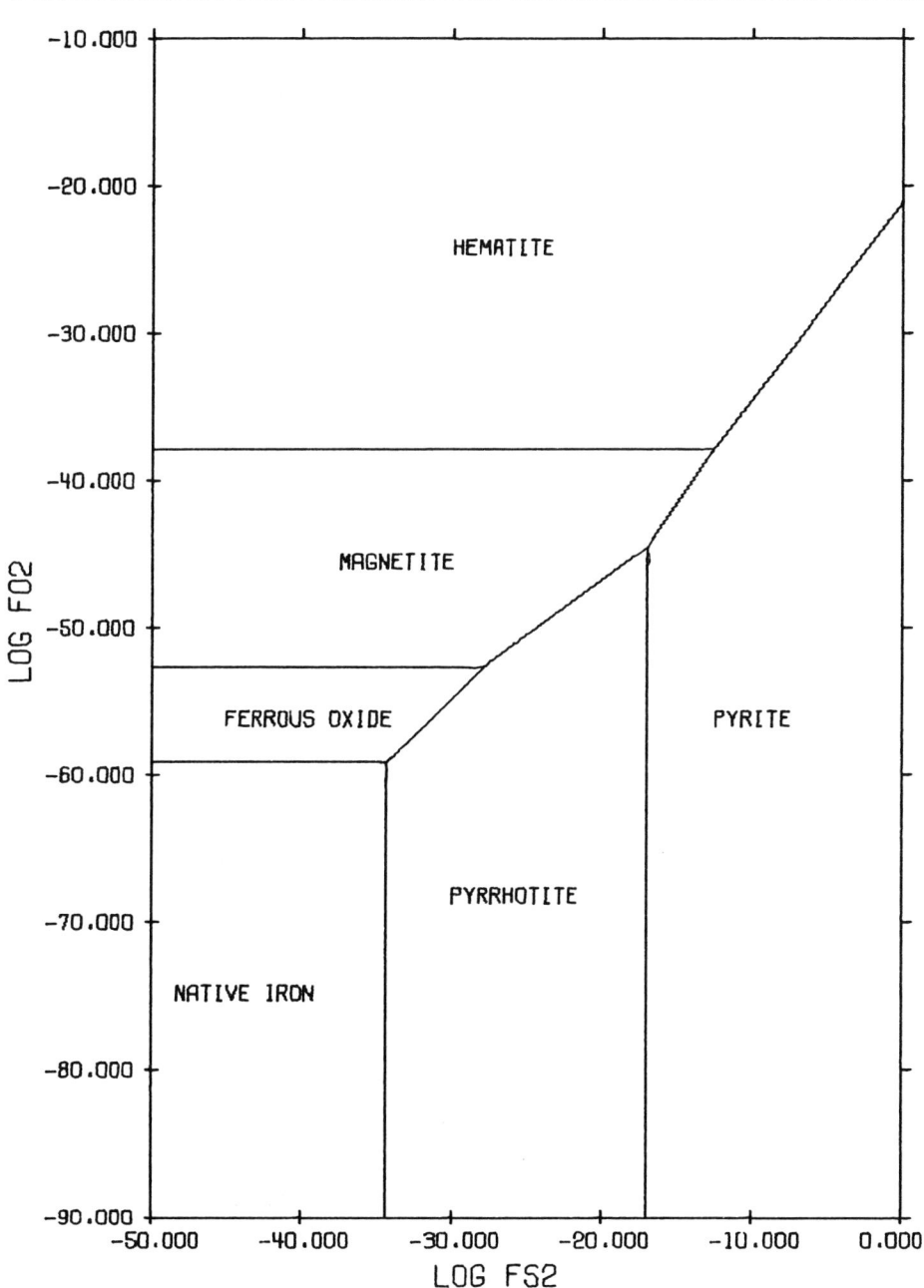

The System HCl—H$_2$O—FeS—H$_2$S—H$_2$SO$_4$ at 200°C.

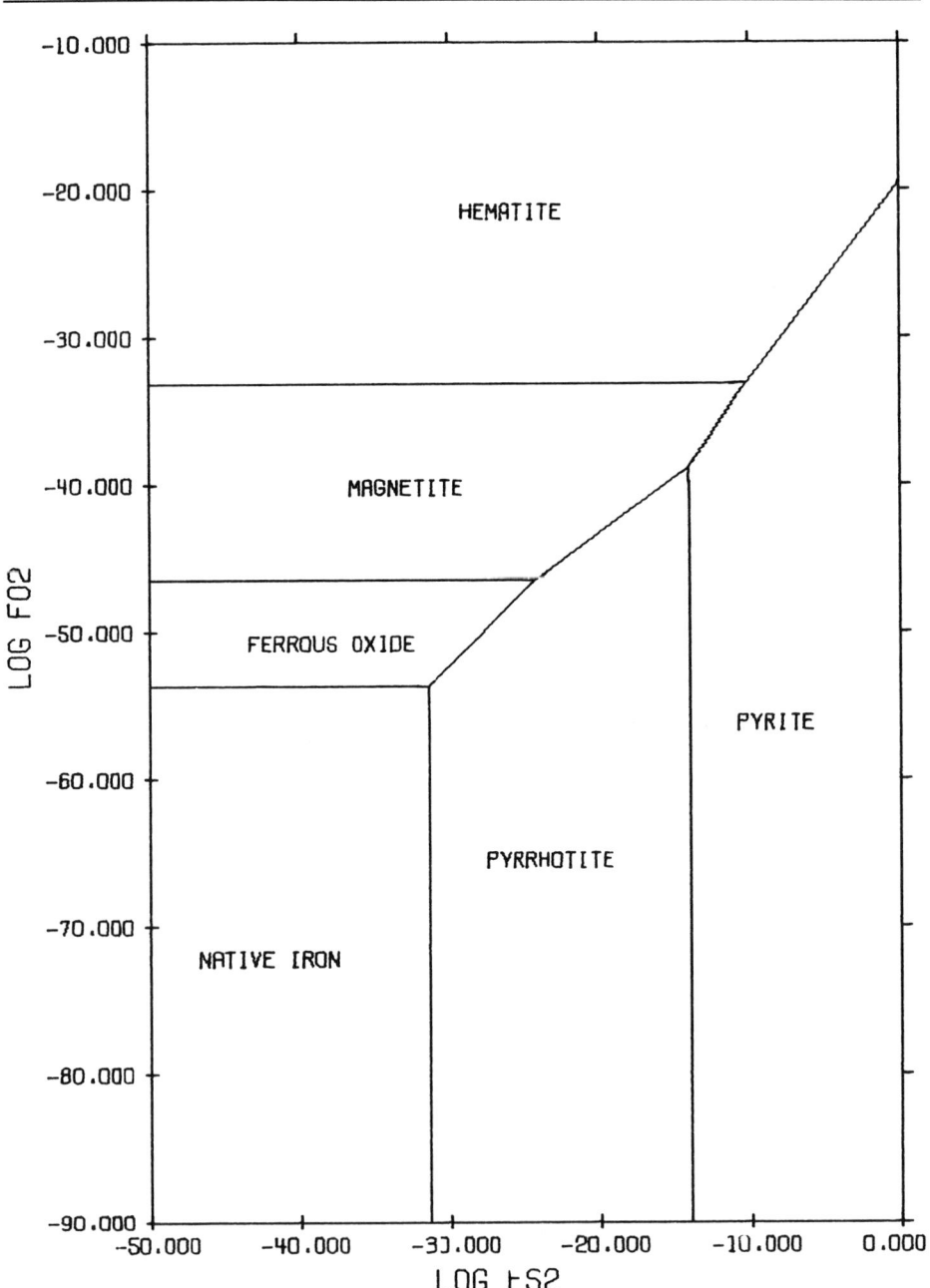

The System HCl—H_2O—FeS—H_2S—H_2SO_4 at 250°C.

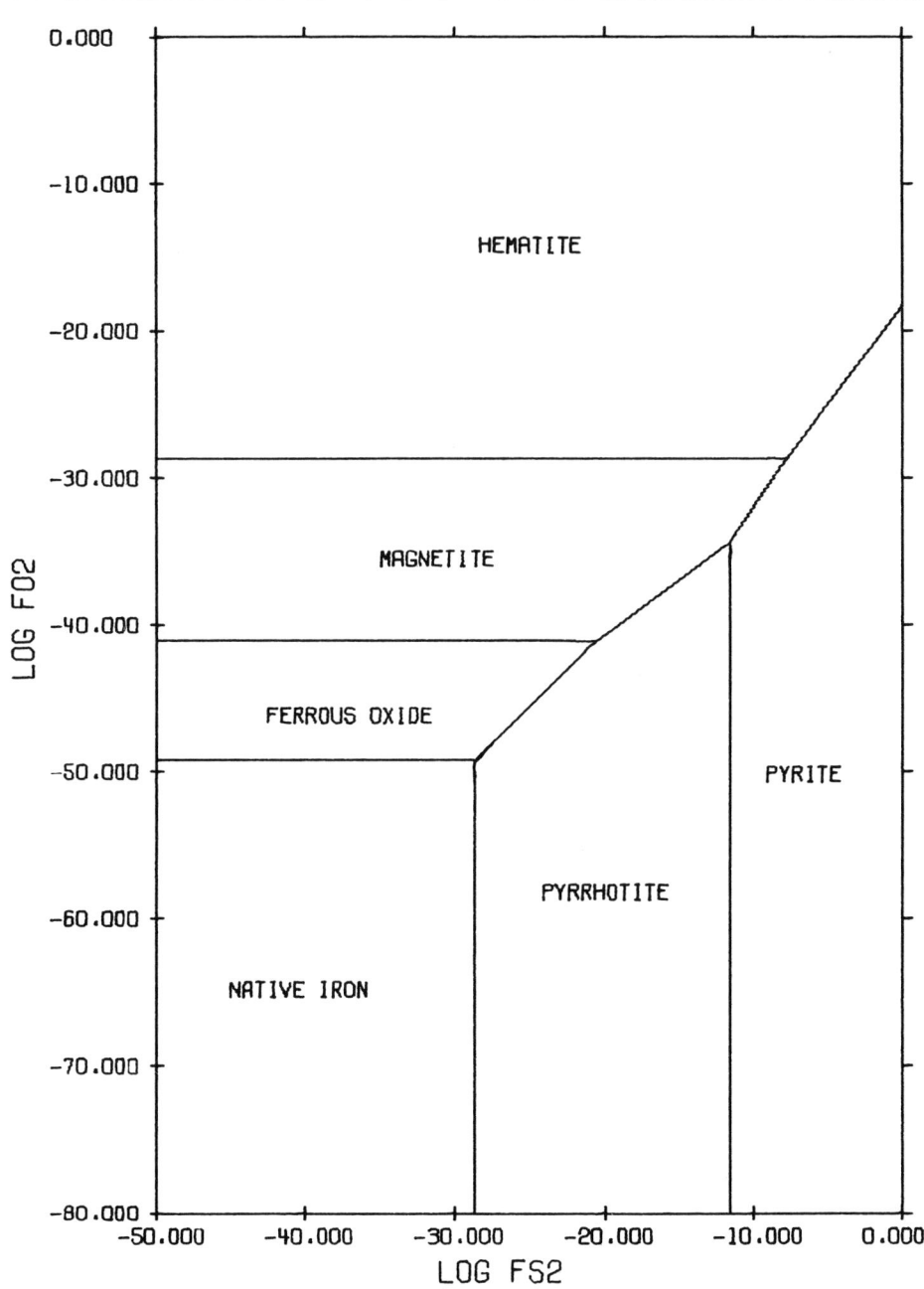

The System HCl—H$_2$O—FeS—H$_2$S—H$_2$SO$_4$ at 300°C.

INDEX OF DIAGRAMS

Index of Diagrams

(Systems arranged alphabetically after HCl—H₂O)

System: HCl—H₂O—	Activity Constraints	Page number by temperature							
		0 °C	25 °C	60 °C	100 °C	150 °C	200 °C	250 °C	300 °C
Ag$_2$S—H$_2$S					208		208		208
Ag$_2$S—H$_2$S—HgS			213			213			213
Ag$_2$S—H$_2$S—PbS					212	212			212
Al$_2$O$_3$—CaO—CO$_2$—K$_2$O—SiO$_2$	quartz saturation	64	65	66	67	68	69	70	71
	amorphous silica saturation	72	73	74	75				
Al$_2$O$_3$—CaO—CO$_2$—MgO—SiO$_2$	quartz saturation		120	121	122	123	124	125	126
	amorphous silica saturation		127	128	129				
	quartz saturation		152		153		154		155
	quartz saturation		156		157		158		159
	gibbsite or corundum saturation		160	161	162	163	164	165	166
Al$_2$O$_3$—CaO—CO$_2$—Na$_2$O—SiO$_2$	quartz saturation	108	109	110	111	112	113	114	115

247

System: HCl—H$_2$O—	Activity Constraints	0 °C	25 °C	60 °C	100 °C	150 °C	200 °C	250 °C	300 °C
Al$_2$O$_3$—CaO—CO$_2$—Na$_2$O—SiO$_2$ (continued)	amorphous silica saturation	116	117	118	119				
Al$_2$O$_3$—CaO—CO$_2$—SiO$_2$			25	26	27				
		100	101	102	103	104	105	106	107
			148		149		150		151
Al$_2$O$_3$—CO$_2$—FeO—K$_2$O—SiO$_2$	quartz saturation		167	168	169	170	171	172	173
	amorphous silica saturation		174	175	176				
Al$_2$O$_3$—CO$_2$—K$_2$O—MgO—SiO$_2$	quartz saturation	40	41	42	43	44	45	46	47
	amorphous silica saturation	48	49	50	51				
Al$_2$O$_3$—K$_2$O—Na$_2$O—SiO$_2$	quartz saturation	28	29	30	31	32	33	34	35
	amorphous silica saturation	36	37	38	39				
Al$_2$O$_3$—K$_2$O—SiO$_2$		16	17	18					
		76	77	78	79	80	81	82	83
					141		142		143
Al$_2$O$_3$—CO$_2$—MgO—Na$_2$O—SiO$_2$	quartz saturation	52	53	54	55	56	57	58	59
	amorphous silica saturation	60	61	62	63				
Al$_2$O$_3$—CO$_2$—MgO—SiO$_2$		22	23	24					
		92	93	94	95	96	97	98	99
			144		145		146		147
Al$_2$O$_3$—Na$_2$O—SiO$_2$		19	20	21					
		84	85	86	87	88	89	90	91
			137		138		139		140

Index of Diagrams

System: HCl—H$_2$O—	Activity Constraints	Page number by temperature							
		0 °C	25 °C	60 °C	100 °C	150 °C	200 °C	250 °C	300 °C
CaO—Al$_2$O$_3$—CO$_2$—K$_2$O—SiO$_2$	quartz saturation	64	65	66	67	68	69	70	71
	amorphous silica saturation	72	73	74	75				
CaO—Al$_2$O$_3$—CO$_2$—MgO—SiO$_2$	quartz saturation		120	121	122	123	124	125	126
	amorphous silica saturation		127	128	129				
	quartz saturation		152		153		154		155
	quartz saturation		156		157		158		159
	gibbsite or corundum saturation		160	161	162	163	164	165	166
CaO—Al$_2$O$_3$—CO$_2$—Na$_2$O—SiO$_2$	quartz saturation	108	109	110	111	112	113	114	115
	amorphous silica saturation	116	117	118	119				
CaO—Al$_2$O$_3$—CO$_2$—SiO$_2$			25	26	27				
		100	101	102	103	104	105	106	107
			148		149		150		151
CaO—CO$_2$—MgO		177	178	179	180	181	182	183	184
CaO—CO$_2$—FeO		187					187		187
CaO—CO$_2$—MgO—SiO$_2$			130	131	132	133	134	135	136
CO$_2$ and other components—see listing under other components									
Cu$_2$S—FeS—H$_2$S—H$_2$SO$_4$	$a_{H_2S} = 10^{-2}$		196		197		198		199
	$a_{H_2S} = 10^{-3}$		200		201		202		203
	$a_{H_2S} = 10^{-4}$		204		205		206		207
Cu$_2$S—H$_2$S—H$_2$SO$_4$			192		193		194		195
			229		229	229			229
			230	231	232	233	234	235	236

System: HCl–H_2O–	Activity Constraints	0 °C	25 °C	60 °C	100 °C	150 °C	200 °C	250 °C	300 °C
FeO–Al_2O_3–CO_2–K_2O–SiO_2	quartz saturation		167	168	169	170	171	172	173
	amorphous silica saturation		174	175	176				
FeO–CaO–CO_2		187					187		187
FeO–CO_2–MgO		186	185			186			186
FeS–Cu_2S–H_2S–H_2SO_4	$a_{H_2S} = 10^{-2}$		196		197		198		199
	$a_{H_2S} = 10^{-3}$		200		201		202		203
	$a_{H_2S} = 10^{-4}$		204		205		206		207
FeS–H_2S–H_2SO_4			188		189		190		191
			228			228	228		228
			237	238	239	240	241	242	243
FeS–H_2S–ZnS	$a_{H_2S} = 10^{-2}$		215		216		217		218
	$a_{H_2S} = 10^{-3}$		219		220		221		222
	$a_{H_2S} = 10^{-4}$		223		224		225		226
H_2S–Ag_2S				208		208			208
H_2S–Ag_2S–HgS			213			213			213
H_2S–Ag_2S–PbS				212		212			212
H_2S–Cu_2S–FeS–H_2SO_4	$a_{H_2S} = 10^{-2}$		196		197		198		199
	$a_{H_2S} = 10^{-3}$		200		201		202		203
	$a_{H_2S} = 10^{-4}$		204		205		206		207
H_2S–Cu_2S–H_2SO_4			192		193		194		195
			229		229	229			229
			230	231	232	233	234	235	236
H_2S–FeS–H_2SO_4			188		189		190		191
			228			228	228		228
			237	238	239	240	241	242	243
H_2S–FeS–ZnS	$a_{H_2S} = 10^{-2}$		215		216		217		218
	$a_{H_2S} = 10^{-3}$		219		220		221		222
	$a_{H_2S} = 10^{-4}$		223		224		225		226
H_2S–HgS					209	209	209	209	
H_2S–PbS			210			210			210
H_2S–ZnS			211			211			211

Index of Diagrams

System: HCl—H$_2$O—	Activity Constraints	0 °C	25 °C	60 °C	100 °C	150 °C	200 °C	250 °C	300 °C
H$_2$SO$_4$—Cu$_2$S—H$_2$S			192		193		194		195
			229		229	229			229
			230	231	232	233	234	235	236
H$_2$SO$_4$—Cu$_2$S—FeS—H$_2$S	$a_{H_2S} = 10^{-2}$		196		197		198		199
	$a_{H_2S} = 10^{-3}$		200		201		202		203
	$a_{H_2S} = 10^{-4}$		204		205		206		207
H$_2$SO$_4$—FeS—H$_2$S			188		189		190		191
			228			228	228		228
			237	238	239	240	241	242	243
HgS—Ag$_2$S—H$_2$S			213			213			213
HgS—H$_2$S					209	209	209	209	
K$_2$O—Al$_2$O$_3$—CaO—CO$_2$—SiO$_2$	quartz saturation	64	65	66	67	68	69	70	71
	amorphous silica saturation	72	73	74	75				
K$_2$O—Al$_2$O$_3$—CO$_2$—FeO—SiO$_2$	quartz saturation		167	168	169	170	171	172	173
	amorphous silica saturation		174	175	176				
K$_2$O—Al$_2$O$_3$—CO$_2$—MgO—SiO$_2$	quartz saturation	40	41	42	43	44	45	46	47
	amorphous silica saturation	48	49	50	51				
K$_2$O—Al$_2$O$_3$—Na$_2$O—SiO$_2$	quartz saturation	28	29	30	31	32	33	34	35
	amorphous silica saturation	36	37	38	39				
K$_2$O—Al$_2$O$_3$—SiO$_2$		16	17	18					
		76	77	78	79	80	81	82	83
					141		142		143

Index of Diagrams

System: HCl—H$_2$O—	Activity Constraints	Page number by temperature							
		0 °C	25 °C	60 °C	100 °C	150 °C	200 °C	250 °C	300 °C
MgO—Al$_2$O$_3$—CaO—CO$_2$—SiO$_2$	quartz saturation		120	121	122	123	124	125	126
	amorphous silica saturation		127	128	129				
	quartz saturation		152		153		154		155
	quartz saturation		156		157		158		159
	gibbsite or corundum saturation		160	161	162	163	164	165	166
MgO—Al$_2$O$_3$—CO$_2$—K$_2$O—SiO$_2$	quartz saturation	40	41	42	43	44	45	46	47
	amorphous silica saturation	48	49	50	51				
MgO—Al$_2$O$_3$—CO$_2$—Na$_2$O—SiO$_2$	quartz saturation	52	53	54	55	56	57	58	59
	amorphous silica saturation	60	61	62	63				
MgO—Al$_2$O$_3$—CO$_2$—SiO$_2$		22	23	24					
		92	93	94	95	96	97	98	99
			144		145		146		147
MgO—CaO—CO$_2$		177	178	179	180	181	182	183	184
MgO—CaO—CO$_2$—SiO$_2$			130	131	132	133	134	135	136
MgO—CO$_2$—FeO		186	185			186			186
Na$_2$O—Al$_2$O$_3$—CaO—CO$_2$—SiO$_2$	quartz saturation	108	109	110	111	112	113	114	115
	amorphous silica saturation	116	117	118	119				
Na$_2$O—Al$_2$O$_3$—K$_2$O—SiO$_2$	quartz saturation	28	29	30	31	32	33	34	35

Index of Diagrams

System: HCl—H₂O—	Activity Constraints	Page number by temperature							
		0 °C	25 °C	60 °C	100 °C	150 °C	200 °C	250 °C	300 °C
Na_2O—Al_2O_3—K_2O—SiO_2 (continued)	amorphous silica saturation	36	37	38	39				
Na_2O—Al_2O_3—CO_2—MgO—SiO_2	quartz saturation	52	53	54	55	56	57	58	59
	amorphous silica saturation	60	61	62	63				
Na_2O—Al_2O_3—SiO_2		19	20	21					
		84	85	86	87	88	89	90	91
			137		138		139		140
PbS—Ag_2S—H_2S					212		212		212
PbS—H_2S				210		210			210
PbS—H_2S—ZnS					214		214		214
SiO_2 and other components—see listing under other components									
ZnS—H_2S				211			211		211
ZnS—H_2S—PbS					214		214		214
ZnS—FeS—H_2S	$a_{H_2S} = 10^{-2}$		215		216		217		218
	$a_{H_2S} = 10^{-3}$		219		220		221		222
	$a_{H_2S} = 10^{-4}$		223		224		225		226